Lecture Notes in Computer Science   11830

More information about this series at http://www.springer.com/series/7410

Sushil Jajodia · George Cybenko ·
Peng Liu · Cliff Wang ·
Michael Wellman (Eds.)

Adversarial
and Uncertain Reasoning
for Adaptive Cyber Defense

Control- and Game-Theoretic Approaches
to Cyber Security

Springer

Editors
Sushil Jajodia (ID)
George Mason University
Fairfax, VA, USA

George Cybenko
Dartmouth College
Hanover, NH, USA

Peng Liu
Pennsylvania State University
University Park, PA, USA

Cliff Wang
Army Research Laboratory
Triangle Park, NC, USA

Michael Wellman
University of Michigan
Ann Arbor, MI, USA

ISSN 0302-9743   ISSN 1611-3349   (electronic)
Lecture Notes in Computer Science
ISBN 978-3-030-30718-9   ISBN 978-3-030-30719-6   (eBook)
https://doi.org/10.1007/978-3-030-30719-6

LNCS Sublibrary: SL4 – Security and Cryptology

Cover illustration: Figure 5 from 'Adaptive Cyber Defenses for Botnet Detection and Mitigation', p. 170

This Springer imprint is published by the registered company Springer Nature Switzerland AG
The registered company address is: Gewerbestrasse 11, 6330 Cham, Switzerland

Preface

Today's cyber defenses are largely static. They are governed by slow deliberative processes involving testing, security patch deployment, and human-in-the-loop monitoring. As a result, adversaries can systematically probe target networks, pre-plan their attacks, and ultimately persist for long times inside compromised networks and hosts.

In response to this situation, researchers in recent years have started to investigate various methods that make networked information systems less homogeneous and less predictable. The basic idea of Adaptation Techniques (AT) is to engineer systems that have homogeneous functionalities but randomized manifestations. Homogeneous functionality allows authorized use of networks and services in predictable, standardized ways while randomized manifestations make it difficult for attackers to engineer exploits remotely.

Examples of AT include concepts such as Moving Target Defenses (MTD) as well as artificial diversity and bio-inspired defenses in order to assess the extent to which they involve system adaption for security and resiliency purposes. Unfortunately, the majority of AT research has been focused on developing specific new techniques as opposed to understanding their overall operational costs, when they are most useful, and what their possible inter-relationships might be. Moreover, the AT approaches assume stationary and stochastic, but non-adversarial, environments. Situations with intelligent peer adversaries operating in and changing a networked environment produce dynamic behaviors that violate these assumptions, potentially defeating these adaptations.

This volume aims to synthesize the recent advances made by a large team of researchers working on the same U.S. Department of Defense Multidisciplinary University Research Initiative (MURI) project during 2013–2019.[1] This project has developed a new class of technologies called Adaptive Cyber Defense (ACD) by building on two active but heretofore separate research areas: Adaptation Techniques and Adversarial Reasoning.

Our research has yielded a rich repertoire of AT methods for introducing diversity and uncertainty into networks, applications, and hosts.[2] Moreover, we have investigated the criteria for deciding where, when, and how to best employ available AT options. Such management decisions are complex due to the performance and security tradeoffs inherent in AT approaches. To address such challenges, this project has harnessed a

---

[1] George Cybenko, Sushil Jajodia, Michael P. Wellman, Peng Liu, "Adversarial and uncertain reasoning for adaptive cyber defense: Building the cyber foundation (invited paper)," *Proc. 10th Int'l. Conf. on Information Systems Security (ICISS), Springer Lecture Notes in Computer Science, Vol. 8880,* Atul Prakash, Rudrapatna Shyamsundar, eds., Hyderabad, India, December 2014, pages 1–8.

[2] See Chapter 1 for a brief summary of advances documented in this book along with the pointers to the relevant literature.

broad array of Adversarial Reasoning (AR) techniques to identify effective and stable strategies for deploying AT options in operational systems. AR combines machine learning, behavioral science, operations research, control theory, and game theory to address the goal of computing effective strategies in dynamic, adversarial environments.

These techniques force adversaries to continually re-assess, re-engineer, and re-launch their cyberattacks. By integrating game-theoretic and control-theoretic analyses for tradeoff analysis, ACD presents adversaries with optimized and dynamically changing attack surfaces and system configurations, thereby significantly increasing the attacker's workloads and decreasing their probabilities of success.

This coherent and focused research effort has yielded: (a) scientific and engineering principles that enable effective ACD, and (b) prototypes and demonstrations of technologies embodying these principles in several real-world scenarios.

We are extremely grateful to the numerous participants for their contributions to the MURI project. In particular, it is a pleasure to acknowledge the authors for their contributions. Special thanks go to the US Army Research Office (ARO) and Alfred Hofmann, Vice-President of Publishing at Springer for their support of this volume summarizing the project. We also wish to thank the Army Research Office for their financial support under the MURI grant W911NF-13-1-0421.

August 2019

Sushil Jajodia
George Cybenko
Peng Liu
Cliff Wang
Michael Wellman

Contents

Overview of Control and Game Theory in Adaptive Cyber Defenses

George Cybenko[1]([✉]), Michael Wellman[2], Peng Liu[3], and Minghui Zhu[3]

[1] Dartmouth, Hanover, USA
gvc@dartmouth.edu
[2] University of Michigan, Michigan, USA
[3] Pennsylvania State University, University Park, USA

Abstract. The purpose of this chapter is to introduce cyber security researchers to key concepts in modern control and game theory that are relevant to Moving Target Defenses and Adaptive Cyber Defense. We begin by observing that there are fundamental differences between control models and game models that are important for security practitioners to understand. Those differences will be illustrated through simple but realistic cyber operations scenarios, especially with respect to the types and amounts of data require for modeling. In addition to modeling differences, there are a variety of ways to think about what constitutes a "solution." Moreover, there are significant differences in the computational and information requirements to compute solutions for various types of Adaptive Cyber Defense problems. This material is presented in the context of the advances documented in this book, the various chapters of which describe advances made in the 2012 ARO ACD MURI.

Keywords: Control Theory · Game Theory ·
Adaptive Cyber Defense · Moving Target Defense ·
Autonomous Cyber Operations

1 Moving Target Defenses (MTD)

The computer systems, software applications, and network technologies that we use today were developed in user and operator contexts that greatly valued standardization, predictability, and availability. Even today, performance and cost-effectiveness remain dominant market drivers. It is only relatively recently that security and resilience (not to be confused with fault tolerance) have become equally desirable properties of cyber systems. As a result, the first generation of cyber security technologies were largely based on system hardening through improved software security engineering [7,21] (to reduce vulnerabilities and attack surfaces) and layering security through defense-in-depth [28,31] (by adding encryption, access controls, firewalls, intrusion detection systems, and malware scanners, for example). These security technologies sought to respect the homogeneity, standardization, and predictability that have been so valued by the market but at the same time increasing security.

© Springer Nature Switzerland AG 2019
S. Jajodia et al. (Eds.): Adaptive Cyber Defense, LNCS 11830, pp. 1–11, 2019.
https://doi.org/10.1007/978-3-030-30719-6_1

Consequently, most of our cyber defenses remain static today. They are governed by slow and deliberative processes such as software testing [40], episodic penetration testing [39], security patch deployment [32], and human-in-the-loop monitoring and analysis of security events [12, 24, 36].

Adversaries benefit greatly from this situation because they can continuously and systematically probe targeted systems with the confidence that those systems will change slowly if at all. Adversaries can afford the time to engineer reliable exploits and pre-plan their attacks because their targets are essentially fixed and almost identical. Moreover, once an attack succeeds, adversaries persist for long times inside compromised networks and hosts because the hosts, networks, and services – largely designed for availability and homogeneity – do not reconfigure, adapt or regenerate except in deterministic ways to support maintenance and uptime requirements. This creates serious information and opportunity asymmetry between IT system defenders and potential attackers [6].

In response to this situation, researchers in recent years have started to investigate a variety of technologies that can make networked information systems less homogeneous and less predictable. Among the terms and concepts used to describe such cyber defense technologies are:

- *Diversity*: Inspired by biological systems [23], cyber diversity is a general concept for introducing robustness and resilience into engineered systems by reducing common failure modes in redundant system components. That is, the goal is to avoid technology "monocultures" [44, 53]. In cyber security systems, this is typically accomplished by introducing software or network variants appropriately [10, 16, 19, 30].
- *Randomization*: One approach to introduce cyber diversity is to randomize specific components of an information system. Such randomization can be done at the low level of a system's address space to defeat certain types of memory-based exploits [43], at the software level by generating multiple software variants through compiler randomization [30], instruction set randomization to defeat injected malware [9], or randomization of a network's address space [26] or protocols [33], to give just a few examples.
- *Moving Target Defenses*: Motivated by the observation that a moving target is harder to hit than a fixed one, the general concept behind Moving Target Defenses in the cyber domain is that an information system that changes dynamically *during its operation* will be more difficult for an attacker to surveil, reverse engineer and ultimately exploit with sufficient degrees of persistence than a fixed target [27]. Randomization and diversity are two ways to implement moving target defenses but not all randomization and diversity techniques necessarily realize moving targets. That is because some implementations of diversity and randomization do not in fact change during execution or system recovery after an attack.

A basic goal of Moving Target techniques is to engineer systems that have homogeneous functionalities but dynamically different manifestations. Homogeneous functionality allows authorized use of networks and services in predictable, standardized ways while randomized manifestations make it difficult for attackers to engineer exploits remotely, let alone parlay one exploit into successful attacks against a multiplicity of hosts or even the same host after reboot. Ideally, each compromise of a system deploying a Moving Target Defense would require the same, significant effort by the attacker who is exploiting the system component in which the Moving Target Defense is deployed.

Although functionality is preserved, it should be noted that there are intrinsic and important tradeoffs between increased security through such means and increased maintenance overhead for managing systems that are less predicable and heterogeneous. Moreover, there are also tradeoffs among the classical security properties of Confidentiality, Integrity and Availability (CIA) when deploying some forms of diversity [20]. For example, having N different and diverse web servers mirroring the same content can increase availability because an attacker has to bring down all N variants, presumably requiring a workfactor about N times higher than bringing down any one web server. On the other hand, the N variants make for a larger attack surface because a breach of any of one of them can compromise confidentiality.

This is but one example of the kinds of tradeoffs that arise when deploying diverse moving targets in an operational environment, namely the possible tradeoffs among security properties valued in the deployment.

In fact, virtually all techniques for increasing security through diversity, randomization and/or moving target defenses involve parameter choices both as individual standalone techniques and especially so when used in combinations [5, 17, 38, 52].

Good or optimal choices for such parameter settings requires modeling the problem, quantifying the model with realistic data and ultimately "solving" the resulting optimization problem. Because the operating environment, mission objectives, mission priorities, attacker behaviors and attacker objectives can all change over time, in fact during exection, moving target deployment solutions might have to be constantly recomputed.

These aspects of Moving Target Defense are the subject of "Adaptive Cyber Defenses" technology addressed in the chapters of this book, and explained in more detail in the following section.

2 Adaptive Cyber Defense: Control and Game Theory for MTD

Research and development in Moving Target Defense has been significant over the past few years.

A 2016 survey paper documented at least 100 different types of Moving Target Defense techniques [14], indicating a significant growth in the number of techniques compared to a 2013 survey [37] that documented 59 different types of Moving Target Defenses. In fact, the development of individual Moving Target Defense Techniques continues at a significant pace today according to a 2018 update to the 2013 survey article [48]. Research on new techniques continues today [2].

The variety of Moving Target Defense techniques together with the variety of options and parameter settings for deploying each individual technique means that there are several types of decisions that an information system operator needs to make to effectively use such techniques. Those decisions include:

- Decisions about which single or combination of MTD's to use;
- Decisions about which MTD parameter settings to use for an individual technique;
- Decisions about which combination of MTD's together with their parameter settings to use (deciding about both of the above simultaneously).

Such decisions are made when MTD's are first deployed and then should be continuously reassessed and updated during deployment seeing as operating and threat conditions change over time. These choices constitute the decision making aspect of the "MTD OODA" (Observe-Orient-Decide-Act) Loop [8,13]. The study of such decisions within the context of MTD's is called Adaptive Cyber Defense (ACD) - the topic of this book.

The rigorous, analytic framework for ACD, namely studying the decision problems arising in MTD-based systems falls within the general scope of Operations Research [51] but more specifically Control Theory and Game Theory. The decision problems are especially challenging when there is inherent uncertainty in the decision-making's operating environment as is typically the case in cyber operations.

The key distinction between Control Theory and Game Theory is the nature of the operating environment and how it is modeled. To illustrate the fundamental difference, consider the following simple but representative MTD cyber defense situation.

In a cloud computing environment, performance of servers and applications degrade over time (due to memory leaks or other inadequate memory management among other reasons, for example). Given availability requirements (such as the average or minimal number of servers available over time) and historical data on performance degradation, it is possible to quantitatively formulate a decision problem regarding schedules for regenerating individual server software. Two fundamentally different modeling frameworks in this scenario are briefly described and compared below.

2.1   Control Theory Models

In this modeling approach, there is a benefit for each time unit that a server is up and fulfilling requests at various rates and there is a time cost for restarting the server with a fresh image. For simplicity of exposition, assume that the server is either working properly or not. During restarts, no requests can be fulfilled because the server is not working. Moreover, there is a probability distribution for the time that the server will fail after a restart. That probability distribution, as well as the value of server uptime and time to restart, are independent of how many restarts have occured and when they occured. Note that if the system is not memoryless, the system operator can be inclined to restart a system even before it fails outright because the cost of downtime is higher than the cost of restarting. This kind of model is common within the cloud server reliability research literature and can be formulated as a control problem [11, 29].

A key aspect of this formulation of the problem is that the operating environment in which the system operates is *non-adversarial* in that the failures are random and independent of each other.

Moreover, control theoretic formulations typically involve computing minima or maxima of objective functions so that the models can be solved using optimization techniques such as dynamic programming.

2.2 Game Theory Models

In the game theory modeling approach, the same costs and benefits for correct server operation hold as in the above control theory model. However, the server failures are no longer solely the result of natural, benign operation but are influenced or even explicitly triggered by rational adversaries (the attackers) who have their own costs and benefits for bringing a server down. The attacker accrues benefit when the server is down but has a cost for launching an attack, successful or not, because some effort is required to exploit a novel vulnerability or to use a new source IP address that is not black-listed.

A key aspect of the game theory formulation of the problem is that the operating environment in which the system operates is decidedly *adversarial* in that the system failures are due to the actions of a rational agent whose objectives are typically at odds with the system operator's objective. As in the above control theory formulation, the system operator can benefit from restarting a system even before it is fulling compromised in an attack because the cost of downtime is higher than the cost of restarting.

The concept of solution to a game theoretic formulation of a problem is typically expressed in terms of equilibria, such as Nash Equilibia. By contrast with control problems, equilibria in games are typically saddle points in the sense that they are maxima for one player and minima for another player.

Such game theory-based models can lead to complex analyses in which there are several open problems [34, 46, 50].

In both control and game theory, the term "policy" refers to the actions the operator takes to change system states (for example, a "restart" action will take the system from the "failed" state to the "normal operation" state for the operator but for an attacker, the "attack" action will take the targeted system from the "normal" state to the "failed" state. Given an objective function and a concept for a "solution" with respect to that objective function, an optimal policy for each actor is a policy that achieves optimal performance for them with respect to their concept of solution and their objective function.

In this context, Adaptive Cyber Defense (ACD) is the application of control and game theory to Moving Target Defenses (MTD). Notwithstanding the above distinctions, both control and game theory as used in Adaptive Cyber Defense involve many common ingredients. We list the ingredients below along with brief descriptions of them as well as pointers to the literature, including chapters in this book, with detailed approaches (Table 1).

Table 1. Adaptive Cyber Defense (ACD) ingredients

Ingredient	Description	Book chapters	Other references
Moving Target Defense Techniques	Adaptive Cyber Defenses involve the deliberate and rational actions that an operator can invoke to protect their systems. Specific actions considered include possible network, operating systems and applications randomizations, diversity and Moving Target Defenses. Possible actions include configuration and parameter selections for individual techniques. In its totality, this is an enormous action space that no enterprise would consider deploying altogether so it is more realistic to consider these techniques individually or in small combinations only	Chapter 7 [15] Chapter 8 [3]	
Moving Target Defense Quantification	In order to effectively use Moving Target Defenses through the application of control and/or game theory, it is necessary to quantify the methods, their effects, their costs as well as the situation picture the operating environment in which they operate. A variety of efforts have investigated both empirical and analytic techniques for such quantifications	Chapter 5 [1] Chapter 10 [42]	[18, 22, 41, 47]
Adaptive Cyber Defense Control Models and Techniques	Decisions about MTD deployment and operation that are made under worst-case and/or stationary operating conditions are typically modeled as control problems and therefore solvable by control techniques	Chapter 2 [35] Chapter 4 [25] Chapter 8 [3] Chapter 9 [4]	
Adaptive Cyber Defense Game Models and Techniques	Decisions about MTD deployment and operation that are made under operating conditions that are adversarial are typically modeled as game problems and therefore solvable by techniques used for solving game models	Chapter 3 [45] Chapter 6 [49]	[54]

3 Chapter Summaries

Chapter 1 - Overview of Control and Game Theory in Adaptive Cyber Defenses. This chapter is an introduction and overview of the structure and motivation for this book.

Chapter 2 - Control-Theoretic Approaches to Cyber Security. This chapter reviews control theoretic formulations of cyber security problems, focusing on state-based approaches and modeling of uncertainty.

Chapter 3 - Game-Theoretic Approaches to Cyber Security. This chapter reviews game theoretic formulations of cyber security problems, focusing on stochastic dynamic games and modeling of asymmetric information in such games.

Chapter 4 - Reinforcement Learning in Adaptive Cyber Defense. This chapter presents reinforcement learning approaches to solving certain control theoretic formulations of zero-day attack situations.

Chapter 5 - Moving Target Defense Quantification. In order to build and solve either control or game theoretic formulations of cyber security problems, it is necessary to quantify various aspects of the attack/defend engagement. This chapter presents a novel approach to such quantifications.

Chapter 6 - Empirical Game-Theoretic Methods. Empirical game theory does not start with a stylized .abstract model of an adversarial encounter, using simulations of such encounters to create increasingly more complex and accurate models and solutions to the underlying game.

Chapter 7 - Adaptive Cyber Defense Techniques for Memory Protection. This chapter describes several memory corruption cyber attacks and develops dynamic adaptive address space layout randomization (ASLR) approaches to defend against novel attacks.

Chapter 8 - Adaptive Cyber Defense Techniques for Botnet Detection and Mitigation. This chapter describes the botnet detection and mitigation problems together with adaptive cyber defense approaches to solving them using both control and game theoretic formulations.

Chapter 9 - Optimizing Alert Management Processes in Cyber Security. This chapter describes the cyber security alert management problem together with control theory based approaches to optimizing tasks and personnel assignments in Cyber Security Operations Centers (CSOC).

Chapter 10 - Online and Scalable Adaptive Cyber Security Defense. This chapter decribes problems related to the online state and parameter estimation and approximation required in certain adaptive defense techniques. The focus is on using recently developed so-called "sketching" techniques that allow approximating various structural and statistical properties of data streams using only limited storage and processing time.

Acknowledgements and Disclaimer. The work presented in this book was support by the Army Research Office under grant W911NF-13-1-0421. The authors of this book and other participants in the Adaptive Cyber Defense project are grateful for the direction and support of Dr. Clifford Wang (U.S. Army Research Office).

The views and opinions expressed in this book are those of the authors and do not necessarily reflect the official policy or position of any agency of the U.S. Government.

References

1. Albanese, M., Connell, W., Venkatesan, S., Cybenko, G.: Moving Target Defense Quantification (chap. 5). Springer, New York (2018)
2. Albanese, M., Huang, D.: MTD 2018: 5th ACM workshop on Moving Target Defense (MTD). In: Proceedings of the 2018 ACM SIGSAC Conference on Computer and Communications Security, pp. 2175–2176. ACM (2018)
3. Albanese, M., Jajodia, S., Venkatesan, S., Cybenko, G., Nguyen, T.: Adaptive Cyber Defenses for Botnet Detection and Mitigation (chap. 8). Springer, New York (2018)
4. Albanese, M., Jajodia, S., Venkatesan, S., Cybenko, G., Nguyen, T.: Adaptive Cyber Defenses for Botnet Detection and Mitigation (chap. 9). Springer, New York (2018)
5. Anderson, N., Mitchell, R., Chen, R.: Parameterizing moving target defenses. In: 2016 8th IFIP International Conference on New Technologies, Mobility and Security (NTMS), pp. 1–6. IEEE (2016)
6. Anderson, R., Moore, T.: The economics of information security. Science **314**(5799), 610–613 (2006)
7. Anderson, R.J.: Security Engineering: A Guide to Building Dependable Distributed Systems. Wiley, Hoboken (2010)
8. Angerman, W.S.: Coming full circle with Boyd's OODA loop ideas: an analysis of innovation diffusion and evolution. Technical report, Air Force Inst Of Tech Wright-Patterson AFB OH School of Engineering and Management (2004)
9. Barrantes, E.G., Ackley, D.H., Forrest, S., Stefanović, D.: Randomized instruction set emulation. ACM Trans. Inf. Syst. Secur. (TISSEC) **8**(1), 3–40 (2005)
10. Baudry, B., Monperrus, M.: The multiple facets of software diversity: Recent developments in year 2000 and beyond. ACM Comput. Surv. (CSUR) **48**(1), 16 (2015)
11. Bertsekas, D.P.: Dynamic Programming and Optimal Control, vol. 1. Athena Scientific Belmont, Belmont (2005)
12. Bhatt, S., Manadhata, P.K., Zomlot, L.: The operational role of security information and event management systems. IEEE Secur. Priv. **5**, 35–41 (2014)
13. Boyd, J.R.: The essence of winning and losing. Unpublished lecture notes 12(23), 123–125 (1996)
14. Cai, G.L., Wang, B.S., Hu, W., Wang, T.Z.: Moving target defense: state of the art and characteristics. Front. Inf. Technol. Electron. Eng. **17**(11), 1122–1153 (2016)
15. Chen, P., et al.: MTD Techniques for Memory Protection against Zero-Day Attacks (chap. 7). Springer, New York (2018)
16. Co, M., et al.: Double Helix and RAVEN: a system for cyber fault tolerance and recovery. In: Proceedings of the 11th Annual Cyber and Information Security Research Conference, p. 17. ACM (2016)
17. Collins, M.P.: A cost-based mechanism for evaluating the effectiveness of moving target defenses. In: Grossklags, J., Walrand, J. (eds.) GameSec 2012. LNCS, vol. 7638, pp. 221–233. Springer, Heidelberg (2012). https://doi.org/10.1007/978-3-642-34266-0_13

18. Connell, W., Albanese, M., Venkatesan, S.: A framework for moving target defense quantification. In: De Capitani di Vimercati, S., Martinelli, F. (eds.) SEC 2017. IAICT, vol. 502, pp. 124–138. Springer, Cham (2017). https://doi.org/10.1007/978-3-319-58469-0_9

19. Cox, B., et al.: N-variant systems: a secretless framework for security through diversity. In: USENIX Security Symposium, pp. 105–120 (2006)

20. Cybenko, G., Hughes, J.: No free lunch in cyber security. In: Proceedings of the First ACM Workshop on Moving Target Defense, pp. 1–12. ACM (2014)

21. Devanbu, P.T., Stubblebine, S.: Software engineering for security: a roadmap. In: Proceedings of the Conference on the Future of Software Engineering, pp. 227–239. ACM (2000)

22. Farris, K.A., Cybenko, G.: Quantification of moving target cyber defenses. In: Sensors, and Command, Control, Communications, and Intelligence (C3I) Technologies for Homeland Security, Defense, and Law Enforcement XIV, vol. 9456, p. 94560L. International Society for Optics and Photonics (2015)

23. Forrest, S., Somayaji, A., Ackley, D.H.: Building diverse computer systems. In: The Sixth Workshop on Hot Topics in Operating Systems, pp. 67–72. IEEE (1997)

24. Ganesan, R., Jajodia, S., Cam, H.: Optimal scheduling of cybersecurity analysts for minimizing risk. ACM Trans. Intell. Syst. Technol. (TIST) (TIST) 8(4), (2017). Article no. 52

25. Hu, Z., Chen, P., Zhu, M., Liu, P.: Reinforcement Learning for Adaptive Cyber Defense against Zero-day Attacks (chap). 4. Springer, New York (2018)

26. Jafarian, J.H., Al-Shaer, E., Duan, Q.: Openflow random host mutation: transparent moving target defense using software defined networking. In: Proceedings of the First Workshop on Hot Topics in Software Defined Networks, pp. 127–132. ACM (2012)

27. Jajodia, S., Ghosh, A.K., Swarup, V., Wang, C., Wang, X.S.: Moving Target Defense: Creating Asymmetric Uncertainty for Cyber Threats, vol. 54. Springer, Cham (2011)

28. Jajodia, S., Noel, S., Kalapa, P., Albanese, M., Williams, J.: Cauldron mission-centric cyber situational awareness with defense in depth. In: IEEE MILCOM, pp. 1339–1344 (2011)

29. Jung, G., Joshi, K.R., Hiltunen, M.A., Schlichting, R.D., Pu, C.: Performance and availability aware regeneration for cloud based multitier applications. In: 2010 IEEE/IFIP International Conference on Dependable Systems and Networks (DSN), pp. 497–506. IEEE (2010)

30. Larsen, P., Homescu, A., Brunthaler, S., Franz, M.: SoK: automated software diversity. In: 2014 IEEE Symposium on Security and Privacy (SP), pp. 276–291. IEEE (2014)

31. Lippmann, R., et al.: Validating and restoring defense in depth using attack graphs. In: IEEE MILCOM, pp. 1–10 (2006)

32. Lippmann, R., Webster, S., Stetson, D.: The effect of identifying vulnerabilities and patching software on the utility of network intrusion detection. In: Wespi, A., Vigna, G., Deri, L. (eds.) RAID 2002. LNCS, vol. 2516, pp. 307–326. Springer, Heidelberg (2002). https://doi.org/10.1007/3-540-36084-0_17

33. MacFarland, D.C., Shue, C.A.: The SDN shuffle: creating a moving-target defense using host-based software-defined networking. In: Proceedings of the Second ACM Workshop on Moving Target Defense, pp. 37–41. ACM (2015)

34. Marden, J.R., Shamma, J.S.: Game theory and control. Annu. Rev. Control Robot. Auton. Syst. 1, 105–134 (2018)

35. Miehling, E., Rasouli, M., Teneketzis, D.: Control-Theoretic Approaches to Dynamic Cyber Security (chap. 2). Springer, New York (2018)
36. Novikova, E., Kotenko, I.: Analytical visualization techniques for security information and event management. In: 2013 21st Euromicro International Conference on Parallel, Distributed and Network-Based Processing (PDP), pp. 519–525. IEEE (2013)
37. Okhravi, H., et al.: Survey of cyber moving target techniques. Techical report, Massachusetts Institute of Technology: Lexington Lincoln Lab (2013)
38. Okhravi, H., Riordan, J., Carter, K.: Quantitative evaluation of dynamic platform techniques as a defensive mechanism. In: Stavrou, A., Bos, H., Portokalidis, G. (eds.) RAID 2014. LNCS, vol. 8688, pp. 405–425. Springer, Cham (2014). https://doi.org/10.1007/978-3-319-11379-1_20
39. Pfleeger, C.P., Pfleeger, S.L., Theofanos, M.F.: A methodology for penetration testing. Comput. Secur. 8(7), 613–620 (1989)
40. Potter, B., McGraw, G.: Software security testing. IEEE Secur. Priv. 2(5), 81–85 (2004)
41. Priest, B.W., Vuksani, E., Wagner, N., Tello, B., Carter, K.M., Streilein, W.W.: Agent-based simulation in support of moving target cyber defense technology development and evaluation. In: Proceedings of the 18th Symposium on Communications & Networking, pp. 16–23. Society for Computer Simulation International (2015)
42. Priest, B.W., Cybenko, G., Liu, P., Singh, S., Albanese, M.: Online and Scalable Adaptive Cyber Defense (chap. 10). Springer, New York (2018)
43. Shacham, H., Page, M., Pfaff, B., Goh, E.J., Modadugu, N., Boneh, D.: On the effectiveness of address-space randomization. In: Proceedings of the 11th ACM Conference on Computer and Communications Security, pp. 298–307. ACM (2004)
44. Stamp, M.: Risks Monoculture. Communications of the ACM 47(3), 120 (2004)
45. Tavafoghi, H., Ouyang, Y., Teneketzis, D., Wellman, M.: Game Theoretic Approaches to Cyber Security: Challenges, Results and Open Problems (chap. 3). Springer, New York (2018)
46. Van Dijk, M., Juels, A., Oprea, A., Rivest, R.L.: Flipit: the game of "stealthy takeover". J. Cryptol. 26(4), 655–713 (2013)
47. Van Leeuwen, B., Stout, W.M., Urias, V.: Operational cost of deploying moving target defenses defensive work factors. In: Military Communications Conference, MILCOM 2015 – 2015 IEEE, pp. 966–971. IEEE (2015)
48. Ward, B.C., et al.: Survey of cyber moving targets, 2nd edn. Technical report, MIT Lincoln Laboratory Lexington United States (2018)
49. Wellman, M.P., Nguyen, T.H., Wright, M.: Empirical Game-Theoretic Methods for Adaptive Cyber-Defense (chap. 6). Springer, New York (2018)
50. Wellman, M.P., Prakash, A.: Empirical game-theoretic analysis of an adaptive cyber-defense scenario (preliminary report). In: Poovendran, R., Saad, W. (eds.) GameSec 2014. LNCS, vol. 8840, pp. 43–58. Springer, Cham (2014). https://doi.org/10.1007/978-3-319-12601-2_3
51. Winston, W.L., Goldberg, J.B.: Operations Research: Applications and Algorithms, vol. 3. Thomson Brooks/Cole, Belmont (2004)
52. Xu, J., Guo, P., Zhao, M., Erbacher, R.F., Zhu, M., Liu, P.: Comparing different moving target defense techniques. In: Proceedings of the First ACM Workshop on Moving Target Defense, pp. 97–107. ACM (2014)

53. Zhang, M., Wang, L., Jajodia, S., Singhal, A., Albanese, M.: Network diversity: a security metric for evaluating the resilience of networks against zero-day attacks. IEEE Trans. Inf. Forensics Secur. **11**(5), 1071–1086 (2016)
54. Zhu, Q., Başar, T.: Game-theoretic approach to feedback-driven multi-stage moving target defense. In: Das, S.K., Nita-Rotaru, C., Kantarcioglu, M. (eds.) GameSec 2013. LNCS, vol. 8252, pp. 246–263. Springer, Cham (2013). https://doi.org/10. 1007/978-3-319-02786-9_15

Control-Theoretic Approaches
to Cyber-Security

Erik Miehling[1]([⊠]), Mohammad Rasouli[2], and Demosthenis Teneketzis[3]

[1] Coordinated Science Lab, University of Illinois at Urbana–Champaign,
Urbana, IL 61801, USA
miehling@illinois.edu
[2] Department of Management Science & Engineering,
Stanford University, Stanford, CA 94305, USA
rasoulim@stanford.edu
[3] Department of Electrical Engineering & Computer Science,
University of Michigan, Ann Arbor, MI 48109, USA
teneket@umich.edu

Abstract. In this chapter, we discuss the control-theoretic approach to cyber-security. Under the control-theoretic approach, the defender prescribes defense actions in response to security alert information that is generated as the attacker progresses through the network. This feedback information is inherently noisy, resulting in the defender being uncertain of the underlying status of the network. Two complementary approaches for handling the defender's uncertainty are discussed. First, we consider the probabilistic case where the defender's uncertainty can be quantified by probability distributions. In this setting, the defender aims to specify defense actions that minimize the expected loss. Second, we study the nondeterministic case where the defender is unable to reason about the relative likelihood of events. The appropriate performance criterion in this setting is minimization of the worst-case damage (minmax). The probabilistic approach gives rise to efficient computational procedures (namely sampling-based approaches) for finding an optimal defense policy, but requires modeling assumptions that may be difficult to justify in real-world cyber-security settings. On the other hand, the nondeterministic approach reduces the modeling burden but results in a significantly harder computational problem.

1 Introduction

The field of control theory studies how one can make a sequence of decisions in order to most efficiently guide, or *control*, a system toward a specified objective subject to some uncertainty regarding the system's evolution. Some examples of problems addressed by control theory include maintaining a system's output at a

This work is partially supported by the Army Research Office under grant W911NF-13-1-0421.

S. Jajodia et al. (Eds.): Adaptive Cyber Defense, LNCS 11830, pp. 12–28, 2019.
https://doi.org/10.1007/978-3-030-30719-6_2

desired set-point in the presence of external disturbances, *e.g.*, an aircraft autopilot system responsible for maintaining speed and altitude in varying weather conditions, or tracking a path or trajectory subject to measurement noise and estimation errors, *e.g.*, an autonomous vehicle's road following algorithm tasked with translating noisy measurements from multiple sensors into real-time steering, acceleration, and braking decisions. Depending on the control environment, the information available for making decisions can take different forms. In some settings, the current status of the system is directly observable and can be used in the decision making process. In others, the uncertainty is not only due to the effect of the control action on the evolution of the system, but also includes the inability to perfectly observe the system's status, requiring control decisions to be made based on noisy observations or measurements. In either setting, sequential control decisions must be made based on new, potentially noisy, information that is revealed as the problem evolves. The precise topic that control theory addresses is the nature of this *feedback loop* – the influence of control decisions on the observable output and the dependency of revealed information on the choice of subsequent control actions – with the end goal of prescribing *optimal* control actions, that is, those that achieve the objective at the lowest operational cost.

In this chapter, we study the role of control theory in cyber-security. In particular, we focus on the (dynamic) defense problem: how a defender can prescribe actions in real-time as a function of a stream of intrusion information in order to interfere with, and potentially mitigate, attacks carried out by one or more adversaries.[1] It is worth emphasizing the defining characteristic of control theory, namely the one-sided nature of the decision-making process. As such, the control-theoretic approach studied in this chapter considers the defender as the *only active decision-maker* in the system.[2] All other decision-making processes that may be present in the system, *e.g.*, actions of the attacker(s) or the behavior of trusted (non-malicious) users, are abstracted into the model of the cyber environment. The one-sided nature of the control theoretic approach is in contrast with the two-sided (or in general, many-sided) decision making environment of game theory which consists of many agents, each possessing different information and (at least partially) conflicting objectives. Game-theoretic tools, specifically how they can be used to address the problems in cyber-security, are discussed in-depth in Chap. 3. While modeling the cyber-security problem as a control problem is an approximation of the true problem, it is a valuable first step for addressing the full complexity of the game-theoretic approach. Indeed, many of the challenges of the cyber-security problem present in the control-theoretic approach also exist in the game-theoretic approach.

---

[1] Such systems are referred to as *intrusion response systems* in the cyber-security literature; see [1] for a review of the area.

[2] In some control settings, the "decision maker" may actually consist of a collection of agents making decisions based on their own localized information in order to achieve some common objective. Such problems still fall within the realm of control theory, due to all agents having an identical objective, but are referred to as *decentralized control problems* or *team problems* [2,3].

The defense problem presents many challenging requirements from both modeling and computational perspectives. The problem is inherently dynamic, evolving over time as a function of the defender's actions and (potentially unobservable) events from the cyber environment. New information is continuously revealed to the defender as the problem evolves, all of which, in general, must be used in the defender's decision making process. The model for the cyber environment, termed the *threat model*, must be sufficiently expressive to describe the complex nature of attacks. In particular, attacks are *progressive*, consisting of multiple stages and involving the combination of many vulnerabilities across multiple network elements, and *persistent*, with attackers continuing to attempt to fulfill their objective, using various attack pathways, until they are successful. The defender, in its attempts to interfere with or mitigate attacks, must be aware of the conflicting effects of its defense decisions on the system. It is faced with an unavoidable tradeoff between security and availability; performing system modifications that lower an attack's chance of success also interfere with the normal functionality and usability of the system by trusted users. Beyond modeling challenges, the defense problem presents significant challenges from a computational perspective. The systems that are targeted by cyber attacks are large-scale, consisting of many hosts, each containing a wide-range of software and operated by a large collection of users. Reasoning about all possible ways such systems can be attacked often leads to a combinatorial explosion in complexity. As a result, scalable algorithms must be developed, often requiring approximations or novel solution techniques (such as sampling methods or system decompositions). One must also ensure that algorithms are able to meet the strict timing requirements of the system by prescribing defense decisions quickly. Oftentimes, defense decisions have a limited window of usefulness; prescribing a defense decision too late can be as ineffective as taking no action at all.

The tools offered by control theory are a natural fit for addressing the aforementioned requirements. First, quantifying the status of the system through assignment of a state allows one to formally describe the evolution of the system's level of security as a function of the defense actions and events from the cyber environment (*e.g.*, the description of the threat model). Furthermore, under the state-based approach, one can define an appropriate cost structure (costs for states and actions) that captures the desired tradeoff between security and availability. Defending the system then amounts to determining actions that ensure the system stays out of undesirable (high-cost) regions of the state space. In general, the defender's decisions must be made based on all available information. The notion of an *information state* from control theory allows for a compression of the available information into a summary that is sufficient for making optimal decisions. Once an appropriate information state for the problem is identified, one can cast the problem of determining the optimal defense policy (the sequence of functions mapping the information state to actions) as a set of sequential optimization problems (via dynamic programming). Computational concerns can then be more directly addressed by investigating approximations to the dynamic programming recursion, leading to approximately optimal defense policies.

In what follows, we discuss the philosophy behind the control theoretic app-roach to cyber-security. First, in Sect. 2, we describe the assignment of a state to quantify the level of security of the system and how this state evolves as a function of the defender's actions and the events from the cyber environment. The defender's lack of perfect information regarding the state, and how this is addressed, is discussed in Sect. 3. Section 4 introduces the notion of defense poli-cies and the computational procedure for obtaining them. Section 5 provides two model instances of the general control-theoretic approach, differing primarily in the assumed nature of the uncertainty in the problem (probabilistic vs. non-deterministic). The general idea of each approach is described, as well as each model's benefits and drawbacks. Concluding remarks are provided in Sect. 6.

2   The State-Based Approach to Cyber-Security

At the heart of any control problem is the notion of a state. The state describes the current operating status of the system, quantifying how the system reacts to the control input and events from the environment, and influencing how the control translates to the observable output. Viewing cyber-security as a control problem first requires that one defines a state that accurately quantifies the level of security of the system. To this end, the state, denoted by $x_t \in \mathscr{S}$ at any given time t, should reflect some aspect of the attacker's current capabilities. For example, the state could represent the permissions that the attacker possesses or its progress (in terms of compromised hosts) toward reaching a specific target host. In Sect. 5, we will define the state in the context of two formal security models; for the current discussion, however, consider the state to be abstract representation of the system's security level.

The next ingredient in the control-theoretic description of cyber-security is the specification of the control, that is, the defender's actions. Defense actions can take a wide variety of forms. One class of such actions is patches. A patch for a vulnerability renders the corresponding exploit(s) ineffective, offering an effec-tive strategy for hardening the system and interfering with the attacker's goal. Unfortunately, the time between discovery of the vulnerability and the instal-lation of a patch, termed the *vulnerability exposure window*, can be upwards of five months [4].[3] As a result, relying solely on patches would inevitably allow systems to be operational while exposed to vulnerabilities. Alternative defen-sive measures that operate on faster time-scale than patches are needed. The defense actions we consider throughout this chapter use known vulnerabilities and security alert information to actively interfere with the attacker's progres-sion. Specifically, a defense action at time t, denoted by $u_t \in \mathscr{A}$, corresponds to system modifications that directly influence the ability of the attacker to induce a state transition, $x_t \to x_{t+1}$. For example, a defense action may disable the precondition of an exploit (such as connectivity between two hosts via a specific port) in order to block the attacker from using the exploit. While not a

---

[3] For a deeper discussion of this issue, see the related topic of vulnerability disclosure policies [5].

permanent solution, these defense strategies can be effective for interrupting an attack, buying useful time for forensic analysis and the development of a patch.

Defense decisions are made based on the predicted evolution of the state under various defense actions. In order to carry out such a prediction, one needs a model for the attacker. The concept of *threat modeling* [6] from the computer security community addresses precisely this task. Informally, threat models describe what the attacker can do given its current capabilities. More specifically, a threat model describes the various ways in which an attacker can infiltrate the system (attack vectors/pathways), what resources it finds valuable (the attacker's objectives), and what sort of security information is generated/detected during an attack (*e.g.*, via intrusion logs). In the context of the control-theoretic approach of this chapter, the threat model describes how the state evolves as a function of defense actions and events from the cyber environment, as well as what observations are generated during this evolution. For example, given a set of attacker capabilities (quantified by the current state) the threat model serves to define what exploits the attacker can attempt and, given the defense action, an updated set of attacker capabilities and any security alerts that may have been generated during the attempt of the exploits. It is important to note that while the control theoretic approach requires a well-defined threat model, it need not be completely known *a priori*. Simultaneous learning of the model and control of the system based on feedback information still falls within the realm of control theory (termed *adaptive control* [7] and *reinforcement learning* [8]).

While the defender's primary objective is to prevent the attacker from reaching its goals, it must also consider the effect of its defense actions on the normal operation of the system. Defenses that are most effective at interfering with the attacker also tend to be most disruptive to the normal operation of the system (*e.g.*, shutting down the email server to block phishing emails). On the other hand, prioritizing system availability unavoidably preserves attack pathways. In short, keeping the attacker away from its goals is largely in conflict with maintaining availability. Quantifying this tradeoff is achieved by assigning costs to both states and defense actions, via a cost function $c(x_t, u_t)$. High costs should be assigned to undesirable states, *e.g.*, the attacker possessing root access on a critical host, as well as to actions that significantly limit availability. Using the threat model, the defender can reason about costs of state-action trajectories which in turn guide the selection of defense actions that achieve the desired security-availability tradeoff.

3   The Defender's Information

A fundamental aspect of the dynamic defense problem is that the defender cannot perfectly observe the attacker's activity, *i.e.*, the events from the cyber environment. Instead the defender receives observations, denoted by $y_t \in \mathscr{O}$, generated as a function of the underlying events. The monitoring devices that generate the observations are inherently noisy. For instance, intrusion detection systems suffer from both missed detections (generating no security alerts when something malicious has occurred) and false alarms (triggering security alerts in the absence of malicious behavior). As a consequence of this imperfect detection, the defender has uncertainty over the true state of the system.

The control-theoretic concepts of *information structure* and *information state*[4] allow one to formalize the defender's lack of perfect information regarding the true state. The information structure of a problem is a formal description of the phrase "who knows what about the system and when" [10]. Under the centralized control theoretic approach of this chapter, the information structure of the problem has a straightforward interpretation; it simply describes the set of variables that the defender knows at any given time. Throughout the chapter, it is assumed that the information structure satisfies *perfect recall*, that is, the defender remembers all of its past observations and defense decisions. In other words, at time t the defender has access to the history $h_t = (u_0, y_1, \ldots, u_{t-1}, y_t) \in (\mathscr{A} \times \mathcal{O})^t$. Given that there is only one decision-maker, the problem is said to have a strictly classical information structure [11]. This allows one to compress the history into a summary, termed an *information state* and denoted by I_t, that has a time-invariant domain $\mathscr{I}$ [10]. The information state is sufficient for making optimal decisions, *i.e.*, basing decisions on the information state, rather than the whole history, is without loss of optimality. Treating the information state as the state of the problem, one can formulate a completely observable decision problem that admits a dynamic programming decomposition. The evolution of the information state is dictated by the new information that is revealed as time progresses (defense actions and observations).

4   Computation of Defense Policies

The defense action at any given time is computed as a function of the defender's current information (given by the information state). Formally, the translation from information states to defense actions is specified by a *defense policy*, denoted by $g = (g_0, g_1, \ldots, g_{T-1})$, where T is the decision horizon (the finite horizon case will be considered in this chapter; however T can also be infinite) and each g_t is a function from the given information state I_t to a distribution over defense actions, that is, $g_t : \mathscr{I} \to \Delta(\mathscr{A})$. Determining the *best* defense policy depends on the defender's model for how events are generated (*i.e.*, how the attacker chooses its actions). As will be discussed in more detail in Sect. 5, the assumed nature of uncertainty in the problem dictates the cost criterion for the problem. For example, if uncertainty is quantified by probability distributions and the defender is risk-neutral, the defender's objective may be to minimize the total expected cost. On the other hand, under nondeterministic uncertainty, an appropriate criterion would be to minimize the worst-case cost, termed the *minmax* criterion. The best defense policy, termed an *optimal* defense policy

---

[4] For a deeper discussion of information structures and information states, see [9].

denoted by $g^* = (g_0^*, g_1^*, \ldots, g_{T-1}^*)$, is a policy that minimizes the corresponding cost criterion.

In general, each defense action has a long-run impact on the evolution of the system. As such, defense decisions cannot be made in isolation; one must balance immediate costs with future costs to ensure that early defense decisions don't result in the system ending up in an undesirable or vulnerable state. Reasoning about sequences of actions is a computationally formidable task, especially when the time horizon, T, is long. Fortunately, results from control theory allow one to sequentially decompose the long-run optimization problem into a collection of simpler subproblems. The sequential decomposition, known as dynamic programming, relies on a concept known as the *principle of optimality* [12,13]. A problem is said to satisfy the principle of optimality if, given a sequence of optimal control actions from time t onward, the remainder of the action sequence from $t + 1$ onward will still be optimal for the problem that starts from the state resulting from the action taken at t. The cost of the remainder of the action sequence from a given state, termed the cost-to-go, is captured by defining a *value function*. The value function represents the best that one can do from the given state. The resulting recursive expression, termed the Bellman (or dynamic programming) equation is solved in the finite horizon case by starting from the final decision time and working backwards, a process termed backward induction. In the infinite horizon case, one must solve a fixed point equation [13]. The optimal policies are recovered from the value functions by finding the action, for a given state, that minimizes the cost criterion.

Dynamic programming is the predominant approach for solving centralized control problems (and thus the dynamic defense problem studied in this chapter); however, it suffers from computational challenges as the problem size grows. The main challenge arises from the need to compute and store the value functions for every possible state. As the state space grows, this procedure becomes increasingly burdensome (referred to as the *curse of dimensionality*). Due to the very large state space in many cyber-security settings, the curse of dimensionality becomes a significant issue for the dynamic defense problem. This problem is further compounded by the fact that the defender possesses imperfect information of the state; the domain of the value functions is thus the set of information states $\mathscr{I}$, an uncountably infinite space. These challenges preclude the computation of optimal actions for every possible (information) state. One must resort to approximations of the dynamic programming recursion, resulting in approximately optimal defense policies. As will be illustrated in the following section, the information state of the problem provides guidance for an appropriate approximation, allowing for scalable and fast computation without significantly impacting decision quality.

5   Some Models from the Literature

There is a large body of research concerning the design of systems that prescribe automated defenses based on real-time intrusion information. Such systems are referred to by various names in the literature: automated intrusion response systems [14], autonomic & self-protecting systems [15,16], and survivable systems [17,18], among others [19,20]. The seminal work of [18] was the first to investigate the design of such a system from a formal, control-theoretic perspective. More recent work has taken a similar approach, developing control-theoretic automated defense systems for completely observable [15,16,20] and partially observable settings [1,19,21–23].

This section will focus on the partially observable setting. In particular, we investigate two complementary approaches to modeling the defender's uncertainty: (1) probabilistic uncertainty, and (2) nondeterministic uncertainty. Probabilistic uncertainty quantifies all uncertainty in the problem via probability distributions. For instance, under a given defense action, the transition from one state to another is assumed to be dictated by probabilities. The second approach, nondeterministic uncertainty, considers a more coarse form of uncertainty where one only knows the possible events and not their specific probabilities. For each setting, the general decision environment and form of the information state is described. To aid in exposition, we draw upon two existing models developed in the literature, namely [1,21] for the probabilistic approach and [22,23] for the nondeterministic approach. In both cases, solving for an optimal defense policy is intractable, requiring solution techniques that yield approximate defense policies. Each section concludes with a general discussion of the benefits and drawbacks of the respective modeling approach.

5.1 Probabilistic Uncertainty

The first approach assumes that the nature of the defender's uncertainty is probabilistic. Under probabilistic uncertainty, the state transitions and the generation of observations are assumed to be dictated by probability distributions. In particular, the state dynamics follow a controlled Markov chain[5] where the control is the defender's action, as illustrated by Fig. 1.

An implicit assumption in this setting is that the underlying distributions characterize, as a function of the defense action, all uncertainty associated with the attacker's behavior. In particular, given a current state $x_t = s_i$ and a defense action $u_t = a$, the transition to the next state $x_{t+1} = s_j$ is given by a fixed conditional probability $p_{ij}^a = \mathbb{P}(X_{t+1} = s_j \mid X_t = s_i, U_t = a)$.[6] Further, given a successor state $x_{t+1} = s_j$ and an action $u_t = a$, an observation $y_{t+1} = o_k$ is generated according to the conditional probability $r_{jk}^a = \mathbb{P}(Y_{t+1} = o_k \mid X_{t+1} = s_j, U_t = a)$.

The above described model is known in the literature as a *partially observable Markov decision process* (POMDP). It is well known that the information state in a POMDP is the conditional probability measure, that is, the probability mass function on the state space $\mathscr{S}$ conditioned on the history

---

[5] This is a special case of a general probabilistic automaton where the dynamics are assumed to be Markovian.

[6] The uppercase notation, X_t, is used to represent a random variable.

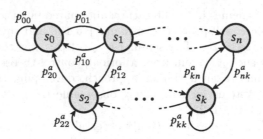

Fig. 1. Under the assumption of probabilistic uncertainty, the state dynamics evolve probabilistically. We represent this evolution as a controlled Markov chain where the control is the defense action.

$h_t = (u_0, y_1, \ldots, u_{t-1}, y_t)$ [24]. The information state (also referred to as the *belief state*) is denoted by $I_t = \pi_t \in \Pi = \Delta(\mathscr{S})$ where $\Delta(\mathscr{S})$ is the probability simplex on the state space (the space of all probability mass functions on $\mathscr{S}$). The belief state π is updated via Bayes rule, as a function of the new information $(u_t, y_{t+1}) = (a, o)$, to $\pi' = (\tau_1(\pi, a, o), \ldots, \tau_n(\pi, a, o))$ where $\tau_j(\pi, a, o) = \sum_i \pi_i p_{ij}^a r_{jk}^a / \sum_i \sum_j \pi_i p_{ij}^a r_{jk}^a$.

Under probabilistic uncertainty, an appropriate performance metric is that of total expected discounted cost. The cost for a given defense policy $g : \Delta(\mathscr{X}) \to \Delta(\mathscr{A})$ is defined as

$$C(g) = \mathbb{E} \left[\sum_{t=0}^{T-1} \beta^t c(x_t, u_t) + \beta^T c(x_T) \right].$$

where $c(\cdot, \cdot)$ is the state-action cost, $c(\cdot)$ is the terminal cost (that only depends on the final state), and $\beta \in [0, 1)$ is a *discount factor* which serves to place more weight on immediate costs compared to later costs. The expectation above is taken with respect to the joint probability distribution on trajectories $(x_0, u_0, \ldots, x_{T-1}, u_{T-1}, x_T)$ as a result of defense policy g. An optimal defense policy g^* is one that minimizes the total expected discounted cost $C(g)$, that is, $g^* = \inf_g C(g)$. Recalling the discussion of Sect. 4, optimal defense policies are computed from the value function. The value function in the probabilistic uncertainty case is defined on the space of beliefs $\Delta(\mathscr{S})$ and is denoted by $V : \Delta(\mathscr{S}) \to \mathbb{R}$. Using the likelihoods encoded by the belief and the probabilities described by the model, one can write the dynamic programming equations, for every $\pi \in \Delta(\mathscr{S})$ and $t = 0, \ldots, T - 1$, as

$$V_t^*(\pi) = \min_{a \in \mathscr{A}} \mathbb{E} \left[c(x, a) + \beta V_{t+1}^*(\tau(\pi, a, y)) \right]$$

$$= \min_{a \in \mathscr{A}} \left\{ \sum_i \pi_i c(s_i, a) + \beta \sum_k \sum_j \sum_i \pi_i p_{ij}^a r_{jk}^a V_{t+1}^*(\tau(\pi, a, o_k)) \right] \right\}$$

with terminal value function $V_T^* = \mathbb{E}\big[c(x)\big]$. The solution of the above equations can, in principle, be obtained via a recursive computational procedure (*i.e.*, value iteration) which, in turn, yields a corresponding optimal defense policy g^*. Unfortunately, due to the scale of real-world cyber-security problems, one must resort to approximate procedures, as will be described later.

To provide context for the probabilistic approach, we review a model from the literature. The automated intrusion response system, developed in [1, 21], models how a defender can optimally interfere with the progression of an adversary through a computer network. The progression of the attacker is described by a directed acyclic graph, termed an *attack graph*, that encodes the relationships between exploit preconditions (attacker capabilities that are needed to attempt the exploit) and postconditions (attacker capabilities that are realized upon success of the exploit). The state of the system at any given time is the set of currently enabled conditions. As the attacker attempts exploits and moves through the network, alerts are triggered via an intrusion detection system. The defender uses these noisy security alerts to construct a belief of the currently enabled conditions. Using the belief, the defender prescribes actions that induce system modifications that block exploits from being carried out. While these system modifications interfere with the progression of the attacker (the evolution of the state), they are also costly, requiring the defender to tradeoff between interfering with the attacker's progression and maintaining system availability.

The novelty of the model developed in [1, 21] is the use of attack graphs to model the active progression of an attacker through a network. Prior work primarily considered attack graphs in the context of offline vulnerability analysis, *e.g.*, determining the minimum number of exploits to patch in order to maximize the number of blocked attack paths [25]. Introducing a state and using the attack graph to model the dynamics of the state process enables one to build a control (defense) problem and compute defense policies that optimally interfere with an attack as it is unfolding.

While one can write down the dynamic programming equations that characterize an optimal policy, offline computation for every possible belief that may be encountered during runtime is intractable. This is primarily due to the scale of real-world attack graphs and the size of the resulting state space. To avoid this challenge (termed the *curse of dimensionality*) we take advantage of the fact that the defender's uncertainty is described by probability distributions. In particular, the defender is able to forecast future possible attack pathways (chains of exploits) by sampling from the model's distributions. By conditioning on its current belief of the attacker's capabilities, the defender can reason about the likelihood (and expected costs) of various state trajectories under different defense actions. This allows the defender to prescribe defense actions that guide the system to low cost regions of the state space and reach outcomes that balance between security and availability. Such an approach is termed an online defense algorithm [1, 26] since one is only concerned with prescribing actions from the current (belief) state. While the online defense algorithm requires one to continue to perform computation during runtime, it is much more scalable than offline approaches, yielding good quality defense policies in large domains. Additional details of the algorithm can be found in [1].

The benefit of taking a probabilistic approach to the cyber-security problem is primarily computational. Quantifying uncertainty via probability distributions enables the application of scalable computational procedures for computing defense policies. In particular, (provably convergent) solutions techniques based on sampling can readily be applied [1, 26]. However, the probabilistic approach raises some concerns in real-world cyber-security settings. The primary concern is the specification of accurate probability parameters for describing the attacker's behavior. The usual justification for knowing the parameters in a general stochastic control setting is that one has learned them from existing data and previous runs of the problem. This is difficult to justify in the context of cyber-security: attacks are targeted and rarely repeated, leading to sparsity of useful attack data. That said, it is not necessary to specify accurate probabilities for the model to have value. The models can still provide useful qualitative insights that are not sensitive to the specific parameter values. For example, the sampling approach can identify, and focus defensive resources on, structural bottlenecks in the attack graph [1]. These structural properties of the problem are largely independent of the specific probability values. A secondary concern is the question of whether the assumed probabilities are informative for future evolution of the system. This requires that the statistics dictating the attacker's behavior do not change in time, *e.g.*, they are stationary, an assumption which may be difficult to justify in practice. This issue will be discussed in more detail in the following section.

One approach for addressing the above concerns is to consider a *set* of models, as is done in [1]. Consideration of a set of models allows one to capture a wide-range of attacker behavior by not only updating the estimate of the attacker's evolution, but also the estimate of the true model. However, considering a large set of models further compounds computational difficulties. The appropriate tradeoff between model expressiveness and computational tractability will likely be guided by the requirements of specific security settings and deserves further research.

5.2   Nondeterministic Uncertainty

The probabilistic approach discussed in Sect. 5.1 is not the only way to reason about uncertainty [27]. A more coarse description of uncertainty, termed *nondeterministic uncertainty*, places no assumptions on how events from the cyber environment are generated. Under nondeterminism, one cannot reason about the probability of events and thus cannot construct likelihoods of individual states. One can only reason about the set of possible states that are consistent with the available information [28–30].[7] Due to the lack of probabilities, the defender cannot differentiate between the set of possible states in the support. As a result,

---

[7] Comparing to the probabilistic approach, one would only keep track of the support of the distribution and not the likelihoods.

the defender adopts a worst-case cost criterion. Assuming the worst-case can be interpreted as the defender preparing for the attacker to perform the most damaging action (or even further, taking the most conservative viewpoint by assuming that the attacker is omniscient and is able to compute and execute this action). Throughout the discussion, we refer to the attacker as *nature*. We adopt this terminology since we are not modeling an explicit strategy for the attacker.[8] Throughout this section, a general model from the literature will be described along with a discussion of its application to a specialized cyber-security model [22, 23].

The general system model consists of a finite set of states, $\mathscr{S} = \{s_0, s_1, \ldots, s_n\}$, where the state transition $x_t \to x_{t+1}$ is due to both the defender's action, $u_t \in \mathscr{A}$, and the event, $w_t \in \mathscr{E}$, from the cyber environment. Formally, we describe the dynamical system as a nondeterministic finite automaton (NFA). For any given state, the transition due to an action-event pair $(u_t, w_t) = (a, e) \in \mathscr{A} \times \mathscr{E}$ is in general nondeterministic, that is, the state may transition to one of a set of states, as illustrated by Fig. 2. The distinguishing feature of the nondeterministic case compared to the probabilistic case is that, in the latter, the defender cannot reason about the relative likelihood of transitioning to various successor states and must treat all successor states, from a given state, as possible.

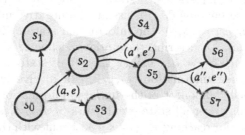

Fig. 2. Under nondeterministic uncertainty, the state dynamics are modeled by a nondeterministic finite automaton. For a given action-event pair, (a, e), state transitions are nondeterministic, meaning a given state can transition to one of a collection of states.

The state dynamics encoded by the nondeterministic finite automaton are described by the function $f_t : \mathscr{S} \times \mathscr{A} \times \mathscr{E} \to \mathscr{S}$, that is, given an action-event pair $(u_t, w_t) = (a, e)$ the state $x_t = s$ follows the update $x_{t+1} = f_t(s, a, e)$. The defender does not perfectly observe the state or nature's events. Instead, it receives an observation y_t generated as a function of the true underlying state

---

[8] Such settings are sometimes referred to as *games against nature* in the literature [31]; however, since no strategy is assumed for the attacker (nature), it is not viewed as an active decision maker, and thus we view the problem in the context of control theory.

and the event, as described by the function $l_t : \mathscr{S} \times \mathscr{E} \to \mathscr{O}$. A slightly more general cost function is considered in this section, namely one that depends on nature's event in addition to the state-action pair, that is, $c : \mathscr{S} \times \mathscr{A} \times \mathscr{E} \to \mathbb{R}_+$. It is assumed that cost functions are bounded above by $\bar{c} < \infty$, that is, $c(s, a, e), c(s) \le \bar{c}$ for all $(s, a, e) \in \mathscr{S} \times \mathscr{A} \times \mathscr{E}$. Define $\mathscr{C} = \left[0, \frac{1-\beta^{T+1}}{1-\beta}\bar{c}\right]$ as the possible range of cumulative costs accrued over the duration of the problem.[9]

The problem of decision-making under nondeterministic uncertainty has been extensively studied in the literature. Early work, see [32–36], has established a duality between probabilistic and nondeterministic uncertainty, proposing *cost measures*, *cost densities*, and *feared values* as analogous concepts to probability measures, probability densities, and expected values. Further connections to robust control and game theory have been established in [34,37]. As illustrated in the literature, the relevant notion in the nondeterministic case is that of cost, rather than probabilities. In particular, one should base control decisions on the (worst-case) cost for reaching each state as opposed to reasoning about their likelihoods.

An appropriate information state in this setting is the *worst-case cost-to-come* statistic of [34,36,37], denoted by $I_t = \theta_t \in \Theta = (\mathscr{C} \cup \{-\infty\})^n$, defined as the maximum possible cost for reaching each state given the current history. That is, for any given time t, the information state θ_t consists of a collection of costs, one for each state, $\theta_t = \{\theta_t(s)\}_{s \in \mathscr{S}}$, where each $\theta_t(s)$ is defined as the maximum cost for reaching state s. If state s is not consistent with the current history then the corresponding $\theta_t(s)$ is assigned a negative infinite value. Given new information $(u_t, y_t) = (a, o)$, the information state θ_t is updated via the rule $\theta_{t+1} = \mu(\theta_t, a, o)$. To describe the update, define $\Omega(s', a, o) := \{(s, e) \in \mathscr{S} \times \mathscr{E} \mid s' = f_t(s, a, e), o = l_t(s, e)\}$ as the set of state-event pairs that are consistent with the new information $(u_t, y_t) = (a, o)$ given that the system is in state $s' \in \mathscr{S}$. Each component of the updated information state, $\theta_{t+1}(s')$, is computed by searching over all state-event pairs $(s, e) \in \Omega(s', a, o)$ in order to find the maximal cost for reaching $x_{t+1} = s'$ consistent with the new information $(u_t, y_t) = (a, o)$. Further details of the information state update in a general setting can be found in Chap. 6 of [37] and Sect. 3 of [36], as well as in the context of cyber-security in [23].

Since one does not have access to probability distributions in the nondeterministic setting, the notion of expected value is no longer relevant. An appropriate cost criterion in this setting is minimization of the worst-case cost. The worst-case cost for a given defense policy $g : (\mathscr{C} \cup \{-\infty\})^n \to \Delta(\mathscr{A})$ is

$$D(g) = \max_z \left[\sum_{t=0}^{T-1} \beta^t c(x_t, u_t, w_t) + \beta^T c(x_T) \right]$$

where the maximization is taken over all feasible trajectories of the form $z = (x_0, u_0, w_0, \dots, x_{T-1}, u_{T-1}, w_{T-1}, x_T)$ as a result of defense policy g. An optimal defense policy is one that minimizes the worst-case cost $D(g)$, that is, $g^* =$

---

[9] For the infinite horizon case, $\mathscr{C} = \left[0, \frac{\bar{c}}{1-\beta}\right]$.

$\inf_g D(g)$. As before, one can construct a dynamic programming recursion on the space of information states in order to recursively compute a value function, denoted by $W : (\mathscr{C} \cup \{-\infty\})^n \to \mathbb{R}$, and a corresponding optimal policy. Defining $\mathscr{O}_t$ as the set of observations that are consistent with the current information at time t,[10] one can write the dynamic programming equations, for each $\theta \in (\mathscr{C} \cup \{-\infty\})^n$ and $t = 0, \ldots, T-1$, as

$$W_t^*(\theta) = \min_{a \in \mathscr{A}} \max_{o \in \mathscr{O}_t} \left[W_{t+1}^*(\mu(\theta, a, o)) \right]$$

with terminal value function $W_T^*(\theta) = \max_{s \in \mathscr{S}} [c(s) + \theta(s)]$. Note that, unlike in the probabilistic case of Sect. 5.1, the cost function is embedded within the definition of the information state itself and does not explicitly appear in the dynamic programming equations.

The computational challenges are more pronounced in the nondeterministic case compared to the probabilistic case. The two main challenges are: (i) complexity of maintaining the information state θ_t, and (ii) solving the dynamic programming equations. To address the first challenge, the model of [22,23] considered a simplified information state in which one only keeps track of the set of states consistent with the current history. That is, the information state θ_t is approximated by the set of states s that have a finite $\theta_t(s)$. The simplification leads to a much simpler information state but comes at the cost of optimality. The second challenge, solving the dynamic programming equations, cannot be addressed by the sampling-based approach outlined in the previous subsection (since we do not have access to probability distributions). Instead, the problem is approximated by (spatially) decomposing the system into a collection of sub-systems. By analyzing the functional dependencies between the state components, one can construct a graph that quantifies the strength of the coupling between states. One can then apply clustering algorithms to partition the graph into sub-systems, each associated with a local defense policy. Allowing defense policies to communicate the necessary security information via messages, the computation of the defense policy can be decomposed into the computation of multiple local defense policies. This improves scalability and permits computation in some moderately-sized settings. Additional details of the decomposition approach can be found in [23].

The main benefit of taking a nondeterministic approach to the defender's uncertainty is the increased modeling flexibility. Reasoning over possible transitions and computing the worst-case includes a wide-range of attacker strategies, even non-stationary behavior. Furthermore, the modeling task is greatly simplified, compared to the probabilistic approach, as one does not need to make claims about which states are more or less likely to be realized. The nondeterministic approach does come with some drawbacks. The main issue is computational – even after the simplification of the information state to describe the set of consistent states, the space of approximate information states is the power set of the state space and thus scales poorly. Furthermore, the defense policies computed under the minmax approach can be overly conservative. Indeed, always

---

[10] In other words, $\mathscr{O}_t$ is the range of the functions $w \mapsto l_t(x_t, w)$.

assuming the worst possible state transitions is a very pessimistic viewpoint. This can be problematic as the attacker may prescribes attacks for the sole purpose of triggering conservative defenses, causing the defender to essentially carry out a denial-of-service attack on itself. Integrating elements of the probabilistic uncertainty approach into the nondeterministic approach can help to alleviate this issue. In particular, considering a range of distributions over which the worst-case is taken (to obtain the *least favorable distribution* [37]) permits one to regulate the degree of pessimism in computing defense policies.

6   Concluding Remarks

Fundamental to the control-theoretic approach is the assumption that the defense problem is one-sided, that is, the defender is the only active decision maker. As such, the threat model serves to absorb the attacker's behavior into the model of the cyber environment. Computational limitations preclude specification of a complete threat model, that is, a full representation of the system (*e.g.*, active services/software, all active users and associated privilege levels, network connectivity, and trust relationships). One must make approximations, specifying threat models that include coarser state (*e.g.*, attacker privilege levels) and observation processes (*e.g.*, noisy security alerts from an intrusion detection system). This unavoidably introduces uncertainty, requiring the defender to estimate the true security status of the system from the observable signals.

Two complementary approaches to handling the defender's uncertainty have been discussed, namely probabilistic uncertainty and nondeterministic uncertainty. Probabilistic uncertainty assumes that the defender's uncertainty can be quantified by probability distributions. While permitting efficient (sampling-based) computational procedures for determining defense policies, taking a probabilistic approach requires some assumptions, *e.g.*, stationarity, that may be difficult to justify in real-world cyber-security settings. Alternatively, the nondeterministic approach leads to both a simpler modeling task (there is no longer a requirement to specify probability parameters) and a more flexible model (allowing for a description of non-stationary behavior). However, these benefits come at a cost of a harder computational problem.

The control-theoretic approach discussed in this chapter provides a foundation for extension to more general settings. One natural extension is to consider the case where the model of the cyber environment is unknown. To address this problem, standard tools from Bayesian adaptive control and reinforcement learning [8] may not be sufficient. The main challenge arises from the need to obtain large quantities of useful attack data. Concepts such as *transfer learning* [38] and *generalization* [39] may be useful for dealing with the sparsity and reproducibility issues in attack data. Another extension is the consideration of more complex threat models, primarily allowing for both the attacker and the defender to be active decision makers. Such a modification results in a game-theoretic interaction where both the attacker and defender optimally respond to each other's actions. The complexities associated with the game-theoretic approach to cyber-security, as well as some foundational results, are presented in the following chapter.

References

1. Miehling, E., Rasouli, M., Teneketzis, D.: A POMDP approach to the dynamic defense of large-scale cyber networks. IEEE Trans. Inf. Forensics Secur. **13**(10), 2490–2505 (2018)
2. Marschak, J., Radner, R.: Economic Theory of Teams. Yale University Press, New Haven (1972)
3. Ho, Y.-C., Kastner, M., Wong, E.: Teams, signaling, and information theory. IEEE Trans. Autom. Control **23**(2), 305–312 (1978)
4. Gorenc, B., Sands, F.: Hacker machine interface: the state of SCADA HMI vulnerabilities. Technical report, Trend Micro Zero Day Initiative Team (2017)
5. Arora, A., Telang, R., Xu, H.: Optimal policy for software vulnerability disclosure. Manage. Sci. **54**(4), 642–656 (2008)
6. Shostack, A.: Threat Modeling: Designing for Security. Wiley, Hoboken (2014)
7. Kumar, P.R., Varaiya, P.: Stochastic Systems: Estimation, Identification, and Adaptive Control. Prentice Hall, Upper Saddle River (1986)
8. Sutton, R.S., Barto, A.G., Williams, R.J.: Reinforcement learning is direct adaptive optimal control. IEEE Control Syst. **12**(2), 19–22 (1992)
9. Mahajan, A., Martins, N.C., Rotkowitz, M.C., Yüksel, S.: Information structures in optimal decentralized control. In: 51st Annual Conference on Decision and Control (CDC), pp. 1291–1306. IEEE (2012)
10. Mahajan, A., Mannan, M.: Decentralized stochastic control. Ann. Oper. Res. **241**(1–2), 109–126 (2016)
11. Schuppen, J.H.: Information structures. In: van Schuppen, J.H., Villa, T. (eds.) Coordination Control of Distributed Systems. LNCIS, vol. 456, pp. 197–204. Springer, Cham (2015). https://doi.org/10.1007/978-3-319-10407-2_24
12. Bellman, R.: Dynamic Programming. Princeton University Press, Princeton (1957)
13. Bertsekas, D.P.: Dynamic Programming and Optimal Control, vol. 1. Athena Scientific, Belmont (1995)
14. Shameli-Sendi, A., Ezzati-Jivan, N., Jabbarifar, M., Dagenais, M.: Intrusion response systems: survey and taxonomy. Int. J. Comput. Sci. Netw. Secur. **12**(1), 1–14 (2012)
15. Iannucci, S., Abdelwahed, S.: A probabilistic approach to autonomic security management. In: IEEE International Conference on Autonomic Computing (ICAC), pp. 157–166. IEEE (2016)
16. S. Iannucci, et al.: A model-integrated approach to designing self-protecting systems. IEEE Trans. Software Eng. (Early Access) (2018)
17. Lewandowski, S.M., Van Hook, D.J., O'Leary, G.C., Haines, J.W., Rossey, L.M.: SARA: Survivable autonomic response architecture. In: DARPA Information Survivability Conference & Exposition II (DISCEX), vol. 1, pp. 77–88. IEEE (2001)
18. Kreidl, O.P., Frazier, T.M.: Feedback control applied to survivability: a host-based autonomic defense system. IEEE Trans. Reliab. **53**(1), 148–166 (2004)
19. Musman, S., Booker, L., Applebaum, A., Edmonds, B.: Steps toward a principled approach to automating cyber responses. In: Artificial Intelligence and Machine Learning for Multi-Domain Operations Applications, vol. 11006, pp. 1–15. International Society for Optics and Photonics (2019)
20. Speicher, P., Steinmetz, M., Hoffmann, J., Backes, M., Künnemann, R.: Towards automated network mitigation analysis. In: Proceedings of the 34th ACM/SIGAPP Symposium on Applied Computing, pp. 1971–1978. ACM, New York (2019)

21. Miehling, E., Rasouli, M., Teneketzis, D.: Optimal defense policies for partially observable spreading processes on Bayesian attack graphs. In: Proceedings of the Second ACM Workshop on Moving Target Defense, pp. 67–76. ACM (2015)
22. Rasouli, M., Miehling, E., Teneketzis, D.: A supervisory control approach to dynamic cyber-security. In: Poovendran, R., Saad, W. (eds.) GameSec 2014. LNCS, vol. 8840, pp. 99–117. Springer, Cham (2014). https://doi.org/10.1007/978-3-319-12601-2_6
23. Rasouli, M., Miehling, E., Teneketzis, D.: A scalable decomposition method for the dynamic defense of cyber networks. In: Rass, S., Schauer, S. (eds.) Game Theory for Security and Risk Management. Springer, Cham (2018). https://doi.org/10.1007/978-3-319-75268-6_4
24. Smallwood, R.D., Sondik, E.J.: The optimal control of partially observable Markov processes over a finite horizon. Oper. Res. **21**(5), 1071–1088 (1973)
25. Albanese, M., Jajodia, S., Noel, S.: Time-efficient and cost-effective network hardening using attack graphs. In: 42nd Annual IEEE/IFIP International Conference on Dependable Systems and Networks (DSN), pp. 1–12. IEEE (2012)
26. Silver, D., Veness, J.: Monte-Carlo planning in large POMDPs. In: Advances in Neural Information Processing Systems, pp. 2164–2172 (2010)
27. Besold, T.R., Garcez, A.A., Stenning, K., van der Torre, L., van Lambalgen, M.: Reasoning in non-probabilistic uncertainty: logic programming and neural-symbolic computing as examples. Minds Mach. **27**(1), 37–77 (2017)
28. Witsenhausen, H.: Sets of possible states of linear systems given perturbed observations. IEEE Trans. Autom. Control **13**(5), 556–558 (1968)
29. Schweppe, F.: Recursive state estimation: unknown but bounded errors and system inputs. IEEE Trans. Autom. Control **13**(1), 22–28 (1968)
30. Bertsekas, D.P.: Control of uncertain systems with a set-membership description of the uncertainty. Technical report, DTIC Document (1971)
31. Milnor, J.: Games against nature. In: Coombs, C.H., Davis, R.L., Thrall, R.M. (eds.) Decision Processes, pp. 49–60. Wiley, Hoboken (1954)
32. Akian, M., Quadrat, J.P., Viot, M.: Bellman processes. In: Cohen, G., Quadrat, J.P. (eds.) 11th International Conference on Analysis and Optimization of Systems Discrete Event Systems. LNCIS, vol. 199, pp. 302–311. Springer, Berlin, Heidelberg (1994). https://doi.org/10.1007/BFb0033561
33. Bernhard, P.: Expected values, feared values, and partial information optimal control. In: Olsder, G.J. (ed.) New Trends in Dynamic Games and Applications. AISDG, vol. 3, pp. 3–24. Birkhäuser Boston, Basel (1995). https://doi.org/10.1007/978-1-4612-4274-1_1
34. Bernhard, P.: A separation theorem for expected value and feared value discrete time control. ESAIM: Control Optimisation Calc. Var. **1**, 191–206 (1996)
35. Akian, M., Quadrat, J.-P., Viot, M.: Duality between probability and optimization. Idempotency **11**, 331–353 (1998)
36. Bernhard, P.: Minimax - or feared value - $L1/L\infty$ control. Theoret. Comput. Sci. **293**(1), 25–44 (2003)
37. Başar, T., Bernhard, P.: H-Infinity Optimal Control and Related Minimax Design Problems: A Dynamic Game Approach. Springer, Cham (2008)
38. Weiss, K., Khoshgoftaar, T.M., Wang, D.: A survey of transfer learning. J. Big data **3**(1), 9 (2016)
39. Oh, J., Singh, S., Lee, H., Kohli, P.: Zero-shot task generalization with multi-task deep reinforcement learning. In: Proceedings of the 34th International Conference on Machine Learning, JMLR, pp. 2661–2670 (2017)

Game Theoretic Approaches to Cyber Security: Challenges, Results, and Open Problems

Hamidreza Tavafoghi[1], Yi Ouyang[2], Demosthenis Teneketzis[3(✉)],
and Michael P. Wellman[4]

[1] Mechanical Engineering, University of California, Berkeley, CA, USA
[2] Industrial Engineering and Operations Research,
University of California, Berkeley, CA, USA
[3] Electrical Engineering and Computer Science, University of Michigan,
Ann Arbor, MI, USA
`teneket@umich.edu`
[4] Computer Science and Engineering, University of Michigan, Ann Arbor, MI, USA

Abstract. We formulate cyber security problems with many strategic attackers and defenders as stochastic dynamic games with asymmetric information. We discuss solution approaches to stochastic dynamic games with asymmetric information and identify the difficulties/challenges associated with these approaches. We present a solution methodology for stochastic dynamic games with asymmetric information that resolves some of these difficulties. Our main results are based on certain key assumptions about the game model. Therefore, our methodology can solve only specific classes of cyber security problems. We identify classes of cyber security problems that our methodology cannot solve and connect these problems to open problems in game theory.

1 Introduction

The high and continually increasing connectivity of modern cyber networks has resulted in significant improvement in the functionality and efficiency of our networked systems, but has also created new entry points for attackers, thus making these systems more vulnerable to intrusion. As noted by Miehling et al. [23], recent events such as information leakage and theft [7], car hacking [11], and denial of service attacks [6], have highlighted this vulnerability. Such vulnerability has become an issue of great concern because (i) the operation of critical infrastructure is increasingly reliant upon (potentially unreliable) networked systems and (ii) cyber attacks are becoming persistent and increasingly sophisticated. As reported by the Department of Homeland Security's Industrial Control Systems Cyber Emergency Response Team (ICS-CERT), attacks on the critical infrastructure sectors (such as energy, communication, manufacturing, transportation, and water systems) have remained persistent over the past few years, with 245 in 2014, 295 in 2015, and 290 in 2016 [28], and many of the

© Springer Nature Switzerland AG 2019
S. Jajodia et al. (Eds.): Adaptive Cyber Defense, LNCS 11830, pp. 29–53, 2019.
https://doi.org/10.1007/978-3-030-30719-6_3

recent intrusions have had the potential to inflict severe and widespread damage (an increasing number of attacks have reached the system's control system layer [28]). Therefore, it is imperative to detect and mitigate cyber attacks so as to ensure secure operation of society's critical systems.

Cyber security is a complex problem. The complexity of the problem stems primarily from the fact that many individuals/agents (attackers, defenders) with different objectives, and different information about the cyber network's structure/topology and security status interact with one another through the network. At each time instant the cyber network's security status and each agent's information depend, in general, on exogenous random events (e.g., random failures in hosts and connections among hosts) and all the agents' strategies; such strategies are not common knowledge [1,41] among all agents. Furthermore, the degree to which each agent achieves his objectives depends on his strategy and *all* the other agents' unknown strategies. Agents can use these these features of the cyber security problem to their advantage. For example, an attacker can take undetectable actions, or detectable actions that do not fully reveal intent; a defender can likewise take actions that are not observable by attackers. Under these conditions, the determination of strategic equilibria—configurations of strategies that leave no agent any incentive to deviate—is a formidable problem.

In this chapter we propose and present a game-theoretic approach to the cyber security problem. In Sect. 2 we explain why the formulation of the cyber security problem as a stochastic dynamic game with asymmetric information provides a reasonable approach to the problem. Then, in the remainder of the chapter we present: the game model that captures the salient features of cyber security problems (Sect. 3); current approaches to stochastic dynamic games with asymmetric information and the difficulties/challenges associated with them (Sect. 4); a new approach/methodology that resolves some of these difficulties along with a discussion of the methodology's key results (Sect. 5); the literature on game theory that is relevant to cyber security problems (Sect. 6); some open problems in game theory that are tightly connected to cyber security (Sect. 7).

The literature on stochastic dynamic games with asymmetric information is rich in deep ideas and is very technical, therefore, it is not easily accessible. Our goal in this chapter is to present and explain, in as plain language as possible, the key ideas behind the approaches to stochastic dynamic games with asymmetric information along with the main results of our approach to these games. For this reason we provide an informal presentation of the approaches to and the results on dynamic games of asymmetric information along with references that formally describe these approaches and results. The results presented in this chapter summarize a series of papers describing our work, motivated by issues in cyber security, on dynamic games with asymmetric information [30,31,36–39].

2 The Cyber Security Problem as a Stochastic Dynamic Game with Asymmetric Information

As pointed out in the introduction, cyber security is a multi-agent problem involving attackers and defenders. The salient features of cyber security problems are: (i) Attacks are progressive in nature. Attackers are using their capabilities to attack and capture computers/hosts. Defenders attempt, through their actions, to retake hosts that are under the attackers' control and to limit the attackers' exploits (e.g., by isolating certain hosts from the rest of the network). As a result of the attackers' and defenders' actions, the security status of the network changes/evolves over time. The evolution is also affected by the occurrence of random events (e.g., random failures in hosts and connections between hosts) in the cyber network. (ii) Each agent has different information about the network's security status. For example: each defender knows in part the network's topology, but does not know the hosts/computers that are under the attackers' control and the information of other defenders; each attacker knows the hosts he has captured, but does not know the network's topology and the hosts captured by other attackers. In addition to their private information, attackers and defenders possess, at any time instant, some common information, consisting of events which they all observe, such as an attack to a group of servers that is detected and the detection is common knowledge [2, 41] among all attackers and defenders (for examples of events that are common information see [6, 7, 11, 28]). (iii) Attackers and defenders are strategic and self-interested: each agent attempts to optimize his own objective rather than a social/agent-wide objective. (iv) Each agent has different objectives. An attacker's objective is to acquire the information he is looking for and to take control of the cyber network. A defender's objective is to protect the information that is stored in the network without compromising too much the network's integrity and availability (a defender can protect his network by turning off the corresponding section of the network, however, such an action would make that section unavailable to its users).

As a result of the above features, the cyber security problem can be formulated as a dynamic game with asymmetric information where the underlying system is stochastic and dynamic. The attackers' and defenders' different objectives along with their strategic behavior lead to a game. The stochastic and dynamic nature of the game is due to the fact that the attackers' and defenders' information changes over time (it increases over time) and the network's security status evolves randomly over time. The fact that attackers and defenders possess, at every time instant, private information (in addition to their common information) results in an asymmetric information structure, thus a game with asymmetric information.

3   The Game Model

We present a model that captures the salient features of cyber security problems discussed in the previous section. We consider a stochastic dynamic system the evolution of which over a time horizon T is affected by the decisions/actions of N strategic agents along with random events that occur in nature. Such a system is described by a stochastic difference equation. Specifically, the system's state at time $t + 1$, X_{t+1}, is a function of its state X_t at time t, the actions A_t^i, $i = 1, 2, \ldots, N$, of the N strategic agents at t, and random events that occur at time t and are statistically independent of the system's state at t and the agents' actions at t. At each time t, each agent has some *noisy/imperfect private information* about the system's evolution up to time t; such information is described by noisy observations of X_t and all the agents' actions $A_{t-1} = (A_{t-1}^i, i = 1, 2, \ldots, N)$ at time $t - 1$. Furthermore, all agents have some *noisy/imperfect common information* about the system's evolution up to t; such information is described by common noisy observations of the system's state X_t and all the agents' actions A_{t-1}. We assume that the system's state, all the agents' private and common observations, and all the agents' actions are finite-valued. All agents have *perfect recall*, that is, at any time t, they remember everything they have observed up to t and all the actions they have taken up to $t - 1$. Denote by P_t^i agent i's private information at t and by C_t the agents' common information at t, $i = 1, 2, \ldots, N$, $t = 1, 2, \ldots, T$. At each time t, agent i's action, $i = 1, 2, \ldots, N$, is generated by g_t^i, his strategy at t; g_t^i is a function of i's private information P_t^i at t and the common information C_t at t. We denote by g^i agent i's *strategy profile* in the T-horizon game; g^i is the collection of strategies g_t^i, $t = 1, 2, \ldots, T$. We term the collection of the agents' strategy profiles g^i, $i = 1, 2, \ldots, N$, the *strategy profile* g in the T-horizon game. At each t, agent i has an instantaneous utility $U_t^i(X_t, A_t)$ that is a function of the system's state X_t and all the agents' actions A_t^i, $i = 1, 2, \ldots, N$. Each strategic agent's objective is to determine his strategy profile so as to maximize the expected sum of his instantaneous utilities from the beginning (time 1) until the end of the game (time T).

The state X_t represents the system's/network's security status at time t. The progressive nature of cyber attacks is captured by the fact that X_t evolves dynamically over time, and its evolution at any time t is affected by the agents' (attackers' and defenders') actions at t and random events that occur in nature at t, such as network failures, and are independent of the agents' actions and the system's security status. Since cyber security systems are networks consisting of a finite number of computers, each computer's security status can be described by one of a finite number of states and each agent can take at any time t a finite number of actions, it is reasonable to assume that the system's state space along with the observation and action spaces are finite. We assume that all agents have *perfect recall*, that is, every agent remembers everything he has seen and every thing he has done. We will discuss the implication of this assumption in the analysis of dynamic games with asymmetric information later in this chapter. We wish to bring to the reader's attention two important features of the above-described model. (1) The instantaneous utility of each agent (hence his overall utility) depends on all agents' actions that are not all perfectly observable and are generated by their respective strategies. Therefore, each agent's choice of strategy must take into account all the other agents' choice of strategies, thus the agents' strategy choices are inter-dependent. (2) Since the dynamic system's evolution over time depends on all the agents' actions through their strategy choices, at any time t, each agent's information, private and common, depends on all agents' strategies up to $t - 1$. These two features of the model result in significant difficulties in the analysis of dynamic games with asymmetric information and in the computation of the appropriate equilibria. We will discuss these difficulties along with ways of overcoming them in the rest of this chapter.

We conclude this section by presenting an example, drawn from [30,31], that we will use throughout the chapter to illustrate the ideas and results we present. Even though the model of the example does not capture all the essential features of cyber security problems, (the current state of the art on stochastic dynamic games with asymmetric information cannot be used to solve the cyber security problem in its full generality), we hope that it will help the reader to understand and appreciate the difficulties/issues that arise in these games along with the key ideas of our approach.

An Example

Consider a game, played over a time horizon T, with N agents that are split into two groups. Group 1 consists of N_1 agents, group 2 consists of N_2 agents, $N_1 + N_2 = N$. The state of the dynamic system at time t is denoted by $X_t = (X_t^1, X_t^2, \ldots, X_t^{N_1})$. The component X_t^n of the state is privately observed by agent n in group 1. The private state X_t^n has uncontrolled Markovian dynamics with given time-invariant matrix of transition probabilities Q^n, $n = 1, 2, \ldots, N_1$. At the beginning of time t, each agent n in group 1 observes X_t^n, $n = 1, 2, \ldots, N^1$ and takes an action A_t^n. The actions $A_t = A_t^n, n = 1, 2, \ldots, N_1$ are announced to all N agents. After this announcement, agent m, $m = 1, 2, \ldots, N_2$, in group 2 makes a decision $D_t^m = (D_t^m(1), D_t^m(2), \ldots, D_t^m(N_1))$. Let $D_t = (D_t^1, D_t^2, \ldots, D_t^{N_2})$ denote the decisions of the agents in group 2 at time t. The decisions D_t are observed by all N agents. After the decisions D_t are made, all agents receive noisy observations $Y_t^1, Y_t^2, \ldots, Y_t^{N_1}$ of the states $X_t^1, X_t^2, \ldots, X_t^{N_1}$, respectively. The utility of agent n, $n = 1, 2, \ldots, N_1$, in group 1 is given by $U_t^{n,1}(A_t, D_t) = (A_t^n - c)(\sum_{i=1}^{N_2} D_t^i(n))$. The utility of agent m, $m = 1, 2, \ldots, N_2$, in group 2 is given by $u_t^m(Y_t, D_t, A_t) = V_t^m(Y_t, D_t) - (\sum_{i=1}^{N_1} D_t^m(i)A_t^i)$, where $V_t^m(Y_t, D_t)$ is a given function of Y_t and D_t, and $Y_t = (Y_t^1, Y_t^2, \ldots, Y_t^{N_1})$.

We assume that the state X_t of the dynamic system, the agents' actions A_t, D_t and the observations Y_t, $t = 1, 2, \ldots, T$ take values in finite spaces. Furthermore, we assume that the Markov state processes X_t^n, $t = 1, 2, \ldots, T, n = 1, 2, \ldots, N_1$, describing the evolution of the dynamic system, and the observations Y_t, conditioned on X_t, $t = 1, 2, \ldots, T$, are all mutually independent.

In this game, the private information of agent n in group 1 at time t, before any action/decision is made at t, is $P_t^n = X_{1:t}^n$, where $X_{1:t}^n = X_1^n, X_2^n, \ldots, X_t^n, n = 1, 2, \ldots, N_1$. Agents in group 2 have no private information at any time. The common information of all N agents at time t is $C_t = A_{1:t-1}, D_{1:t-1}, Y_{1:t-1}$,

The action/decision of agent n, $n = 1, 2, \ldots, N$, at time t is a function g_t^n of his private information P_t^n at t and the common information C_t at t. The functions $g_t^n, t = 1, 2, \ldots, T$ define agent n's strategy g^n in the game. The collection of strategies $g^n, n = 1, 2, \ldots, N$, define the strategy profile played in the game.

4   Current Approaches to Dynamic Games with Asymmetric Information and the Associated Challenges

We provide an informal presentation of the key ideas underlying current approaches to dynamic games with asymmetric information along with the challenges associated with these approaches. For a formal presentation of current approaches to dynamic games with asymmetric information we refer the reader to [9, 25, 29].

The fundamental difficulties in the analysis of stochastic dynamic games with asymmetric information arise from the fact that the agents involved in the game (e.g. attackers and defenders) are strategic, they possess private information, their private and common information increase over time, their strategy choices are inter-dependent, and each agent's information depends, in general, on the strategies employed by all other agents.

To address these difficulties, classical approaches to dynamic games with asymmetric information proceed by taking into account the following considerations. At every instant of time, each agent has to form: (i) An appraisal about the current state of the (stochastic dynamic) system (e.g., the current security status of the network) and the other agents' private information; such an appraisal is about the history/past of the game. (ii) An appraisal about how other agents will play in the future so as to evaluate the performance of his strategy choices; such an appraisal is about the future of the game. Consider any agent, say agent i; given the other agents' strategies, at each time t, agent i can utilize his information (private and common) at t along with (i) all other agents' past strategies up to time $t-1$ and (ii) all other agents' future strategies from t up to T to form these appraisals about the history/past and the future of the game, respectively.

Since each agent has his own objective, each agent's strategy is his own private information, thus, it is not known to other agents. Therefore, each agent has to form a prediction about the other agents' strategies. According to Nash's idea/model, all agents have a "common prediction about the strategy profile" played in the game (as stated in Sect. 3, a strategy profile describes all agents' strategies at all times). Such a prediction strategy profile does not necessarily coincide with the actual strategy profile that is being played in the game. Thus, it is possible that an agent's strategy, say agent i's strategy, is different from the prediction strategy profile. Denote by $g^* := (g^{*1}, g^{*2}, \ldots, g^{*N})$ the common prediction about the agents' strategy profile, where g^{*i} is the strategy prediction profile for agent i (his strategy from time 1 up to time T); denote by $g := (g^1, g^2, \ldots, g^N)$ the actual strategy profile, that is, the strategy profile that is being played by the agents, i.e. $g^{*i} \neq g^i$, in the game. Below we discuss the implications of agent i's deviation from the prediction strategy profile on all agents' behavior. For that matter, we first consider agent i who may want to deviate from his predicted strategy, then we examine the response of an agent who faces such a deviation, and finally we discuss how an agent determines his optimal strategy at each time t for all realizations of his information.

When agent i chooses a strategy, he needs to know how other agents will play/react for any choice that is different from the prediction strategy profile g^{*i}. Since a deviation from agent i may generate information (*e.g.* an observation) that is not expected when the prediction strategy profile g^* is played by all agents, (that is, information that has zero probability according to g^*), the prediction strategy profile g^* has to determine how agents will act for all possible realizations of their information, even those realizations that have zero probability according to it. Then, using g^*, agent i can form an appraisal about the future of the game for any choice of his own strategy and can evaluate the performance of that strategy.

By the same rationale, when agent i chooses any strategy g^i, he needs to determine that strategy for all possible realizations of his information, even those that have zero probability according to the prediction strategy profile $g^{*1}, g^{*2}, \ldots, g^{*(i-1)}, g^{*(i+1)}, \ldots, g^{*N}$. This is because some other agent j, different from i, may deviate from the prediction strategy profile g^{*j}, therefore, agent i must foresee such a possible deviation and must determine his response (according to g^i) to these deviations.

To determine his optimal strategy for all realizations of his information (those that have positive or zero probability under the prediction strategy profile g^*), at each time t, an agent, say agent i, needs to form an appraisal about the history of the game at t along with an appraisal about the future of the game under the assumption that all other agents follow the prediction strategy profile g^*. To form an appraisal about the history of the game at t, agent i proceeds as follows. For all realizations of his information up to and including t that have positive probability under g^*, he uses Bayes' rule to form this appraisal. For any realization of information up to and including t that has zero probability under g^*, agent i cannot rely on the strategy prediction g^* up to time $t - 1$ and use Bayes' rule to form this appraisal. The realization of information of zero probability under g^* tells agent i that his original prediction up to time $t - 1$ is not completely correct, consequently, he needs to revise his original strategy prediction up to $t - 1$ and to form a revised appraisal about the history of the game at t. Therefore, agent i must determine how to revise/form his appraisal about the history of the game at t for all realizations of his information that have zero probability under the prediction strategy profile g^*.

In the game theoretic literature [9] the above considerations are formalized as follows. Each agent's appraisals, say agent i's appraisals, about the history and future of the game are captured by an *assessment* which consists of a *strategy prediction profile* g^* (that is common to all agents) and a *belief* μ_t^i, for each time t, on the system's state and the other agents' private information at t, based on agent i's information at t; for each t, the realization of such a belief is, in general, agent i's private information. The collection $\mu := (\mu_t^i, i = 1, 2, \ldots, N, t = 1, 2, \ldots, T)$ of all agents' beliefs at all times is called a *belief system* μ. At any time t and for any realization of agent i's information at t (such a realization may have positive or zero probability under the strategy prediction profile g^*) agent i's belief at t determines his appraisal about the history of the game at t;

agent i's appraisal at t about the future of the game is determined by his belief μ_t^i at t and the prediction strategy profile $g_{t:T}^* = (g_t^*, g_{t+1}^*, \ldots, g_T^*)$ from t until T (the end of the game).

Based on the definition of assessment, game theorists extended the concept of Nash equilibrium to dynamic games with asymmetric information. An equilibrium of the dynamic game is defined as a common assessment among the agents that satisfies the following conditions under the assumption that agents are rational. (1) Agent i, $i = 1, 2, \ldots, N$, chooses his strategy to maximize his total expected utility in all continuation games (*i.e.* the game's continuation from t until T for all $t = 1, 2, \ldots, T-1$) given the assessment about the game. Consequently, the prediction about agent i's strategy that other agents hold must be a maximizer of agent i's total expected utility under the assessment about the game. (2) For all times t, any agent i's belief at t that is based on a realization of information that has positive probability under the common assessment must be equal to the conditional probability, under the strategy prediction profile of the common assessment, of the system's state and the other agents' private information at t; this conditional probability for agent i is determined via Bayes' rule when all other agents play according to the common assessment's prediction strategy profile. When the realization of agent i's information at t has zero probability under the prediction strategy profile of the common assessment, his belief at t cannot be determined via Bayes' rule and must be revised. The revised belief must satisfy a certain set of *reasonable* conditions so as to be compatible with agent i's rationality. Game theorists have proposed various sets of conditions (see [9, 25, 29]) to capture the notion of reasonable beliefs that are compatible with the agents' rationality. Different sets of conditions for off-equilibrium beliefs, that is, beliefs along off the equilibrium paths of the game (i.e. paths of zero probability under the common strategy prediction component of the assessment) result in different equilibrium/solution concepts (such as perfect Bayesian equilibria, sequential equilibria, perfect equilibria, proper equilibria, persistent equilibria, *etc*) that have been proposed for dynamic games with asymmetric information (see [9, 25, 29] for the definition and meaning of all these equilibrium concepts).

Perfect Bayesian Equilibrium (PBE) [9, 42], is an equilibrium concept that has been widely accepted as an appropriate solution concept for dynamic games with asymmetric information. A PBE is defined as an assessment (a strategy prediction g^* and a belief system μ) that satisfies the *sequential rationality* and *consistency* conditions. The sequential rationality and consistency conditions for dynamic games with asymmetric information where the underlying system is dynamic are formally defined in [36]. Here we verbally describe these conditions.

Sequential Rationality. *Consider any agent at any time, say agent i at time t. Given agent i's information (private and common) at t, his belief at t according to the assessment, and the prediction of all other agents' strategies from t until T according to the assessment, agent i's strategies that maximize his (total) expected utility from t until T are the same as his corresponding strategies from t until T in the assessment.*

Sequential rationality requires that the common prediction g^{*i} about agent i's strategy must be an optimal strategy choice for him since it is common knowledge that he is a rational agent. We note that the sequential rationality condition is more restrictive than the optimality condition for Bayesian Nash Equilibrium (BNE) which requires that the optimality condition in italics above should hold only for $t = 1$. By the sequential rationality condition we require the optimality of the strategy prediction g^* even along paths of the game's evolution that are off-equilibrium (*i.e.* paths that have zero probability under g^*), thus, we rule out the possibility of *non-credible threats*. Consider for example an agent who threatens to play an action that is suboptimal for himself upon the realization of a history that has zero probability under the strategy prediction g^* of the assessment. Such a non-credible threat is ruled out by sequential rationality (hence by PBE) but is not ruled out by BNE. Sequential rationality gives rise to a set of conditions that the strategy prediction g^* must satisfy given the belief system μ of the assessment. As discussed above, the belief system μ of the assessment should also be compatible with g^*. The compatibility between the strategy prediction component of the assessment and the belief system component of the assessment is captured by the consistency condition.

Consistency. *The consistency condition requires that along all equilibrium paths (that is, game histories that are realized when all agents play an equilibrium strategy profile) the agents' beliefs should be updated/evolve according to Bayes' rule. Along all other paths of the game's evolution, the consistency condition requires that if the information received by an agent i at time t has zero probability under the assessment, agent i's belief at t must be revised in a "reasonable" manner.*

The work in [36] presents a set of "reasonable" conditions for revising an agent's beliefs along off-equilibrium paths of the game's evolution.

Even though the definition of PBE provides a general formalization of outcomes that are *rationalizable* (that is, consistent with agents' rationality) under some strategy profile and belief system, computation of PBEs is a formidable task. There are two major challenges in computing a PBE (g^*, μ). First, there is an inter-temporal coupling between the agents' strategy prediction g^* and the belief system μ. As discussed before, according to the consistency requirement, the belief system μ must satisfy a set of conditions given a strategy prediction g^* (see [36,39]). On the other hand, sequential rationality dictates that a strategy prediction g^* must satisfy a set of optimality conditions given the belief system μ (see [36,39]). Therefore, there is a circular dependency between a strategy prediction g^* and a belief system μ. For example, by sequential rationality, agent i's strategy g_t^{i*} at time t depends on the agents' future strategies $g_{t:T}^*$, (where $t : T$ denotes the time interval $t, t+1, t+2, \ldots, T$), and on the agents' past strategies $g_{1:t-1}^*$ indirectly through the consistency condition for μ_t^i. As a result, one has to determine the strategy prediction g^* and the belief system μ simultaneously for the whole time horizon so as to satisfy all the sequential rationality and consistency conditions; consequently, one cannot sequentially decompose over time the computation of PBEs. Second, since the agents' are assumed to have perfect recall their information (private and common) increases over time, thus, their strategies have a growing domain over time; this feature of the agents' strategies further complicates the computation of PBEs.

We continue discussing the example we introduced at the end of Sect. 3. Here we illustrate the concepts introduced in this section along with the difficulties that arise in the determination of equilibrium strategies in dynamic games with asymmetric information.

An Example (continued)
An assessment of the game is described by a strategy prediction profile $g^* = (g^{*1}, g^{*2}, \ldots, g^{*N})$, that is common to all agents, and a belief system $\mu_t^i, i = 1, 2, \ldots, N, t = 1, 2, \ldots, T$. The component $g^{*i} = (g_1^{*i}, g_2^{*i}, \ldots, g_T^{*i})$ describes the strategy player i is predicted to play in the game. The strategy g^{*i} may be different from the actual strategy g^i player i plays in the game. The component μ_t^i describes agent i's belief at time t about the state of the system X_t and the private information P_t^j of all agents j other than i at time t, conditioned on agent i's private information P_t^i and the common information C_t (the private and common information for all agents at all times have been specified at the beginning of this example, at the end of Sect. 3).

At any time t,the beliefs μ_t^i, $i = 1, 2, \ldots, N$, depend on $g_{1:t-1}$, the strategies of all agents up to time $t - 1$. At each time t, when an agent, say agent i, determines his best strategy from t up to time T according to the sequential rationality requirement, he takes into account the strategy prediction profile for all other agents form t up to T and his belief μ_t^i. Therefore, the (equilibrium) strategies and beliefs of all agents are interdependent over time and must all be determined simultaneously for the whole duration of the game. This fact highlights one of the major difficulties associated with the current approaches to stochastic dynamic games with asymmetric information.

As pointed out at the beginning of this example, (end of Sect. 3), the private information of each agent i in group 1 at time t is $P_t^i = X_{1:t}^i, i = 1, 2, \ldots, N_1$. The common information of all agents at time t is $C_t = A_{1:t-1}, D_{1:t-1}, Y_{1:t-1}$. Consequently, the domains of private and common information increase with time, therefore, for large time horizons T, the computation of equilibria becomes a formidable task. This fact highlights another difficulty associated with the current approaches to stochastic dynamic games with asymmetric information.

In the next section we present an approach to dynamic games with asymmetric information that addresses and partly resolves the above challenges.

5 A Sufficient Information Approach to Stochastic Dynamic Games with Asymmetric Information

The definition of PBE requires agents to keep track of all the information they acquire over time and to form beliefs about the private information of all other agents. In this section we show that agents do not need to keep track of all their past information to reach an equilibrium; at any time t, they can take into account only a subset of the information available at t that is *relevant* to the continuation of the game, and ignore the rest of it. Such a selection of the relevant information is motivated by computational and philosophical reasons: the resulting strategies are simpler and the corresponding PBE are easier to compute; furthermore, the simpler strategies proposed in this section offer a more plausible prediction of the outcome of the interactions among strategic agents in cyber security games where the underlying system is dynamic (due to the progressive nature of attacks) and there is significant asymmetry in the information possessed by the agents (attackers and defenders).

The above discussion motivates the approach to dynamic games with asymmetric information that we present in this section. The key steps of our approach are as follows.

1. We present conditions sufficient to guarantee that all the agents involved in the game can compress their private information in a mutually consistent manner. Such a mutually consistent compression leads to the notion/concept of *sufficient private information*.
2. Using the notion of sufficient private information, we present a compression of the common information. Such a compression of the common information leads to the concept of *sufficient common information*.
3. Using the notions of sufficient private information and sufficient common information we define a set of strategies termed *sufficient information based strategies* (SIB strategies), and a set of PBE termed *Sufficient Information Based Perfect Bayesian Equilibria* (SIB-PBE). We show that the set of sufficient information based strategies is closed under the best response correspondence. Thus, we establish the following result: if all agents except one, say agent i, play sufficient information based strategies, then there exists a sufficient information based strategy for agent i that is a best response to those strategies. The implication of this result is that one can restrict attention to SIB strategies to determine SIB-PBE.
4. Using the sufficient common information as an information state we provide a sequential decomposition of stochastic dynamic games with asymmetric information. Such a decomposition leads to an algorithm for the determination of SIB-PBE. We identify instances of games where SIB-PBE exist.

5.1 Sufficient Private Information

Let C_t denote the agents' common information at time t, and let P_t^i denote agent i's private information at t, $i = 1, 2, \ldots, N$.

Sufficient Private Information. We say that the collection $S_t = (S_t^i, i = 1, 2, \ldots, N)$, (where each S_t^i a function of C_t and P_t^i), is sufficient private information for the N agents if the following conditions are satisfied for all agents and for all times: (i) Each S_t^i can be updated recursively, that is, S_t^i can be determined from S_{t-1}^i and the new information agent i acquires at time t. (ii) S_t, C_t and the agents' actions A_t at t provide the information sufficient to statistically determine the sufficient private information S_{t+1} and the common information

C_{t+1}. (iii) Agent i's expected utility at t conditioned on his private information P_t^i, the common information C_t, and the agents' actions A_t at t, is the same as his expected utility at t conditioned on his sufficient private information S_t^i, the common information C_t and the agents' actions A_t at t. Furthermore both expected conditional utilities at t are independent of agent i's strategy. (iv) The information provided by agent i's private sufficient information S_t^i and the common information C_t is sufficient for agent i to statistically determine/predict the sufficient private information of all other agents. Furthermore, this statistical determination/prediction is independent of agent i's strategy.

The above discussion provides an informal presentation of conditions (i)-(iv); a formal/mathematical description of these conditions can be found in [36,38,39].

We now provide an intuitive interpretation of the above conditions. Condition (ii) requires that that agents' sufficient private information must be rich enough so that, combined with their common information and actions at any time t, it leads to the same prediction of the sufficient private information and the new common information at $t+1$ as the one that would be obtained if agents used all their information at t. Condition (iii) is similar in spirit to one of the requirements defining an information state in centralized stochastic control [18]. The essence of conditions (ii) and (iii) is that sufficient private information must be a component of a statistic that is sufficient for decision making purposes. Therefore, sufficient private information must be updated recursively (condition (i)). The essence of condition (iv) is the following: the agents' sufficient private information must be defined by a mutually consistent compression of all the agents' private information. Such a compression must not entail any loss of information, as far as an agent's ability to statistically predict the other agents' sufficient private information is concerned; furthermore, such a private information compression must be robust to agents' possible unilateral deviations from the strategy prediction g^*.

We would like to point out that the above conditions do not uniquely determine the agents' sufficient private information. These conditions may lead to many solutions including the trivial one $S_t^i = P_t^i$ for all agents i, $i = 1, 2, \ldots, N$. Therefore, an important question is: is there a minimal sufficient private information for the agents? The existence of a minimal sufficient private information for all agents is currently an important open problem.

5.2 Sufficient Common Information

Based on the characterization of sufficient private information, we introduce the notion of sufficient common information which at any time t is a statistic/compressed version of the common information C_t at t.

Sufficient Common Information. We define the agents' sufficient common information at any time t, denoted by Π_t, to be the agents' belief about the dynamic system's state X_t at t, and all the agents' sufficient private information S_t at t, conditioned on the common information C_t.

We call Π_t the *Sufficient Information Based (SIB) belief at t*. The agents' SIB belief at t is computable by all agents, thus, it is common information [1,41] among all agents. The SIB belief Π_t is recursively updated according to a *SIB update rule* ψ_t. Specifically, Π_{t+1} is determined by Π_t and the common information that becomes available at t, that is $Z_t = C_t \backslash C_{(t-1)}$, according to ψ_t. If the realization of the information Z_t has non-zero probability according to the strategy prediction g^*, then ψ_t updates Π_t according to Bayes' rule; if the realization of Z_t has zero probability according to g^*, then Π_{t+1} is updated according to ψ_t in a *reasonable* manner that is consistent with the agents' rationality (see for example [36]). We denote by $\Pi^\psi = (\Pi_t^\psi, t = 1, 2, \ldots, T)$ the sequence of SIB beliefs generated by the update rule $\psi := (\psi_t, t = 1, 2, \ldots, T)$.

The above discussion provides an informal presentation of the concept of sufficient common information. For a formal/mathematical description of sufficient common information and its update we refer the reader to [36, 38, 39].

5.3 Sufficient Information-Based Strategies and Sufficient Information-Based Perfect Bayesian Equilibria

The combination of sufficient private information $(S_t^i, i = 1, 2, \ldots, N, t = 1, 2, \ldots, T)$ and sufficient common information Π_t, $t = 1, 2, \ldots, T$, provides a mutually consistent compression of the agents' private and common information, respectively. Using this information compression we define a class of strategies σ_t^i that are based on S_t^i and Π_t for each agent i at each time t. We call σ_t^i a *Sufficient Information Based (SIB) strategy* for agent i at time t. A collection of SIB strategies σ_t^i, $i = 1, 2, \ldots, N$, $t = 1, 2, \ldots, T$, is termed a *SIB strategy profile* σ. We note that at any time t the set of SIB strategies is a subset of all possible strategies agents can choose at t by using all of their private and common information at t. SIB strategies are simpler than general strategies because they have a smaller domain than general strategies as they are based on compressed versions of the agents' private and common information at any time t. We further note that if the dimensionality of S_t^i, agent i's sufficient private information, $i = 1, 2, \ldots, N$, remains fixed over time then the domain of SIB strategies is time-invariant. In Sect. 5.5 we present instances of dynamic games with asymmetric information where the domain of SIB strategies is time-invariant.

Based on the concept of SIB strategies we introduce the concept of *Sufficient Information Based-Perfect Bayesian Equilibrium (SIB-PBE)* that is informally described as follows.

Sufficient Information Based-Perfect Bayesian Equilibrium (SIB-PBE). A SIB-PBE is a PBE in which all agents play SIB strategies.

For a formal definition of SIB-PBE we need to consider, as in the case of PBE, a SIB assessment that consists of a SIB strategy prediction profile σ and a SIB belief system μ^ψ, and to define the sequential rationality and consistency conditions that σ and μ^ψ must satisfy so that they should specify a SIB-PBE. A formal definition of SIB-PBE can be found in [36].

The class of SIB assessments needed to formally define a SIB-PBE imposes two additional restrictions/requirements on the agents' strategies and beliefs as compared to the general class of assessments presented in Sect. 4. First, SIB assessments require that each agent i, $i = 1, 2, \ldots, N$, must play a SIB strategy σ^i instead of a general strategy g^i. Second, SIB assessments require that at every time t each agent i must form a belief about the status of the game using only the SIB belief Π_t along with his sufficient private information S_t^i (instead of a general belief μ_t^i that is based on his private information P_t^i and the common information C_t). Such restrictions generate the following strategic concerns. First, a strategic agent does not have to restrict his choice to SIB strategies, he may deviate from a SIB strategy σ^i to a non-SIB strategy g^i if such a deviation is profitable for him. Second, at any time t, a strategic agent does not have to limit himself to forming a belief about the status of the game by using only the SIB belief Π_t and sufficient private information S_t^i; he may want to form a belief using all of his private information and all the common information if such a belief enables him to improve his overall expected utility. These concerns are addressed by the results of our methodology that appear in the next section.

5.4   Main Results

The results we present in this section address the difficulties associated with current approaches to dynamic games with asymmetric information, specifically: the inter-dependence over time between strategies and beliefs (Sect. 4); the growing domain of the agents' strategies (Sect. 4); and the strategic concerns arising from restricting attention to SIB strategies and SIB beliefs (Sect. 5.3). These results have been derived under the following key assumption, the meaning of which we discuss in the following subsection.

Key Assumption. At any time t, $t = 1, 2, \ldots, T$, and for any sequence of all the agents' actions up to time $t - 1$ the following conditions are satisfied: (C1) Every possible value x_t of the system state X_t can be realized with positive probability. (C2) For every agent, every possible value of his private observations can be realized with positive probability.

   We present an informal statement of the four main results of the sufficient information approach to dynamic games with asymmetric information. A formal statement of conditions (C1) and (C2) along with a formal statement of the four main results and their proofs can be found in [36].

Result 1. At any time t, every agent i's private belief about the state of the dynamic system and the the private information of all other agents is independent of his own strategy.

Result 2. If every agent $j \neq i$ plays a SIB strategy σ^j, then there exists a SIB strategy σ^i for agent i that is a best response to the strategies $(\sigma^j, j \neq i)$.

   Results 1 and 2 address the strategic concerns created by focusing on SIB strategies and SIB beliefs, and, in part, the growing domain of the agents' strategies. Result 1 shows that no agent can alter his private belief about the state of the dynamic system and all the other agents' private information by deviating from the predicted strategy profile. Thus, when all agents $j \neq i$ play according to a SIB assessment (σ^*, μ^ψ) (where the belief system μ^ψ is determined by the update rule ψ described in Sect. 5.2), agent i cannot mislead these agents by playing a strategy g^i different from σ^{i*}, thus, creating dual beliefs (one belief that is based on the SIB assessment (σ^{i*}, μ^ψ) the functional form of which is known to all agents, and another belief that is based on his private strategy g^i that is only known to him) which he can use to his advantage. Result 2 shows that when all agents play SIB strategies, no agent can profit by deviating from his SIB strategy to a non-SIB strategy. Therefore, we can restrict attention to SIB strategies and attempt to determine PBE within the class of SIB assessments, i.e. SIB-PBE. As pointed out above, SIB strategies are simpler than general strategies (which, at any time, are functions of all the private and common information available to an agent at that time) because they are based on compressed information. However, SIB strategies do not have, in general, a time-invariant domain. Nevertheless, there are several instances in practice where SIB strategies have a time-invariant domain [30,31,37].

Using Results 1 and 2 we can obtain a sequential decomposition of stochastic dynamic games with asymmetric information. Such a decomposition is described by the following result.

Result 3. SIB-PBE can be determined by the solution of N coupled dynamic programs (one for each agent). These dynamic programs determine sequentially (moving backwards in time) SIB-PBE via the solution (*i.e.* the Bayesian Nash equilibria) of a series of T static Bayesian games that have the following form. For the game at time T, and for any realization π_t, $s_T^i = (s_T^i, i = 1, 2, \ldots, N)$ of the SIB belief Π_T and the sufficient information $S_T = (S_T^i, i = 1, 2, \ldots, N)$, respectively, agent i's utility is the expectation of his original utility U_T^i (see Sect. 3) conditioned on the π_T and s_T. For the game at time t, $t = 1, 2, \ldots, T - 1$, and for any realization π_t, $s_t = (s_t^i, i = 1, 2, \ldots, N)$ of the SIB belief Π_t and the sufficient private information $S_t = (S_t^i, i = 1, 2, \ldots, N)$, respectively, agent i's utility is the sum of two terms: (i) the expectation of his original utility U_t^i conditioned on π_t and s_t; and (ii) his expected payoff from time t+1 until time T, due to the continuation of the game, conditioned on π_t and s_t. The second term of the above sum is a function of the SIB belief Π_{t+1} (which, according to Sect. 5.2, is recursively determined form π_t and the new common information Z_{t+1} acquired at $t + 1$) and the sufficient private information S_{t+1} (which, according to Sect. 5.1, is recursively determined from s_t and the new information the agents acquire at $t + 1$).

Result 3 shows that for finite horizon stochastic dynamic games with asymmetric information our approach resolves the difficulty due to the interdependence over time between strategies and beliefs (discussed in Sect. 5.4) by providing a systematic method for determining the components of SIB-PBE one at a time, starting at time T and sequentially moving backwards in time. The N coupled dynamic programs provide an algorithm for determining SIB-PBE.

Results 1 and 2 hold for both finite and infinite horizon games. Under certain additional assumptions, Result 3 can be extended to infinite horizon games (see [36,39]).

Result 4. (i) SIB-PBE exist for zero-sum games. (ii) For nonzero-sum games there exists at least one SIB-PBE (σ^*, μ^ψ), if the following condition is satisfied. There exists sufficient information $S_{1:T}^{1:N}$ such that the update rule ψ is independent of the strategy profile σ^*.

The independence condition of Result 4 is not satisfied for all cyber security games. In [36] we present several classes of dynamic games with asymmetric information where the condition of Result 4(ii) is satisfied.

We illustrate the results of our approach to stochastic dynamic games with asymmetric information through the example introduced in Sect. 3.

An Example (continued)
For the example introduced in Sect. 3, at any time t the sufficient private information of agent i in group 1 is $S_t^i = X_t^i$. As pointed out earlier, all agents in Group 2 have no private information, thus, no sufficient private information. The sufficient common information for all agents at time t is the belief on the system state X_t conditioned on the common information $C_t = A_{1:t-1}, D_{1:t-1}, Y_{1:t-1}$. Note that the sufficient private information of each agent, and the sufficient common information for all agents have time invariant domains.

Using the sufficient common information as an information state, we can show [30, 31] that PBE assessments can be determined sequentially in time by a backward induction algorithm.

Therefore, for the game of the example introduced in Sect. 3, our methodology resolves the key difficulties (discussed in Sect. 4) that are associated with previous approaches to dynamic games with asymmetric information. That is, it breaks the interdependence over time between strategies and beliefs (through the sequential decomposition of the game) and discovers, for each agent and each time, sufficient private information and sufficient common information that have time invariant domains.

5.5   Discussion of the Main Results

Our main results show that the mutually consistent compression of the agents' information (private and common) leads to SIB strategies, SIB beliefs, and SIB-PBE which have several desirable features. Specifically, SIB strategies are simpler than general strategies, and SIB beliefs, which are common knowledge among all agents, can serve as information states in the sequential decomposition of stochastic dynamic games with asymmetric information. In general, the set of SIB-PBE of a dynamic game is a subset of all PBE of the game. This is because in a dynamic game agents can incorporate their past irrelevant observations into their future decisions so as to create rewards (respectively, punishments) that incentivize them to play (respectively, not to play) specific actions over time. By compressing the agents' private and common information, we do not capture such punishment/reward schemes that are based on past irrelevant observations. An example of such a situation appears in [36, 39] within the context of a repeated game, where the set of PBE that can not be captured as SIB-PBE are the ones that utilize payoff-irrelevant information to create reward/punishment schemes in the continuation game.

We would like to note that in dynamic games where the agents' equilibrium payoffs are unique, we can restrict attention to SIB-PBE because the above-described punishment/reward schemes do not lead to additional equilibrium payoffs. One class of such games is the class of zero-sum games (*e.g.*, attacker-defender games within the context of cyber security where the defender's only concern is the network's security). In a zero-sum game agents have completely opposite interests, therefore, it is not rational for them to cooperate on the formation of such punishment/reward schemes; we refer the interested reader to [36,39] for more discussion and the proof of existence of SIB-PBEs in zero-sum games.

Even though it is true that, in general, the set of PBE of a dynamic game is larger than the set of SIB-PBE, in our opinion there are reasons on why in a highly dynamic environments, such as the the environment of cyber security problems, SIB-PBE are more plausible to arise as an outcome of the game.

First, we argue that in a highly dynamic environment with significant information asymmetries among agents, the creation/formation of reward/punishment schemes that utilize the agents' payoff-irrelevant information requires prior complex agreements among the agents. These complex agreements are more likely to occur in games where the underlying system is not highly dynamic (as in repeated games [19]) and there is no much information asymmetry among agents. In a highly dynamic environment with significant information asymmetries among agents (as in cyber security games) the formation of such complex agreements becomes less likely for the following reasons. First, in these environments each agent's individual decision making process is described by a complex Partially Observable Markov Decision Process (POMDP); thus, strategic agents are less likely to form a prior common agreement (that depends on the solution of their POMDPs) in addition to solving their individual POMDPs. Second, as the information asymmetry among agents increases, reward/punishment schemes that utilize payoff-irrelevant information require an increasingly complex agreement that is sensitive and not robust to changes in the assumptions on the information structure of the game. An example illustrating the lack of robustness of these agreements to changes in the information structure of the game is provided in [36,39]. The author of [24] provides a general result on the robustness of the above mentioned reward/punishment schemes in repeated games; he shows that the set of equilibria that are robust to changes in the game's information structure that affect only payoff-irrelevant signals do not include the set of equilibria that utilize the reward/punishment schemes described above.

Second, the proposed solution concept SIB-PBE can be viewed as a generalization/extension of Markov Perfect Equilibrium (MPE) [21] to dynamic games with asymmetric information. Therefore, a similar set of rationales that support the notion of MPE also applies to the notion of SIB-PBE as follows. First, the the set of SIB assessments, as presented in [36,39], describes the simplest form of strategies capturing the agents' behavior that is consistent with the agents' rationality. Second, the class of SIB-PBE captures the idea that "bygones are bygones", which also underlies the requirement of subgame perfection in equilibrium concepts for dynamic games. That is, the agents' strategies in two continuation games that differ only in the agents' information about payoff-irrelevant events must be identical. Third, SIB assessments embody the principle that "minor changes in the past should have minor effects". This implies that any perturbation in the specification of the game or in the agents' past strategies that are irrelevant to the continuation game should not change drastically the outcome of the continuation game.

We would like to emphasize that the key assumption of Sect. 5.4 is essential in establishing the assertions of the main results of the approach presented in this section. Condition (C1) says that there is enough exogenous uncertainty (i.e random uncontrollable events) in the system's evolution so that at each time t all states in the system's state space can be reached with positive probability; condition (C2) says that no agent can infer perfectly another agent's actions based only on his private observations; equivalently, condition (C2) says that any deviation that is detected by a certain agent is also simultaneously detected by all the other agents. We believe that within the context of cyber security problems these conditions are fairly reasonable. For example, when the system/network is heavily used there is a high likelihood that random failures induced by the heavy load can potentially lead to one of many security states. Furthermore, the information agents receive from their own (private) sensors can be very noisy, thus they are not able to perfectly detect other agents' actions. Nevertheless, there are instances of cyber security games with many players/agents (attackers and defendants) where an agent's deviation may be detected by a subset of the rest of the agents (this subset of agents use their private information to detect the deviation). The methodology presented in this section cannot address these instances. We present some ideas on how to address these instances in Sect. 7.

Even though there are instances of dynamic games with asymmetric information where the domain of SIB strategies is time-invariant, e.g. [31,37], the methodology for information compression presented in this section does not always result in sufficient private information the domain of which is time-invariant. Thus, our methodology does not completely resolve the difficulty arising from the growing domain of the agents' strategies in dynamic games with asymmetric information. In Sect. 7, we present some ideas on how to address this difficulty.

We conclude our discussion by pointing out that the main results 1–3 can be obtained if we replace the key assumption of Sect. 5.4 with another one where each agent's actions are always observable by all other agents. However, such an assumption is not realistic for cyber security problems.

6 Relevant Literature

The literature on dynamic games with asymmetric information can be divided into two categories: (1) games where the underlying system is static (repeated games); and (2) games where the underlying system is dynamic. There are significant philosophical differences between the approaches to games in the above categories.

Dynamic games where the underlying system is static (repeated games) arise primarily in economic problems where the environment does not change with time or evolves very slowly over time. Works on (discounted) repeated games study primarily their asymptotic properties, specifically their properties when the horizon is infinite and agents are sufficiently patient (that is the discount factor is close to 1). In repeated games agents play a stage (static) game repeatedly over time. The main objectives of the literature on these games are: (i) to analyze situations where the agents can form self-enforcing punishment/reward mechanisms so as to create additional equilibria that improve the payoffs they obtain by playing an equilibrium of the stage game over time; and (ii) to characterize the payoffs corresponding to all the equilibria of the repeated game.

Dynamic games where the underlying system is dynamic arise in engineering problems where the environment evolves rapidly over time. For example, in cyber security, the progressive nature of cyber attacks results in a rapidly changing environment, this is why the underlying system is modeled by a stochastic difference equation (see Sect. 3). The work existing on games with asymmetric information where the underlying system is dynamic does not restrict attention only to situations where the horizon is infinite and agents are sufficiently patient. The literature addresses situations where the decision problem for each agent, in the absence of interactions with other agents (*i.e.* assuming fixed strategies for the other agents), is a POMDP. Therefore, the determination of a set of equilibrium strategies is a complex problem. Consequently, it is unlikely that the agents seek equilibria that result from the formation of self-enforcing punishment/reward mechanisms that are similar to those used in infinitely repeated games. Existing approaches to and results on dynamic games with asymmetric information where the underlying system is dynamic demonstrate that the equilibria of these games have the same features as the equilibria determined by our approach (see Sect. 5.5). For this reason, in this section we will provide a detailed description of the literature on stochastic dynamic games with asymmetric information where the underlying system is dynamic. At the end of the section we will provide a few key references on dynamic games with asymmetric information where the underlying system is static.

Stochastic dynamic games with asymmetric information where the underlying system is dynamic can be classified into two categories, zero-sum and nonzero-sum. Cyber security problems are usually modeled as non zero sum games because the attackers' and defenders' objectives are not exactly the opposite of each other (see Sect. 3). For this reason, first we will briefly review the literature on zero-sum dynamic games and then we will provide a more detailed discussion of the literature on non zero-sum games.

The works in [3,10,16,17,32] consider dynamic zero-sum games with asymmetric information. The authors of [3,32] study two-player games with Markovian dynamics and lack of information on one side (that is, one player/agent who has perfect knowledge of the game that is being played and one player who has partial/incomplete knowledge of the game that is being played). The authors of [10,16] study two-player zero-sum games with Markovian dynamics and lack of information on both sides (that is, both players possess only partial/incomplete knowledge of the game that is being played). We would like to point out that the authors of [3,10,16,32] consider models with specific Markovian dynamics where each agent observes perfectly a local state that evolves independently of all other local states conditioned on the agents' observable actions. Thus, even if one attempted to formulate cyber security games as dynamic zero-sum games with asymmetric information, the results of the above mentioned papers could not provide any answers or insights because in cyber security games the agents' actions are not, in general, observable, agents have imperfect (noisy) observations of the system's/network's security status, and the game's information structure (who knows what and when) is considerably more complex than that of the above mentioned references. One instance of zero-sum stochastic dynamic games where the agents' actions are not observable is analyzed in [26]. The authors of [26] consider zero-sum games with asymmetric information where the agents, in addition to having private information, share, at each time instant, some common information, and they play pure strategies. They prove that if the set of saddle point equilibria of the above games is non-empty, then the (minmax) value of these games is the same as the value of the (symmetric) games where the agents' only information is their common information. They provide an algorithm for determining the value of the symmetric information games.

The literature on stochastic dynamic non zero-sum games with asymmetric information, where the underlying system is dynamic, addresses mostly situations where, in addition to their private information, all agents have some common information (see [5,12,15,27,30,31,33,35–37,39,40]). References [5,15,35], consider infinite horizon discounted games where the underlying system is a controlled Markov chain. The approach taken in [5,15,35] is based on the philosophy and ideas used to analyze infinitely repeated games. In the work reported in [15] the system's state is perfectly observed by all agents at all times, and each agent's actions are his private information (hidden actions); attention is restricted to *Perfect Pubic Equilibria (PPE)*, that is, equilibria that result in when agents play only common information-based strategies. The authors of [15] characterize, under certain assumptions that appear in [15], the set of the agents' payoffs that correspond to all PPE when all agents are sufficiently patient, that is, the discount factor δ approaches 1. The authors of [5] consider games where at each time all agents observe perfectly each others' actions but each agent has imperfect private information about the system's state. They consider PBE as a solution/equilibrium concept, and characterize, under certain assumptions that are explicitly stated in [5], the set of the agents' payoffs corresponding to all PBE of the game when all agents are sufficiently patient. Sugaya [35] analyzes instances of games where each agent has imperfect private information about the system's state and private monitoring of the other agents' actions; furthermore, he assumes that agents communicate with one another via perfect and public cheap talk. He adopts PBE as the equilibrium/solution concept and character-izes, under certain assumptions that are explicitly stated in [35], the agents' payoffs that correspond to all PBEs of the game when the agents are sufficiently patient. References [12,27,30,31,33,36,37,39,40] analyze finite and/or infinite horizon discounted dynamic games. In all of these references, the agents' common information is used as an instrument for coordination of the agents' strategies. In [12,27,30,31,33,37,40], the *Common Information Based (CIB) belief* (the belief on the dynamic state state at time t, and all the agents' private information at t, based on the agents' common information at t, $t = 1, 2, \ldots, T$) is an information state/sufficient statistic for decision making for each agent at t. In [36,38,39] the *SIB belief* Π_t, $t = 1, 2, \ldots, T$, defined in Sect. 5.2, is an information state for decision making for each agent at time t. In the game instances investigated in [12,27] the CIB belief is independent of the agents' strategies; in such a situation, assessments (defined in Sect. 5.4) can be described simply by the agents' strat-egy prediction (defined in Sect. 5.4), and an appropriate equilibrium concept is *Common Information Based Markov Perfect Equilibrium* that was introduced in [27]. In the game instances investigated in [30,31,33,40], the CIB beliefs depend on the agents' strategies, the agents' actions are always perfectly observable, and the agents' (private) beliefs (defined in Sect. 5.4) are common knowledge [1,41] among all agents. An appropriate equilibrium concept for these instances of games is *Common Information Based-Perfect Bayesian Equilibrium (CIB-PBE)* that was introduced in [30,31]. In the game instances investigated in [36,37,39] the agents' SIB beliefs depend on their strategies, the agents' actions are not observable, and the agents' (private) beliefs are their own private information. An appropriate solution concept for these game instances is SIB-PBE that was introduced in [36,39] and presented in Sect. 5.3. Since cyber security games have asymmetric information, unobservable actions, and the domain of the agents' strategies' grows with time, the work of [35] along with the methodology and results reported in [36,39] and informally presented in Sect. 5 is the literature that is the most relevant to these games.

Infinitely repeated games have been extensively studied, primarily by economists. There is a rich literature available on these games; the book by Mailath and Samuelson, [19], presents the main results on this topic until its publication date. In this chapter we briefly discuss this literature, because some of the ideas and philosophy behind the development of key results for this class of games played a significant role in the development of key results for dynamic games with asymmetric information where the underlying system is dynamic [5,15,35]. Infinitely repeated games can be divided into two categories, zero-sum and non-zero sum.

Infinitely repeated zero-sum games with asymmetric (incomplete) informa-tion were initially studied by Aumann et al. [2]; an excellent survey and discus-sion of results on this class of games can be found in [44].

Infinitely repeated non-zero sum games with asymmetric information can be classified into three categories: games with perfect public monitoring, in which the agents observe perfectly each others' actions, and Nash equilibrium or perfect equilibrium as a solution concept (see [19, 20] and references therein); games with imperfect public monitoring, in which the agents can observe public noisy signals about the action profile and focus on perfect equilibria where each agent's continuation strategy depends only on past public signals (see [8, 19] and references therein); and games with imperfect private monitoring, in which players observe private noisy signals about other players' actions, and sequential equilibrium as a solution concept (see [22] for two-player games and [34] for many-player games, and references therein). In all of the above categories the authors consider infinitely repeated discounted games and characterize the set of equilibrium payoffs corresponding to all equilibria in the limit as the discount factor approaches one.

7   Conclusion

We have argued that cyber security problems are stochastic dynamic games with asymmetric information where the underlying system is stochastic and dynamic. We presented current approaches to analyzing dynamic games with asymmetric information along with the currently available literature and the challenges/difficulties associated with these approaches. As we pointed out in Sect. 4, two major difficulties are the interdependence over time between strategy prediction and beliefs, and the increasing domain of the agents' strategies. We presented a "sufficient information approach" (Sect. 5) which breaks the interdependence over time between strategy prediction and beliefs, leads to a sequential decomposition of the dynamic game and specifies an algorithm for determining the SIB-PBE of the game; we also identified instances of games where the sufficient information approach results in a time-invariant domain of the agents' strategies. The results of the sufficient information approach were developed under a key assumption, stated in Sect. 5.4, which in essence says that any deviation by one agent is either not detected or it is detected simultaneously by all other agents and the detection is based on the agents' common information.

In cyber security problems the domain of the agents' strategies increases, in general, with time. Furthermore, a deviation from one agent may not be detected at all, or it may be detected at different times by different agents. These two features of cyber security games cannot be captured by the approach presented in Sect. 5. In the rest of this section we present some ideas on how to address them, and we identify open problems in dynamic games with asymmetric information that are tightly connected to cyber security games.

First, consider the situation where the agents' sufficient private information increases with time. In this case assume that each agent has finite memory which he updates at each time instant; specifically, assume that at any time t part of each agent's memory is used to store his private information and another part is used to store his SIB belief about the system state and all the agents' (including himself) private information. At time $t + 1$, each agent's private information is determined by an update rule which combines his private information at t and the new information he receives at $t + 1$; similarly, the SIB belief at time $t+1$ is formed by an update rule which combines the SIB belief at t and the new common information received at $t+1$. Under these constraints, the objective is to determine decision strategies (that are based on the agents' private information and the SIB belief), private information update rules, and common information update rules that are in equilibrium.

Next, consider the situation where the key assumption of Sect. 5.4 is relaxed. In this case the challenge is to create public monitoring structures/mechanisms that allow each agent to detect deviations from other agents. Within the context of infinite horizon discounted games (with discount factors close to 1) such monitoring structures are presented (i) in [35] for games where the underlying system is dynamic and is described by a controlled Markov chain, the agents' actions are hidden (unobservable) and the agents' private state observations are imperfect (noisy), and (ii) in [34] for repeated games with an information structure similar to that of [35]. These monitoring mechanisms are described by "review phases" the duration of which is chosen appropriately so that at the end of each phase the law of large numbers should hold with high probability, therefore, allowing agents to detect each others' deviations (see [34]). Such ingenious monitoring structures work well for infinite horizon games but can not be used in finite horizon games. The discovery of monitoring structures that allow agents to detect each others' deviations in finite horizon games where the key assumption of Sect. 5.4 is relaxed and the information structure is similar to that of [35] is a challenging and important open problem that is closely connected to cyber security games.

To alleviate the difficulties arising when the key assumption of Sect. 5.4 is relaxed and public monitoring mechanisms are not in place we can focus on *belief-free equilibria*. An equilibrium is belief-free if, after each history profile, each agent's continuation strategy is optimal independently of his beliefs' of the other agents' history profiles. Game theorists have analyzed and solved repeated infinite horizon discounted games with private imperfect state information, observable actions, and belief-free equilibrium as the solution concept (see [4,13,14,43] and references therein). The analysis and solution of games where the underlying system is dynamic, the agents' private state observations are imperfect (noisy), actions are hidden, and the solution concept is belief-free equilibrium, is an important class of open problems. Such problems are tightly connected to cyber security as they capture several important key features of cyber security games.

Acknowledgments. This work was supported in part by the NSF grants CNS-1238962, ARO-MURI grant W911NF-13-1-0421, and ARO grant W911NF-17-1-0232.

References

1. Aumann, R.: Agreeing to disagree. Ann. Stat. **4**, 1236–1239 (1976)
2. Aumann, R., Maschler, M., Stearns, R.: Repeated Games with Incomplete Information. MIT Press, Cambridge (1995)

3. Cardaliaguet, P., Rainer, C., Rosenberg, D., Vieille, N.: Markov games with frequent actions and incomplete information-the limit case. Math. Oper. Res. **41**(1), 49–71 (2015)
4. Ely, J., Hörner, J., Olszewski, W.: Belief-free equilibria in repeated games. Econometrica **73**(2), 377–415 (2005)
5. Escobar, J., Toikka, J.: Efficiency in games with Markovian private information. Econometrica **81**(5), 1887–1934 (2013)
6. Etherington, D., Conger, K.: Large DDos attacks cause outages at Twitter, spotify, and other sites. TechCrunch, Np, 21 (2016)
7. Finkle, J., Skariachan, D.: Target cyber breach hits 40 million payment cards at holiday peak. http://www.reuters.com/article/us-target-breach-idUSBRE9BH1GX20131219. Accessed 09 Sept 2016
8. Fudenberg, D., Levine, D., Maskin, E.: The folk theorem with imperfect public information. Econometrica **62**(5), 997 (1994)
9. Fudenberg, D., Tirole, J.: Game theory. Cambridge, Massachusetts, vol. 393, no. 12, p. 80 (1991)
10. Gensbittel, F., Renault, J.: The value of Markov chain games with incomplete information on both sides. Math. Oper. Res. **40**(4), 820–841 (2015)
11. Greenberg, A.: Hackers remotely kill a jeep on the highway? With me in it, Wired. https://www.wired.com/2015/07/hackers-remotely-kill-jeep-highway. Accessed 15 Dec 2016
12. Gupta, A., Nayyar, A., Langbort, C., Başar, T.: Common information based Markov perfect equilibria for linear-Gaussian games with asymmetric information. SIAM J. Control Optim. **52**(5), 3228–3260 (2014)
13. Hörner, J., Lovo, S.: Belief-free equilibria in games with incomplete information. Econometrica **77**(2), 453–487 (2009)
14. Hörner, J., Lovo, S., Tomala, T.: Belief-free equilibria in games with incomplete information: characterization and existence. J. Econ. Theory **146**(5), 1770–1795 (2011)
15. Hörner, J., Sugaya, T., Takahashi, S., Vieille, N.: Recursive methods in discounted stochastic games: an algorithm for $\delta \to 1$ and a folk theorem. Econometrica **79**(4), 1277–1318 (2011)
16. Li, L., Langbort, C., Shamma, J.: Solving two-player zero-sum repeated Bayesian games. arXiv preprint arXiv:1703.01957 (2017)
17. Li, L., Shamma, J.: LP formulation of asymmetric zero-sum stochastic games. In: 2014 IEEE 53rd Annual Conference on Decision and Control (CDC), pp. 1930–1935. IEEE (2014)
18. Mahajan, A., Mannan, M.: Decentralized stochastic control. Ann. Oper. Res. **241**(1–2), 109–126 (2016)
19. Mailath, G., Samuelson, L.: Repeated Games and Reputations. Oxford University Press, Oxford (2006)
20. Maskin, E., Fudenberg, D.: The folk theorem in repeated games with discounting or with incomplete information. Econometrica **53**(3), 533–554 (1986)
21. Maskin, E., Tirole, J.: Markov perfect equilibrium: I. Observable actions. J. Econ. Theory **100**(2), 191–219 (2001)
22. Matsushima, H.: Repeated games with private monitoring: two players. Econometrica **72**(3), 823–852 (2004)
23. Miehling, E., Rasouli, M., Teneketzis, D.: A POMDP approach to the dynamic defense of large-scale cyber networks. IEEE Trans. Inf. Forensics Secur. **13**(10), 2490–2505 (2018)

24. Miller, D.: Robust collusion with private information. Rev. Econ. Stud. **79**(2), 778–811 (2012)
25. Myerson, R.: Game Theory. Harvard University Press, Cambridge (2013)
26. Nayyar, A., Gupta, A.: Information structures and values in zero-sum stochastic games. In: 2017 American Control Conference (ACC), pp. 3658–3663. IEEE (2017)
27. Nayyar, A., Gupta, A., Langbort, C., Başar, T.: Common information based Markov perfect equilibria for stochastic games with asymmetric information: finite games. IEEE Trans. Autom. Control **59**(3), 555–570 (2014)
28. Department of Homeland Security. Industrial control systems cyber emergency response team (ICS-CERT). https://ics-cert.us-cert.gov
29. Osborne, M., Rubinstein, A.: A Course in Game Theory. MIT Press, Cambridge (1994)
30. Ouyang, Y., Tavafoghi, H., Teneketzis, D.: Dynamic oligopoly games with private Markovian dynamics. In: 54th IEEE Conference on Decision and Control (CDC) (2015)
31. Ouyang, Y., Tavafoghi, H., Teneketzis, D.: Dynamic games with asymmetric information: common information based perfect Bayesian equilibria and sequential decomposition. IEEE Trans. Autom. Control **62**(1), 222–237 (2017)
32. Renault, J.: The value of Markov chain games with lack of information on one side. Math. Oper. Res. **31**(3), 490–512 (2006)
33. Sinha, A., Anastasopoulos, A.: Structured perfect Bayesian equilibrium in infinite horizon dynamic games with asymmetric information. In: American Control Conference (2016)
34. Sugaya, T.: Folk theorem in repeated games with private monitoring. Working Paper (2019)
35. Sugaya, T.: Folk theorem in stochastic games with private state and private monitoring. Working Paper (2012)
36. Tavafoghi, H.: On design and analysis of cyber-physical systems with strategic agents. Ph.D. thesis, University of Michigan (2017)
37. Tavafoghi, H., Ouyang, Y., Teneketzis, D.: On stochastic dynamic games with delayed sharing information structure. In: 2016 IEEE 55th Conference on Decision and Control (CDC), pp. 7002–7009. IEEE (2016)
38. Tavafoghi, H., Ouyang, Y., Teneketzis, D.: A unified approach to dynamic decision problems with asymmetric information-part i: non-strategic agents. Submitted to IEEE Transactions on Automatic Control (2018). arXiv:1812.01130
39. Tavafoghi, H., Ouyang, Y., Teneketzis, D.: A unified approach to dynamic decision problems with asymmetric information-part ii: strategic agents. Submitted to IEEE Transactions on Automatic Control (2018). arXiv:1812.01132
40. Vasal, D., Anastasopoulos, A.: Signaling equilibria for dynamic LQG games with asymmetric information. In: 55th IEEE Conference on Decision and Control (CDC), pp. 6901–6908. IEEE (2016)
41. Washburn, R.B., Teneketzis, D.: Asymptotic agreement among communicating decisionmakers. Stoch.: Int. J. Prob. Stoch. Processes **13**(1–2), 103–129 (1984)
42. Watson, J.: Perfect Bayesian equilibrium: general definitions and illustrations. Working Paper (2016)
43. Yamamoto, Y.: A limit characterization of belief-free equilibrium payoffs in repeated games. J. Econ. Theory **144**(2), 802–824 (2009)
44. Zamir, S.: Repeated games of incomplete information: zero-sum. Handb. Game Theory **1**, 109–154 (1992)

Reinforcement Learning for Adaptive Cyber Defense Against Zero-Day Attacks

Zhisheng Hu[1], Ping Chen[2], Minghui Zhu[1(✉)], and Peng Liu[3]

[1] School of Electrical Engineering and Computer Science,
Pennsylvania State University, University Park, PA 16802, USA
{zxh128,muz16}@psu.edu
[2] JD.com, Beijing 10111, China
chenping19851@hotmail.com
[3] College of Information Sciences and Technology, Pennsylvania State University,
University Park, PA 16802, USA
pliu@ist.psu.edu

Abstract. In this chapter, we leverage reinforcement learning as a unified framework to design effective adaptive cyber defenses against zero-day attacks. Reinforcement learning is an integration of control theory and machine learning. A salient feature of reinforcement learning is that it does not require the defender to know critical information of zero-day attacks (e.g., their attack targets, and the locations of the vulnerabilities). This information is difficult, if not impossible, for the defender to gather in advance. The reinforcement learning based schemes are applied to defeat three classes of attacks: strategic attacks where the interactions between an attacker and a defender are modeled as a non-cooperative game; non-strategic random attacks where the attacker chooses its actions by following a predetermined probability distribution; and attacks depicted by Bayesian attack graphs where the attacker exploits combinations of multiple known or zero-day vulnerabilities to compromise machines in a network.

1 Emerging Adaptive Cyber Defense

A vulnerability is a flaw or weakness that could be exploited to violate system security policies [1]. Based on vulnerability window [2], vulnerabilities can be classified as normal vulnerabilities and zero-day vulnerabilities. Normal vulnerabilities are disclosed to the public and the security patches are published by the software vendor before any attacker could craft workable exploits. Zero-day vulnerabilities can be actively exploited before software engineers develop any patch to fix the vulnerabilities. As a result, zero-day vulnerabilities are particularly dangerous for information and communications technology (ICT) systems and might cause serious and lasting damage. According to [3], 24–55% of HTTPS servers in the Alexa Top 1 Million were initially vulnerable to Heartbleed [4], including GitHub, Stack Overflow, Imgur and 3% of HTTPS sites in the Alexa Top 1 Million remained vulnerable almost two months after the disclosure of

© Springer Nature Switzerland AG 2019
S. Jajodia et al. (Eds.): Adaptive Cyber Defense, LNCS 11830, pp. 54–93, 2019.
https://doi.org/10.1007/978-3-030-30719-6_4

Heartbleed. In addition, the number of zero-day vulnerabilities is trending up. According to the 2016 Internet security threat report [5], 24 new zero-day vulnerabilities were found in 2014, while the number increased to 54 in 2015.

Traditional defenses are effective against normal attacks but powerless for zero-day attacks. In fact, traditional defenses against zero-day attacks are mainly governed by slow and deliberative processes such as testing and patching. The threat report [6] pointed out that it took on average 59 days to release available patches for the top five zero-day vulnerabilities in 2014. The vacancy of effective defenses and the attackers' asymmetric time advantage versus the defenders' time to detect and respond motivate the recent study of moving target defense (MTD) [7]. MTD techniques dynamically and proactively reconfigure deployed defenses so as to increase uncertainty and complexity for attackers during vulnerability windows. Existing MTD techniques mainly utilize the following two proactive defense ideas: (1) leveraging software and platform diversity [8–12] to make zero-day vulnerabilities less exploitable; and (2) dynamically changing defense postures [13–17] to make defenses less predictable. For example, papers [8,10,11] randomize the implementations of programs to introduce uncertainty in the target. The randomization forces attackers to probe each system individually and substantially raises the bar on exploitation.

Existing MTD techniques require manual efforts to determine which system configurations should be deployed based on engineering experience. As a result, they suffer from two main limitations. First, manual selection could be very time-consuming. For example, the defense technique in [17] requires experienced security analysts to find out the security-sensitive data structures among all the data structures before data structure randomization. To achieve it, the analysts need to carefully go through all the authentication information and function pointers for a specific program. Secondly, manually selected configurations might not be able to achieve optimal cost-effectiveness. Also in [17], the authors admit that it is difficult to balance between security and efficiency. Their technique cannot guarantee that the randomized data structures are the attacker's real targets and the system could suffer as high as 28.8% runtime overhead even only when 20% of data structures are selected to be randomized.

To address the limitations of MTD, a new concept, adaptive cyber defense (ACD) [18], has been introduced. ACD integrates MTD with rigorous methodologies (e.g., game theory, machine learning and control theory) to answer the crucial question: how to optimally deploy available MTD techniques? However, the question is very challenging to answer especially when facing zero-day attacks. As discussed, the defender cannot access critical information of zero-day attacks (e.g., their attack targets, and the locations of the vulnerabilities). The limited information prevents the defender from choosing optimal defense actions. As a result, the defender might waste lots of resources on reconfiguring the irrelevant parts of the ICT systems and still leave the ICT systems vulnerable. In this chapter, we will leverage reinforcement learning [19,20] as a unified framework to address the challenges caused by limited prior information on zero-day attacks.

2 Reinforcement Learning for Adaptive Cyber Defense

Reinforcement learning is a systematic methodology to tackle scenarios where a decision-maker interacts with unknown environments. The idea behind reinforcement learning is that the decision-maker learns through continuously interacting with the environment and receiving feedbacks for chosen actions. The idea is inspired by human's nature. Consider a new born baby who sees a fireplace in the room. The baby approaches the fireplace, and he/she gets warmer. He/she receives a positive feedback of the action of approaching the fireplace. When the baby touches the fire, he/she gets hurt, receiving a negative feedback of the action of touching the fireplace. More generally, the decision-maker starts from an initial state in the environment (e.g., the initial location of the baby in the room), and takes an action (e.g., approach, stay still or depart). The decision-maker transitions to a new state, and receives some feedbacks (e.g., warm, cold or hurt at current time) from the environment. Reinforcement learning outputs a loop of state, action and feedbacks so that the decision-maker can achieve maximum cumulative feedback (e.g., total time that the baby feels warm).

In reinforcement learning, the decision-maker on one hand selects actions on basis of its past experience (exploitation), and on the other hand tries new choices (exploration). Essentially, reinforcement learning is trial-and-error learning. The decision-maker receives feedbacks, which evaluate deployed actions, and seeks to learn how to select the best action to accomplish a given mission. The decision-maker only needs to access induced feedbacks but is not required to know all thefactors which determine feedbacks. So reinforcement learning is particularly well suited to problems where the decision-maker interacts with a partially unknown environment but can evaluate its actions via repeated interactions. This intriguing feature well matches the needs of ACD because observable feedbacks can be generated by the ICT systems due to some preliminary and universal defenses deployed in the systems. For example, when defending Heartbleed attacks, the defender cannot predict which part of the memory is going to be over-read by the attacker. However, when attack requests read beyond buffer boundaries and touch guard pages [21], segmentation faults are triggered [22]. Segmentation faults can be observed by the defender and used to evaluate the cost-effectiveness of previously deployed defenses. On the other hand, reinforcement learning requires continuous interactions to provide adequate learning experience. When applied to ACD, reinforcement learning is limited to scenarios where the attacker probes victim systems for a large number of times during a relatively long period. Heartbleed attack and data structure manipulation attack (DSMA) [17] are two examples of continuous attacks. In particular, Heartbleed attack typically requires hundreds of thousands of probing (buffer over-read) requests and lasts hours to steal sensitive information like secret keys. And DSMA typically requires millions of attempts and lasts hours or even days to execute desired functions.

With the above insights, this chapter aims to design, analyze and evaluate effective ACD schemes by integrating reinforcement learning with MTD techniques. Figure 1 shows the generic architecture of reinforcement learning based ACD. The remaining of the chapter is organized as follows. Section 3 introduces an reinforcement learning algorithm against zero-day strategic attacks. The introduced algorithm allows the players to identify Nash equilibrium where each player only uses its own deployed actions and its received utility values in recent history. Section 4 introduces an adaptive defense against zero-day nonstrategic random attacks. The introduced algorithm can guarantee that the regret is upper bounded by a logarithmic function of the number of defense cycles no matter what probability distribution the attacker follows. Section 5 introduces reinforcement learning algorithms on Bayesian attack graphs. The simulation results confirm that our algorithms enable the defender to identify effective defense policies when utility functions are unknown.

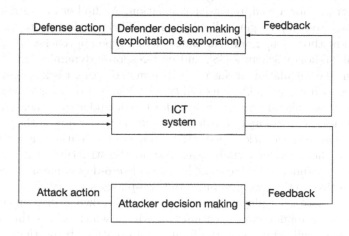

Fig. 1. Reinforcement learning based ACD.

3 Game Theoretic Reinforcement Learning Against Strategic Attacks

(Non-cooperative) game theory provides a mathematically rigorous framework for multiple players to reason about each other and make choices out of their own interest, without considering the overall outcomes of the game (e.g., rock-paper-scissor). The notion of Nash equilibrium [23] is a widely-used solution concept and describes the stable outcome where none can benefit from unilateral deviations. The interactions between an attacker and a defender can be naturally modeled as a non-cooperative game [24–26] in which the defender aims to minimize security risk while the attacker aims to maximize damage. And the solution concept Nash equilibrium provides important guidance of how to deploy defenses. For examples, [27] proposed a stochastic routing framework to make packets take random paths from a source to a destination and increase the complexity of connection eavesdropping attacks. [28] developed a deceptive routing game which is used to design fake routes in wireless communication networks and prevent attackers from jamming the legitimate routes. [29] proposed a framework to design optimal policy of randomizing the IP address space in order to avoid detection of the real nodes.

Numerical algorithms are synthesized for the players to identify Nash equilibria via repeated interactions. Multi-player games can be categorized into discrete games and continuous games. In a discrete (resp. continuous) game, each player has a finite (resp. an infinite) number of action candidates. As for discrete games, learning algorithms include best-response dynamics, better-response dynamics, factitious play, regret matching, logit-based dynamics and replicator dynamics. Please refer to [30–33] for detailed discussion. As an important class of continuous games, generalized Nash games were first formulated in [34], and see survey paper [35] for a comprehensive exposition. A number of algorithms have been proposed to compute generalized Nash equilibria, including, to name a few, ODE-based methods [36], nonlinear Gauss-Seidel-type approaches [37], iterative primal-dual Tikhonov schemes [38], and best-response dynamics [39].

However, the calculation of Nash equilibrium in existing work is mainly offline and requires each player to know its utility function, its opponents' action space. In cyber security, this information is hard for the defender to gather when facing zero-day attacks. For example, the defender may not be aware of the actions the attacker's because the attack targets of zero-day vulnerabilities are not available. Additionally, the defender usually cannot know the structure of its own or the attacker's utility function because the impacts of zero-day vulnerabilities are not available. In this section, we generalize the two-player security game to a multi-player game where each player aims to maximize its own utility function. And then we design a reinforcement learning algorithm which allows the players to compute Nash equilibrium with the limited information. In particular, at each iteration, each player, on one hand, exploits successful actions in recent history via comparing received utility values, and on the other hand, randomly explores any feasible action with a certain exploration rate.

3.1   Game Theoretic Model and Problem Formulation

As mentioned, the interactions between a strategic attacker and a defender can be modeled as a non-cooperative game where two players seek for conflicting objectives. To be more generic, we present the following model that characterizes the interactions of N players in a non-cooperative game. When the player number is 2, it will reduce to the non-cooperative game between the attacker and the defender. The model is demonstrated in Fig. 2 and each component in the figure will be discussed in the following paragraphs.

Players. We consider N players $\mathcal{V} \triangleq \{1, \cdots, N\}$ and each player has a finite set of actions. Let $\mathcal{A}_i$ denote the action set of player i and $a^i \in \mathcal{A}_i$ denote an action of player i. Denote $\mathcal{S} \triangleq \mathcal{A}_1 \times \cdots \times \mathcal{A}_N$ as the Cartesian product of the action sets, where $s \triangleq (a^1, \cdots, a^N) \in \mathcal{S}$ is denoted as an action profile of the players.

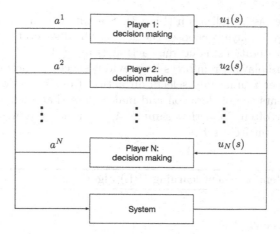

Fig. 2. Game model

Utility. Under the influence of an action profile, the system generates a utility value for each player. The utility function for player $i \in \mathcal{V}$ is defined as $u_i : \mathcal{S} \to \mathbb{R}$. At the end of iteration t, the utility value $u_i(t) = u_i(s(t))$ is measured and sent to player i.

Informational Constraint. Each player does not know the other players' action sets or their deployed actions. Besides, each player is unaware of the structure of its own or the others' utility functions. At iteration t, each player only knows its deployed actions and its received utility values in the past; i.e., $a^i(0), \cdots, a^i(t-1), u_i(0), \cdots, u_i(t-1)$.

Problem Statement. As mentioned, Nash equilibrium describes the stable outcome of the game and provides important guidance of how to deploy defenses. However, the calculation of Nash equilibrium usually requires the defender to know the deployed actions of the attacker and its own utility function. This information is hard for the defender to gather when facing zero-day attacks. We aim to synthesize a learning algorithm under which the action profiles of the two players converge to Nash equilibria. We also quantify the convergence rate of the proposed algorithm in contrast to asymptotic convergence in existing work.

3.2 Learning Algorithm: The RL Algorithm

We propose a learning algorithm called the RL algorithm under the above informational constraint, where each player updates its actions only based on its previous actions and its received utility values. On the one hand, each player chooses the most successful action in recent history. It represents the exploitation phase. However, the exploitation is not sufficient to guarantee that the player can choose the best action given others'. So on the other hand, the player uniformly chooses one action from its action set. It represents the exploration phase. At iterations $t = 0$ and $t = 1$, each player uniformly chooses one action from its action set as initialization (Line 3). Starting from iteration $t = 2$, with probability $1 - \tilde{\epsilon}_i(t)$, player i chooses the action which generates a higher utility value in last two iterations as current action (Line 8–13). This represents the exploitation where player i reinforces its previous successful actions. With probability $\tilde{\epsilon}_i(t)$, player i uniformly selects an action from its action set $\mathcal{A}_i$ (Line 14). This represents the exploration and makes sure that each action profile is selected infinitely often. Note that sample($\mathcal{A}_i$) in Line 14 represents uniformly choosing one element from set $\mathcal{A}_i$.

Algorithm 1. Reinforcement learning (RL) algorithm

1: **while** $0 \leq t \leq 1$ **do**
2:    **for** $i \in \mathcal{V}$ **do**
3:       $a^i(t) \leftarrow$ sample($\mathcal{A}_i$);
4:    **end for**
5: **end while**
6: **while** $t \geq 2$ **do**
7:    **for** $i \in \mathcal{V}$ **do**
8:       With prob. $(1 - \tilde{\epsilon}_i(t))$,
9:       **if** $u_i(t-1) \geq u_i(t-2)$ **then**
10:      $a^i(t) = a^i(t-1)$;
11:      **else**
12:      $a^i(t) = a^i(t-2)$;
13:      **end if**
14:      With prob. $\tilde{\epsilon}_i(t)$, $a^i(t) \leftarrow$ sample($\mathcal{A}_i$);
15:    **end for**
16: **end while**

3.3   Convergence Analysis

We will present the analytical results of the RL algorithm. In particular, we prove the convergence of the RL algorithm to the set of pure Nash equilibria when the interactions of the players consist of a weakly acyclic game. And we quantify the convergence rate of the proposed algorithm. But first let us introduce the notations and assumptions used throughout the section. Denote by $|\mathcal{V}|$ the cardinality of player set, $|\mathcal{A}_i|$ the cardinality of action set of player i and $|\mathcal{A}|_\infty \triangleq \max_{i \in \mathcal{V}} |\mathcal{A}_i|$ the maximum cardinality among all action sets. The exploration rate for player i at iteration t is decomposed into two parts; i.e., $\tilde{\epsilon}_i(t) \triangleq \epsilon_i(t) + e_i(t) \in (0,1]$, where $\epsilon_i(t) = \gamma_i \epsilon^c(t)$, $\gamma_i > 0$, $\epsilon^c(t)$ is common for all the players and $e_i(t)$ represents the exploration deviation. Define $e(t) \triangleq (e_1(t), \cdots, e_N(t))^T$, $\tilde{\epsilon}(t) \triangleq (\tilde{\epsilon}_1(t), \cdots, \tilde{\epsilon}_N(t))^T$ and $\epsilon(t) \triangleq (\epsilon_1(t), \cdots, \epsilon_N(t))^T$. And we define $e_r(t) \triangleq ||e(t)||_\infty^N / \prod_{i=1}^{N} \tilde{\epsilon}_i(t)$. Here we denote by $|| \cdot ||_\infty$ the infinity norm of a vector. In addition, we also use $|| \cdot ||$ to represent the L^1-norm of a vector, and $||P||$ to represent the 1-norm of a matrix.

Assumption M-1. (1). For each $i \in \mathcal{V}$, $\epsilon_i(t) \in (0,1]$ is non-negative, strictly decreasing, and $\lim_{t \to \infty} \epsilon_i(t) = 0$. (2). The sequences $\{\prod_{i=1}^{N} \epsilon_i(t)\}$ and $\{\prod_{i=1}^{N} \tilde{\epsilon}_i(t)\}$ are not summable. (3). $\lim_{t \to \infty} e_r(t) = 0$.

Assumption 3-1 indicates that the players can choose heterogeneous exploration rates. The exploration rates diminish slowly enough and their deviations decrease in faster rates than the common part. In the paper [40], it is assumed that exploration rates $\epsilon_i(t)$ are identical for all i, diminishing and not summable. Assumption 3-1 allows for heterogeneous exploration rates and includes homogeneous exploration rates in the paper [40] as a special case. Actually, papers [41,42] adopt heterogenous step-sizes for distributed optimization and game theory. They impose similar assumptions on the step-sizes.

Markov Chain Induced by the RL Algorithm. Denote by $\mathcal{Z} \triangleq \mathcal{S} \times \mathcal{S}$ the state space, where each state $z(t) \triangleq (s(t), s(t+1))$ consists of the action profiles at iteration t and the next iteration. And denote by $diag(\mathcal{S} \times \mathcal{S}) \triangleq \{(s,s)|s \in \mathcal{S}\}$ the diagonal space of $\mathcal{Z}$. By the definition of $z(t)$, the sequence $\{z(t)\}_{t \geq 0}$ forms a time-inhomogeneous Markov chain, denoted by $\mathcal{M}$. We define $P^{\tilde{\epsilon}(t)}$ as the transition matrix of Markov chain $\mathcal{M}$ at iteration t, where each entry $P^{\tilde{\epsilon}(t)}(z', z)$ represents the transition probability from state z' to z. Besides, denote by $\pi(t)$ the distribution on $\mathcal{Z}$ at iteration t.

Now we consider the interactions of the players consist of a weakly acyclic game. A game is called to be weakly acyclic if from every action profile, there exists a finite best-response improvement path leading from the action profile to a pure Nash equilibrium.

Definition 1. An action profile $s_* \triangleq (a_*^1, \cdots, a_*^i, \cdots, a_*^N)$ is a pure Nash equilibrium if $\forall i \in \mathcal{V}, \forall a^i \in \mathcal{A}_i$, $u_i(s_*) \geq u_i(a^i, a_*^{-i})$.

We will show the convergence of the RL algorithm to the set of pure Nash equilibria when the game is weakly acyclic. But let us continue introducing the notations for analysis.

z-tree of Time-Homogenous Markov Chain $\mathcal{M}^{\tilde{\epsilon}}$. Given any two distinct states z' and z of Markov chain $\mathcal{M}^{\tilde{\epsilon}}$, consider all paths starting from z' and ending at z. Denote by $p_{z'z}$ the largest probability among all possible paths from z' to z. A path might contain intermediate states $z_1, \cdots, z_k$ ($k = 0$ means there is no intermediate state) between z' and z. So $p_{z'z}$ is the product of $P^{\tilde{\epsilon}}(z', z_1), P^{\tilde{\epsilon}}(z_1, z_2), \cdots, P^{\tilde{\epsilon}}(z_k, z)$. We define graph $\mathcal{G}(\tilde{\epsilon})$ where each vertex of $\mathcal{G}(\tilde{\epsilon})$ is a state z of Markov chain $\mathcal{M}^{\tilde{\epsilon}}$ and the probability on edge (z', z) is $p_{z'z}$. A z-tree on $\mathcal{G}(\tilde{\epsilon})$ is a spanning tree rooted at z such that from every vertex $z' \neq z$, there is a unique path from z' to z. Denote by $G_{\tilde{\epsilon}}(z)$ the set of all z-trees on $\mathcal{G}(\tilde{\epsilon})$. The total probability of a z-tree is the product of the probabilities of its edges. The *stochastic potential* of the state z is the largest total probability among all z-trees in $G_{\tilde{\epsilon}}(z)$. Let $\Lambda(\tilde{\epsilon})$ be the states which have maximum stochastic potential for a particular $\tilde{\epsilon} \in (0,1]^N$. Denote the limit set $\Lambda^* \triangleq \lim_{\tilde{\epsilon} \to 0} \Lambda(\tilde{\epsilon})$. And the elements in Λ^* are referred to as *stochastically stable states*.

Remark 1. The above notions are inspired by the resistance trees theory [43]. However, the above notions are defined for any $\tilde{\epsilon} \in (0, 1]$ instead of $\tilde{\epsilon} \to 0$ in the resistance trees theory. This allows us to characterize the transient performance of the RL algorithm.   □

Denote the set of pure Nash equilibria of the game Γ as $\mathcal{N}(\Gamma)$ and $diag(\mathcal{N}(\Gamma) \times \mathcal{N}(\Gamma)) \triangleq \{(s,s) | s \in \mathcal{N}(\Gamma)\}$. The following corollary implies that the action profiles converge to $\mathcal{N}(\Gamma)$ with probability one.

The following theorem is the main analytical result of this section. It shows that the state $z(t)$ converges to the set of stochastically stable state with probability one. Moreover, the convergence rate is quantified using the distance between $\pi(t)$ and the limiting distribution π^*; i.e., $D(t) \triangleq ||\pi(t) - \pi^*||$. The formal proof of Theorem 1 can be found in [44].

Theorem 1. *Let Assumption 3-1 hold and the N-player game Γ be a weakly acyclic game, the following properties hold for the RL algorithm:*

(P1) $\lim_{t \to \infty} Pr\{z(t) \in \Lambda^*\} = 1$ *and* $\Lambda^* \subseteq diag(\mathcal{N}(\Gamma) \times \mathcal{N}(\Gamma))$;

(P2) there exist positive integer t_{min} and positive constant C such that for any $t^* > t_{min}$ and $t \geq t^* + 1$, the following is true:

$$D(t) \leq C(||\epsilon(t^*)||_\infty + ||\epsilon(t)||_\infty + e_r(t^*)$$

$$+ \exp(-\sum_{\tau=t^*}^{t-1} \prod_{i=1}^{N} \epsilon_i(\tau)|\mathcal{A}_i|) + \exp(-\sum_{\tau=t^*}^{t-1} \prod_{i=1}^{N} \tilde{\epsilon}_i(\tau)|\mathcal{A}_i|)). \qquad (1)$$

3.4   Evaluation

We evaluate the RL algorithm on a real-world cyber security scenario. The server containing several zero-day security vulnerabilities. A zero-day attack happens once that a software/hardware vulnerability is exploited by the attacker before software the engineers develop any patch to fix the vulnerability. Here the number of players is 2 and the attacker is equipped with a set of zero-day attack scripts denoted as $\mathcal{A}$ while the defender is equipped with a set of platforms denoted as $\mathcal{D}$. The defender uses a defensive technique called dynamic platforms [12], which changes the properties of the server such that it is harder for the attacker to succeed. In particular, the defender periodically restarts the server, chooses one platform from $\mathcal{D}$ and deploys it on the server each time it restarts the server. The action $d(t)$ is the platform deployed at iteration t. The attacker periodically chooses one of the attack scripts to attack the server. Notice that the attack period is often smaller than the defense period because the defender cannot restart the server too frequently due to the resource consumption of restarting the server. In fact, the defense period is usually a multiple of the attack period. The attack action at iteration t, denoted as $a(t)$, is a subset of the attack scripts and the attack action set $\mathcal{A}$ includes all possible subsets. The order of choosing the attack scripts in one iteration does not matter.

Once an attack action succeeds, the attacker can control the server for a certain amount of time. And every time the server restarts, the defender takes over the control of the server. And here we assume the time consumed by the attack scripts to succeed and the time consumed by restarting the server are negligible compared with the length of an iteration. The goal of both the attacker and defender is to gain longer control time of the server. The utility of the defender $u_d(d(t), a(t))$ is the fraction of the time controlled by the defender during iteration t and the utility of the attacker $u_a(d(t), a(t))$ is the fraction of the time controlled by the attacker. Notice that $u_d(d(t), a(t)) + u_a(d(t), a(t)) = 1$.

Evaluation Setup. In this section, we use Matlab simulations to evaluate the performance of our algorithm based on real-world platform settings, attack scripts and server control data [12, 45]. The total number of defense actions is 5; i.e., the defender has five different platforms: Fedora 11 on x86, Gentoo 9 on x86, Debian 6 on x86, FreeBSD 9 on x86, and CentOS 6.3 on x86. The attacker has two zero-day attack scripts: TCP MAXSEG exploit, and Socket Pairs exploit. The defense period is set to be ten times as large as the attack period; i.e., during one iteration, the attacker launches 10 attack scripts. Since the time consumed by the attack scripts to succeed is negligible, one attack script enables the attacker control $\frac{1}{10}$ of the iteration if it succeeds. The total number of attack actions is 11; i.e., $a_1 = (0, 10), a_2 = (1, 9), \cdots, a_{11} = (10, 0)$, where $a(t) = (0, 10)$ means the attacker launches 0 TCP MAXSEG exploit and 10 Socket Pairs exploits at iteration t.

Real-World Utility Values. Based on the evaluation setup and the real-world attack scripts, we first replay different attack actions on different platforms to get the utility table for the defender and the attacker. The results are shown in Table 1, where the defender is the row player and the attacker is the column player. In each cell, the first number represents the utility value to the defender, and the second number represents the utility value to the attacker.

Nash Equilibrium. By Proposition 1 in [46], we know any 2-player finite game and its any sub-game (any game constructed by restricting the set of actions to a subset of the set of actions in the original game) has at least one pure Nash equilibrium is a weakly acyclic game. From Table 1, we can see any sub-game has at least one pure Nash equilibrium. Now we want to calculate the pure Nash equilibrium (equilibria). From Table 1, we can see if the defense strategy is d_4 (deploying Debian 6) or d_5 (deploying FreeBSD 9), then the utility of the defender is 1 (the utility of the attacker is 0) not matter what action the attacker uses. From Definition 1, we know the combinations of any attacker action and defense action d_4 or d_5 are pure Nash equilibria.

Simulation Results. Based on Table 1, we simulate the interactions of the defender and attacker in Matlab. We choose the exploration rates $\epsilon_d(t) = \epsilon_a(t) = \frac{1}{11t^{1/2}}$. The exploration deviations are chosen as $e_d(t) = \frac{1}{110t^2}$ and $e_a(t) = \frac{1}{110t^2}$. The duration of each simulation (from the attack begins till the attack ends)

Table 1. Utility table

Defense actions	Attack actions					
	(0,10)	(1,9)	(2,8)	(3,7)	(4,6)	(5,5)
d_1(Fedora 11)	0,1.0	0.1,0.9	0.2,0.8	0.3,0.7	0.4,0.6	0.5,0.5
d_2(Gentoo 9)	1.0,0	0.9,0.1	0.8,0.2	0.7,0.3	0.6,0.4	0.5,0.5
d_3(CentOS 6.3)	0,1.0	0.1,0.9	0.2,0.8	0.3,0.7	0.4,0.6	0.5,0.5
d_4(Debian 6)	1.0,0	1.0,0	1.0,0	1.0,0	1.0,0	1.0,0
d_5(FreeBSD 9)	1.0,0	1.0,0	1.0,0	1.0,0	1.0,0	1.0,0
Defense actions	Attack actions					
	(6,4)	(7,3)	(8,2)	(9,1)	(10,0)	
d_1(Fedora 11)	0.6,0.4	0.7,0.3	0.8,0.2	0.9,0.1	1.0,0	
d_2(Gentoo 9)	0.4,0.6	0.3,0.7	0.2,0.8	0.1,0.9	0,1.0	
d_3(CentOS 6.3)	0.6,0.4	0.7,0.3	0.8,0.2	0.9,0.1	1.0,0	
d_4(Debian 6)	1.0,0	1.0,0	1.0,0	1.0,0	1.0,0	
d_5(FreeBSD 9)	1.0,0	1.0,0	1.0,0	1.0,0	1.0,0	

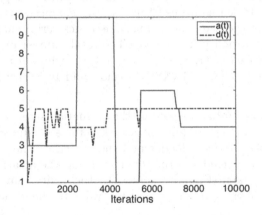

Fig. 3. Trajectories of attack $a(t)$ and defense $d(t)$ with diminishing exploration rate $\epsilon_d(t) = \epsilon_a(t) = \frac{1}{11t^{1/2}}$.

is 10,000 iterations and we repeat 100 identical simulations. Figure 3 shows the trajectories of the defense and attack actions in one certain simulation. And for each simulation, we record the defense action at each iteration. Then at each iteration t, we have 100 chosen defense actions and we use the number of each defense action over 100 as the probability of choosing such defense action at t. The result in Fig. 3 suggests that the defense action converges to the set $\{d_4, d_5\}$. Notice that the combinations of any attacker action and defense action d_4 or d_5 are pure Nash equilibria. Then the simulation results confirm that the convergence of the action profiles to the set of pure Nash equilibria.

4 Multi-arm Bandit Against Non-strategic Random Attacks

In Sect. 3, the attacker is assumed to be a rational decision-maker and willing to play the game and employ the action at Nash equilibrium. However, Nash equilibrium has several limitations in terms of practical implementations in security. First, if the attacker does not want to employ the action at Nash equilibrium, then the defender may be able to get a higher utility value by deviating from its action at Nash equilibrium. Second, in the game, the attacker is restricted to stationary actions. The assumption is easily violated if the attack is unwilling to settle at any stationary action. Third, the game between the defender and the attacker is a two-player finite matrix game. For this class of games, mixed Nash equilibrium always exists but pure Nash equilibrium may not [23]. Notice that mixed Nash equilibrium could be difficult to implement in practice. And based on our observation, not all attackers are rational and willing to stick to Nash equilibrium.

In this section, we investigate a complementary class of attacks which are not strategic and instead follow predetermined but unknown probability distributions. Zero-day continuous buffer over-flow attacks, including over-read attacks (e.g., Heartbleed attacks) and over-write attacks (e.g., non-control data attacks). The attackers cannot easily get some useful feedback from the system. All they can do is to keep over-reading/over-writing the memory until some desired information is obtained/crafted. The attackers have no explicit utility maximizing goal. This class of attacks is characterized by the salient feature that a short period of time is insufficient for the attacker to achieve its goal, and it probes victim systems for a large number of times during a relatively long period. For example, DSMAs, one kind of data-oriented attacks that focus on data structures, typically require millions of attempts and last hours or even days. Another observation is that the attack frequency is usually fixed for a specific attack script.

We use multi-arm bandit [47] to model the interactions between the defender and such attackers. Then an extension of the upper confidence bound (UCB) algorithm [48] is proposed to practically defeat the attacks by considering utility errors and delays. In our algorithm, the defender on one hand, exploits successful actions in history based on their induced average utility values, and on the other hand, explores the actions that are not chosen sufficiently often.

4.1 System Model and Problem Formulation

Multi-arm bandit provides a mathematically tractable model by extending the observations of these two classes of attacks. The model involves a server system and two decision-makers: a defender and an attacker. Each decision-maker has a set of available actions. Their actions jointly influence the security state of the server system. Given any pair of defense and attack actions, the server system may return some feedbacks to the defender. The feedbacks are used by the defender to evaluate the cost-effectiveness of previously deployed defense actions.

This evaluation will guide the defender to deploy future defense actions. Notice that the attacker is not necessarily aware of the feedbacks or unnecessarily changes its actions based on the feedbacks

4.1.1 Defender

The defender has a set of defense actions $\mathcal{D} \triangleq \{d_1, \cdots, d_n\}$. The defender periodically adjusts defense actions on the basis of the feedbacks provided by the server. The uniform time interval between two adjustments is denoted as a defense cycle. In this chapter, we consider only one attacker with a sequence of attack requests whose duration is denoted by a finite time horizon T and the time horizon T is a multiple of defense cycles: $T \triangleq \{t_1, \cdots, t_N\}$. Then for any defense cycle $t \in T$, we denote $a(t) \in \mathcal{A}$ as the attacker's action.

4.1.2 Attacker

The attacker has a set of attack actions, $\mathcal{A} \triangleq \{a_1, \cdots, a_m\}$, but does not necessarily have a utility-maximizing goal. Instead, we extend the attacker's behavior based on observations of existing DMSA scripts. The attacker associates each defense action $d \in \mathcal{D}$ with a predetermined probability distribution DA_d, observes the defender's current action $d(t)$, and then chooses its action $a(t)$ according to the probability distribution $DA_{d(t)}$.

4.1.3 Utility

Given a pair of defense and attack actions, the server system can generate some observable feedbacks. Then we define the feedback as utility in the form of $u = W_r r - W_c c$, where $r : \mathcal{D} \times \mathcal{A} \to \mathbb{R}$ (e.g., the number of failed attack tires), is the effectiveness and $c : \mathcal{D} \to \mathbb{R}$ (e.g., the overhead induced by the defense action) is the cost. And the constant weights W_r and W_c are chosen according to the preference of the defender on security and efficiency. And for defense cycle t, we denote $u(t) = u(d(t), a(t))$. The specific utility with respect to specific attacks will be discussed later. The utility values in all defense cycles are bounded. More formally, $\forall t \in T$, $u(t) \in [u^-, u^+]$, where u^- and u^+ are the lower and upper bounds. Additionally, the bounds are available to the defenders. The server might not be able to generate the utility value at the end of each defense cycle due to heavy server load. And the communication between the defender and the server may suffer from transmission delays [49]. Therefore, at the end of defense cycle t, the defender receives several utility values of previous defense cycles; i.e., $\boldsymbol{u}(t) = \left[u(t - \mathcal{T}_1) \ u(t - \mathcal{T}_2) \ \cdots \ u(t - \mathcal{T}_{h(t)}) \right]^T$, where $\mathcal{T}_1, \cdots, \mathcal{T}_{h(t)}$ are the nonnegative delays, and $h(t)$ is the number of the utility values received by the defender at t. The utility value of each defense cycle once is received once. The utility delays are uniformly upper bounded by $\mathcal{T}$; i.e., $\forall t \in T, \forall i \in \{1, \cdots, h(t)\}, \mathcal{T}_i \leq \mathcal{T}$. We also consider that the received utility values may contain estimation or transmission errors. The utility error for each defense cycle t is defined as $\epsilon(t)$. Then for any defense cycle t, the defender receives several utility values with errors $\boldsymbol{u}'(t) = \boldsymbol{u}(t) + \left[\epsilon(t - \mathcal{T}_1) \cdots \epsilon(t - \mathcal{T}_{h(t)})\right]^T$. Similarly, the utility errors are uniformly upper bounded by ϵ, i.e., $\forall t \in T, |\epsilon(t)| \leq \epsilon$.

The extension of the attacker's behavior and the effect of the utility are formally described the following assumption.

Assumption 4-1. The utility value in defense cycle t, $u(t)$, is also a random variable conditional on $d(t)$. That is, defense actions $\{d_1, \cdots, d_n\}$ can be characterized by utility distributions $\{v_{d_1}, \cdots, v_{d_n}\}$ respectively. And $\{\mu_{d_1}, \cdots, \mu_{d_n}\}$ are the associated expected utility values.

4.1.4   Regret

The regret of a defense algorithm φ is the distance between the aggregate utility value the defender receives if the zero-day attack is known and the aggregate utility value the defender receives if φ is used to defend against the same attack with limited information. Define an *optimal* action: $d^* \in \mathcal{D}$ with the highest *expected* utility value i.e., $d^* \triangleq \arg\max\limits_{d \in \mathcal{D}} \mu_d$. Then the regret is formally defined

as: $R_\varphi(N) = \sum\limits_{t=1}^{N} \mathbb{E}U(d^*, R_a(d^*), \boldsymbol{R}_{SA}(d^*)) - \sum\limits_{t=1}^{N} \mathbb{E}\mathbf{1}^T \boldsymbol{u}'(t)$, where $\sum\limits_{t=1}^{N} \mathbb{E}\mathbf{1}^T \boldsymbol{u}'(t) = $

$\sum\limits_{t=1}^{N} \sum\limits_{j=1}^{h(t)} \mathbb{E}\left(u(t - \mathcal{T}_j) + \epsilon(t - \mathcal{T}_j)\right)$, $\mathbf{1} = \mathbf{1}_{h(t)}$ is an $h(t)$ dimensional column vector, $u(t)$ is the error-free utility value. Then the defender aims to specify a strategy φ to minimize the regret $R_\varphi(N)$ without knowing DA_d.

4.2   Learning Algorithm: The UCB-Z Algorithm

We propose an adaptive defense algorithm, named the UCB-Z (UCB zero-day) algorithm, to guide the defender how to periodically update its actions. An attractive feature of the algorithm is that it only requires the previous defenses and their induced utility values. Further, the cost-effectiveness of the algorithm shows that it can provide worst-case cost-effectiveness guarantees among all possible unknown attacks. The UCB-Z algorithm is an extension of the UCB algorithm in Multi-armed Bandit problems [47,50]. For the ease of presentation, $T = \{t_1, \cdots, t_N\}$ will be referred to as $T = \{1, \cdots, N\}$ in the rest of the section. Before we introduce the steps of the algorithm, let us introduce a set of notations:

- $\mathbf{1}_{\{\Pi\}}$ is an indicator function: $\mathbf{1}_{\{\Pi\}} = 1$ if Π is true and $\mathbf{1}_{\{\Pi\}} = 0$ if Π is false.
- $T_d(t) = \sum\limits_{\tau=1}^{t-1} \mathbf{1}_{\{d(\tau)=d\}}$ is the number of times defense action d has been chosen by the defense cycle $t - 1$.
- $\forall d \in \mathcal{D}, \bar{\mu}'_d(t) = \frac{1}{T_d(t)} \sum\limits_{\tau=1}^{t-1} (\mathbf{1}^T \boldsymbol{u}'(\tau)\mathbf{1}_{\{d(\tau)=d\}})$ represents the *empirical average utility* the defender actually receives by choosing defense action d by the end of defense cycle $t - 1$.
- $\forall d \in \mathcal{D}, I_d(t) = \left(\bar{\mu}'_d(t) + (u^+ - u^-)\sqrt{\frac{2\ln(t)}{T_d(t)}}\right)$ represents the *upper confidence index* of action d at the beginning of defense cycle t.

Algorithm 2. The UCB-Z Algorithm

1: **for** $d \in \mathcal{D}$ **do**
2:     $T_d(1) = 0$;
3:     $\bar{\mu}'_d(1) = 0$;
4: **end for**
5: **for** $t = 1; t \leq N; t++$ **do**
6:     **for** $d \in \mathcal{D}$ **do**
7:         **if** $T_d(t) == 0$ **then**
8:             $I_d(t) = +\infty$;
9:         **else**
10:             $I_d(t) = \bar{\mu}'_d(t) + (u^+ - u^-)\sqrt{\frac{2\ln(t)}{T_d(t)}}$;
11:         **end if**
12:     **end for**
13:     $d(t) = \arg\max_{d \in \mathcal{D}} I_d(t)$;
14:     $T_{d(t)}(t+1) = T_{d(t)}(t) + 1$;
15:     **for** $d \in \mathcal{D} \setminus \{d(t)\}$ **do**
16:         $T_d(t+1) = T_d(t)$
17:     **end for**
18:     Defender receives $\boldsymbol{u}'(t)$;
19:     **for** $d \in \mathcal{D}$ **do**
20:         $\bar{\mu}'_d(t+1) = \frac{1}{T_d(t+1)} \sum_{\tau=1}^{t} (\mathbf{1}^T \boldsymbol{u}'(\tau)\mathbf{1}_{\{d(\tau)=d\}})$;
21:     **end for**
22: **end for**

The true cost-effectiveness of a defense action can be reflected by its expected utility value. And by the law of large number, we know the expected utility value is determined by the empirical average utility value if a defense action is chosen enough times. The UCB-Z algorithm is using average utility value (the first term of $I_d(t)$) to evaluate how well a defense works. To ensure each defense action can be chosen enough times, the algorithm adds a penalty term (the second term of $I_d(t)$). In particular, the penalty term for a particular defense action d explodes if the action is not chosen sufficiently often. For this case, $I_d(t)$ becomes the largest and action d is chosen again.

At the beginning of the defense cycle t, the index $I_d(t)$ of each defense action (Line 6 to 12) is updated. Especially, for the actions that have never been chosen before, their indices are set highest value so that they will be chosen with highest priorities (Line 7 to 8). For the actions that have been chosen before, their indices are updated on the basis of the empirical average utility values (Line 9 to 11). Then the new action $d(t)$ with the largest index (Line 13) is chosen the numbers of times each defense action is updated (Lines 14 to 17). At the end of the defense cycle t, the utility values $\boldsymbol{u}'(t)$ are received (Line 18) and then the empirical average utility values of all defense actions are updated. (Line 19 to 21).

4.3 Cost-Effectiveness Analysis

The common method to evaluate the defense ability is using test-beds and benchmarks (e.g., SPEC 2006 [51], metasploit [52]). However, test-beds and benchmarks typically can only explore some of the possible (attack, workload, defense) scenarios, and the cost-effectiveness could vary a lot under different testing settings. In addition, it may take a long time (e.g., hours) for the evaluation of a single scenario. This is not acceptable for time-critical missions. To overcome the shortcomings, we derive an upper bound on the worst-case regret of the UCB-Z algorithm against all possible attack distributions. More specifically, if the defender chooses defense actions by following the UCB-Z algorithm, the regret cannot be larger than the derived upper bound no matter what (unknown) distribution the attacker follows. This is formalized in Theorem 2.

Theorem 2. *Under the* **Assumption 4-1 in Sect. 4.1.3, for any set of utility distributions** $\{v_{d_1}, \cdots, v_{d_n}\}$**, the regret of the UCB-Z algorithm after** N **defense cycles is upper bounded in the following way:**

$$R_{UCB-Z} \leq \mathcal{T}\mu_{d*} + \sum_{d:0<\Delta_d \leq 2\epsilon} 3\epsilon(N-\mathcal{T})$$

$$+ 8(u^+ - u^-)^2 \sum_{d:\Delta_d > 2\epsilon} \frac{(\Delta_d + \epsilon)\ln(N-\mathcal{T}+1)}{(\Delta_d - 2\epsilon)^2}$$

$$+ \left(1 + \frac{2}{(u^+ - u^-)^2 \frac{8\ln(N-\mathcal{T}+1)}{(\Delta_d-2\epsilon)^2} - 1}\right) \sum_{d:\Delta_d > 2\epsilon} (\Delta_d + \epsilon),$$

where $\Delta_d \triangleq \mu_{d*} - \mu_d \geq 0$ **is the suboptimal parameter of defense action** $d \in \mathcal{D}$**.**

In [47], LAI AND ROBBINS proved that when the utility distributions are Bernoulli, for any strategy φ, the *Regret* is a lower bounded asymptotically. That is, $\lim_{N\to\infty} R_\varphi(N) \geq \sum_{d:\mu_d<\mu_{d*}} \frac{\Delta_d \ln(N)}{D(v_d||v_{d*})}$, where $D(v_d||v_{d*}) \triangleq \sum_{v_d} v_d \ln \frac{v_d}{v_{d*}}$ is the Kullback-Leibler divergence [53] between the utility distribution v_d of any suboptimal defense action and the utility distribution v_{d*} of the optimal defense action. When the utility distributions are Bernoulli, then the regret is always larger than the logarithmical bound no matter what strategy φ the defender follows.

The upper bound stated in Theorem 2 consists of four terms. The first term $\mathcal{T}\mu_{d*}$ is the regret brought by the utility delays, which is a constant. The second term is the regret brought by choosing the suboptimal defense actions whose expected utility values are ϵ-close to the optimal one's. If $\epsilon < \min_{d:\mu_d<\mu_{d*}} \frac{\Delta_d}{2}$ (e.g., $\epsilon = 0$), this term is 0. The third term $8(u^+ - u^-)^2 \sum_{d:\Delta_d>2\epsilon} \frac{(\Delta_d+\epsilon)\ln(N-\mathcal{T}+1)}{(\Delta_d-2\epsilon)^2}$ increases logarithmically in N. The fourth term is the regret brought by choosing the suboptimal defense actions after they are chosen enough times, and this term will decrease to 1 if N increases to infinity. If the utility errors are very small (e.g., $\epsilon = 0$), the upper bound in Theorem 2 is a logarithmic upper bound with respect to N and increases as slow as the lower bound when the utility distributions are Bernoulli. Therefore, the logarithmic upper bound in Theorem 2 is the best possible upper bound which holds for any utility distribution.

4.4   Co-design with DSMAs

4.4.1   DSMAs
The UCB-Z algorithm may not be directly implemented in current servers. We need to modify the server (security engineering). We will use DSMAs as the concrete prototype (shown in Fig. 4) to illustrate the server modification and show the experiment results.

4.4.2   Attacker
In this section, we investigate the scenario when the attacker exploits DSMAs. DSMAs [17] iteratively exploit a vulnerability to manipulate specific fields of one or more data structures. Such attacks require millions of attempts to attack one or multiple specific data structures, which typically span hours or even days. In particular, each attack action is a combination of several attack scripts. One attack script targets some data structure types (a fixed script targets fixed data structure types). Even when the target fields is protected by static data structure layout randomization (DSLR) or adaptive DSLR, the attacker can guess the offset of the fields by using brute force attacks.

Based on the real-world DSMA attack scripts [54], we observe two mathematical features of the real-world attack: (1) the attack frequency is fixed; (2) the combination of attack scripts is fixed. The attacker cannot avoid being detected since it cannot ensure trigger no segmentation faults. So to increase the probability of modifying correct data structures, the attacker would like to change

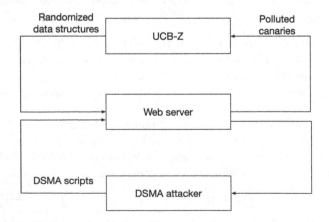

Fig. 4. The concrete system with DSMAs

its attack actions. Using probability distributions is a reasonable choice for the attacker to adjust its actions. We will also extend feature (2) by assuming that the attacker associates each defense action $d \in \mathcal{D}$ with a predetermined probability distribution DA_d and updates attack actions (e.g., adjusting the attack script combination).

4.4.3   Defense

Our defense technique against DSMAs is combining DSLR and canary detection. In particular, each defense action randomizes a set of data structures at runtime. In order to generate feedbacks, some 32-bit specific values called canaries are inserted into the fields of all data structures. Once a canary is polluted, it indicates that a potential modification of the fields in a data structure is discovered.

4.4.4   Cost-Effectiveness Utility

Specific to DSMAs, the most direct quantification of the effectiveness is the number of failed DSMAs during a defense cycle. As mentioned, a polluted canary indicates a discovered potential modification. Therefore, we take a bold move by regarding a polluted canary as one failed DSMA and use the number of polluted canaries to quantify the effectiveness.

If adaptive DSLR does not incur any cost, the best defense is randomizing all feasible data structures. However the study of SALADS [17] shows that the performance overhead is proportional to the number of total randomized data structures. So we use the number of randomized data structures to quantify the cost.

4.4.5   Server Modification

To implement UCB-Z Algorithm, the server need to be modified to provide to "offer" two functions. First, it should provide adaptive DSLR. Second, it should provide canary detection to provide utility values. Based on our evaluation (see Sect. 4.5), the memory allocator introduces low runtime overhead and acceptable memory overhead.

Adaptive DSLR. We modify the server so that it can generate data structure self-randomization (DSSR) binary. The DSSR binary maintains the metadata for all the data structure instances, including the base addresses and offsets of the fields in data structures. The DSSR binary handles randomization with a dynamic whitelist. The data structures in the dynamic whitelist are randomized at the beginning of each defense cycle. To properly access the randomized data structure layout, the read/write operations of data structures are replaced with a set of DSSR statements [17]. What's more, all the definitions of the data structures are randomized at compile-time, and padding bytes are inserted into the data structures [55].

Canary Detection. The canary detection is executed based on the memory forensic analysis. A separate thread scans the canaries in the randomized data structures and compares current values with a random canary value [56]. The random canary is chosen at the beginning of each defense cycle through /dev/urandom. To quickly pinpoint the canaries, we maintain an array to record their addresses along with their binary states. Only canaries with state 1 need to be checked. After DSSR statements complete the data structure access, the canary detection checks whether the data structure type is in the whitelist, and then DSSR statements will update the array for the corresponding canaries.

4.5   Evaluation

We will use real experiments to evaluate our adaptive defense when facing DSMAs. The evaluation focuses on the effectiveness and overhead of our defense.

Experiment Environment. The experiments are conducted on an Intel(R) Core(TM) i5 machine with 4 GB memory running Red Hat Linux 7.3 with Linux kernel version 2.4.18 and we use Apache (1.1.1) compiled with OpenSSL (0.9.6d) and Glibc (2.2.2) as the vulnerable server. The programs contain 348, 132, and 2329 data structure types, respectively. In the following experiments, we partition the data structures into five groups equally. We choose the length of the defense cycle as 5 s. And we measure that the upper bound of utility delays is 5 s and the upper bound of utility errors is 1 under our experiment setting.

Real-World DSMA Scripts. The attacker has four real-world DSMA scripts [54,57–59]. The attack action set consists of the 4 DSMA scripts. The first script overwrites a data structure ssl_session_st and malloc_chunk. The second script overflows the pw_uid in data structure passwd to do privilege escalation. The third script pollutes data structure malloc_chunk. The fourth script overflows data structure timeval. All the four DSMA scripts use brute force method to guess the layouts of the target data structures.

Effectiveness. To evaluate how effective our adaptive defense is in real world, we compare our defense with static DSLR against the four attack scripts. Static DSLR is a defense against DSMA attacks at compile-time, once the program is loaded into the memory, the layout of the data structure is fixed. Defense results are shown in Table 2. The results demonstrate that in two hours, static DSLR cannot defend against DSMA with brute force attacks. However, our adaptive defense is robust enough to prevent such attacks.

Runtime Overhead. To evaluate the runtime overhead introduced by adaptive DSLR, we compare the modified web server Apache (1.1.1) with original one compiled with GNU GCC. The defense cycles are 1/5/10 s and in each defense cycle. As Table 3 shown, the average runtime overheads are 6.4%, 3.9%, 2.3% on average.

Memory Overhead. We also evaluate the memory overhead. We compare the memory usage of the modified web server with the original web server. The memory overhead is 2.1% on average, which is mainly introduced by the paddings and canaries.

Table 2. Effectiveness (from [60], page 8)

Attack script	Data structure	Static DSLR	Adaptive defense
CVE-1999-0071 [59]	timeval	×	√
CVE-2001-0144 [57]	passwd	×	√
CVE-2002-0656 [54]	ssl_session_st	×	√
	malloc_chunk		
CVE-2015-0235 [58]	malloc_chunk	×	√

Table 3. Runtime overhead

Apache web server compiled with GNU GCC	Modified system	
	Defense cycle	Overhead
3756.2 req/s	10 s	3669.8 req/s (2.3%)
	5 s	3609.7 req/s (3.9%)
	1 s	3515.7 req/s (6.4%)

4.6 Validation of Mathematical Predictions

We will also validate the upper bound in Theorem 2. First we simulate the interactions in Matlab, and then compare the regret, aggregate utility with the ones in Apache experiment.

Matlab Simulation. The Matlab simulations remain the same features as the vulnerable apache web server in terms of data structure types and instances. And the attack scripts in simulations remain the same features as CVEs in terms of targets and attack frequency.

Comparison between Matlab Simulations and Apache Experiments. We test the UCB-Z algorithm in both Matlab simulation and Apache environment with the same DSMA which lasts 2 h (1440 defense cycles). The attack is repeated 20 times and the average regret over 20 repetitions are collected and compared with the upper bound in Theorem 2. Figure 5(a) presents the comparison among the Matlab regret, Apache regret, and the UCB-Z upper bound. The comparison between the aggregate utility of the UCB-Z in Matlab simulation and Apache experiment is also shown in Fig. 5(b).

Implications. From Fig. 5(a), we can see that the both Matlab regret and Apache regret are lower than the upper bound. This validates the correctness of the upper bound in Theorem 2, which provides the worst-case cost-effectiveness of our adaptive defense. From Fig. 5(b), we can see the curve of the aggregate utility in all Matlab simulation is close to the curve of Apache experiment. This indicates that our mathematical model has high fidelity of the real-world defense system. We also compare the time consumption of the Matlab simulation and the Apache experiment. Under our experiment environment, the Matlab simulation

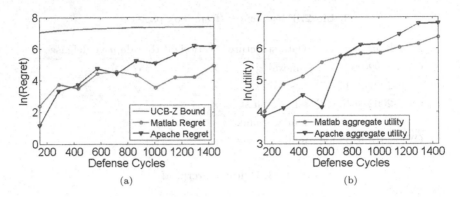

Fig. 5. (a) shows the comparison among regrets of UCB-Z in Matlab simulation, Apache experiment and the mathematical upper bound; (b) compares the aggregate utility in Matlab simulation and Apache experiment during 2-h attack

consumes 0.4 s while the experiment in Apache consume 2 h. We can see that Matlab simulation can save 4 orders of magnitudes of time when we evaluate a defense against the same attack.

5   Reinforcement Learning on Bayesian Attack Graphs

In Sects. 3 and 4, the attacker manipulates single zero-day vulnerability (or single class of vulnerabilities) and targets single ICT system. In this section, we consider that an intelligent attacker exploits combinations of multiple known or zero-day vulnerabilities to compromise machines in a network. In particular, we consider the ACD problem in a computer network where an intelligent attacker exploits combinations of multiple known or zero-day vulnerabilities to compromise machines in the network. There are two main challenges for the defender: (1) partial observability: the defender can only observe the states of a subset of the machines due to limited detection capabilities; (2) unknown utility functions: the defender is only able to access some utility values; i.e., feedbacks, after some defense actions are taken, but is not able to know the utility functions; i.e., the mapping from actions and environment to the feedbacks because the locations and the impacts of zero-day vulnerabilities are not available. Partially observable Markov decision process (POMDP) [61] has been proposed to address the first challenge. And a simple version of the ACD problem has been studied in [62]. This chapter uses *Bayesian attack graphs* (BAGs) [63–65] to describe how an intelligent attacker exploits combinations of vulnerabilities to compromise the network. In addition, the ACD problem on a BAG is modeled as a POMDP problem. POMDP demonstrates the decision making process where the decision-maker is able to interact with the environment, receive observations that partially reflect system states. By analyzing about the consequences of actions and observations on the environment, the decision-maker can estimate the system state using a probability distribution, which is called belief, over the possible system states. Then the decision-maker selects actions on basis of the beliefs instead of unavailable system states. We propose reinforcement learning algorithms to solve POMDP problems with unknown utility functions.

5.1   ACD on Bayesian Attack Graphs

In this subsection, we consider an ACD problem where the attacker exploits multiple vulnerabilities to compromise computer network and the defender tries to defend a computer network against external intrusions. In particular, we first use BAGs to model how the attacker attacks the network when there is no defender. After that, we introduce defense and model the interactions among the attacker, network and defender.

5.2   Problem Formulation of POMDP

In last subsection, we introduce an ACD problem on BAGs. Essentially, the problem is an instance of POMDP problems. In this section, we present the problem formulation of generic POMDP problems. We consider a computer network which consists of multiple machines and each machine has a vulnerability. The attacker wants to compromise the network by exploiting reachable vulnerabilities. But each exploit can only succeed with a certain probability. As first introduced in [63], a BAG can depict the interactions between the attacker and network, which is formally defined as follows:

Definition 2. A BAG is defined as a tuple $G = (\mathcal{N}, \mathcal{E}, \mathcal{P})$. $\mathcal{N} = \{1, ..., K\}$ is the set of machines. $\mathcal{E}$ is the set of directed edges, where each edge is an exploit and $(i, j) \in \mathcal{E}$ if machine i can be compromised through machine j. $\mathcal{P}$ is the set of exploit probabilities associated with the edges, where $\rho_{ij} \in \mathcal{P}$ represents the likelihood that exploit (i, j) can succeed; i.e., how likely the attacker can successfully compromise machine j after he/she compromises machine i.

If $(i, j) \in \mathcal{E}$, node i is called an in-neighbor of node j and node j is called an out-neighbor of node i. The nodes in a BAG can be classified into two categories: leaf nodes $\mathcal{N}_L \subseteq \mathcal{N}$ and non-leaf nodes. Note that leaf nodes do not have any in-neighbor. And the probability of a leaf node $l \in \mathcal{N}_L$ is compromised is denoted by ρ_l. For any non-leaf node i, their in-neighbors are denoted by $\bar{D}_i \triangleq \{j \in \mathcal{N} | (j, i) \in \mathcal{E}\}$. We refer NIST's Common Vulnerability Scoring System (CVSS) [66] as a way to capture the principal characteristics of a vulnerability and produces a numerical score reflecting its severity. Here we use the exploitability metrics in CVSS to calculate the value of each ρ_{ij} and ρ_l.

System State. The state of machine $i \in \mathcal{N}$ at time t is either compromised (value 1) by the attacker or not (value 0); i.e., $s_t^i \in \{0, 1\}, \forall i \in \mathcal{N}$. Combing the states of the machines, The system state at time step t is denoted by the tuple $s_t = (s_t^1, \cdots, s_t^K) \in \mathcal{S} = \{0, 1\}^K$. An example of BAGs is shown in Fig. 6. In the example, the state of BAG is $s = (1, 0, 0, 0)$ where only machine 1 is compromised.

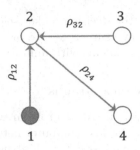

Fig. 6. An example of BAGs. (from [67], page 101)

Attacker's Action. At the very beginning, the attacker selects a subset of leaf nodes as entry points. Once some leaf nodes are compromised, the attacker may use them as stepping stones to exploit their out-neighbors. Since the defense is absent, once a machine was compromised, it remains compromised. The attacker stops when all machines are compromised.

The state transitions are autonomous given attacker's initial actions because we do not consider defense for the time being. In real world, machine $i \in \mathcal{N} \setminus \mathcal{N}_L$ can be compromised at t under one of two conditions: all of its in-neighbors are compromised or at least one of its in-neighbors compromised at $t-1$. The choice of the condition depends on the type of vulnerability in machine i. Machine i is an *And-machine* if it can be compromised only after all of the machines in $\bar{D}_i$ are compromised. Otherwise, Machine i is an *Or-machine*. Next, we consider there is a defender in the network, which will make the state transitions a little more complicated.

Defender's Actions. The defender's actions contains two parts: detection and reimage. Formally, at time step t, the defender's action is $a_t = (a_t^r, a_t^d) \in \mathcal{A} \subset \mathscr{P}(\mathcal{N}) \times \mathscr{P}(\mathcal{N})$, where a_t^r are the reimaged machines, a_t^d are the monitored machines and $\mathscr{P}(\mathcal{N})$ is the power set of $\mathcal{N}$.

Concern About High False Positives. High false positives is a common problem in Intrusion Detection System (IDS) [68]. However, in this chapter, the detection is assumed to be implemented by manual analysis. Therefore, our detection does not include any false positive. On the other hand, the defender can only detect a subset of the machines in the network due to limited resources.

State Transition. State transitions consist of a Markovian process which can be modeled by conditional probability $P(s_t|s_{t-1}, a_{t-1})$. Here $P(s_t^j = 1|s_{t-1}, a_{t-1})$ represents the probability that machine j is compromised at step t when defense a_{t-1} is taken at state s_{t-1}. For an And-machine which was not compromised nor reimaged, it can only be compromised after all of its in-neighbors are compromised. And for an Or-machine which was not compromised nor reimaged, it can be compromised if at least one of its in-neighbors is compromised. If the machine was compromised at the previous step, it remains compromised if it was not reimaged. In all other cases, the probability of this machine being compromised is 0. The state transition can be written as follows. If $i \in \mathcal{N}$ is an And-machine:

$$P(s_t^i = 1|s_{t-1}, a_{t-1}) =$$

$$\begin{cases} \prod_{j \in \bar{D}_i} \rho_{ji} & \text{if } s_{t-1}^i = 0, s_{t-1}^j = 1, \forall j \in \bar{D}_i \text{ and } i \notin a_{t-1}^r, \\ 1 & \text{if } s_{t-1}^i = 1 \text{ and } i \notin a_{t-1}^r, \\ 0 & \text{otherwise.} \end{cases}$$

And if $i \in \mathcal{N}$ is an Or-machine:

$$P(s_t^i = 1|s_{t-1}, a_{t-1}) =$$

$$\begin{cases} 1 - \prod_{\{j \in \bar{D}_i | s_{t-1}^j = 1\}} (1 - \rho_{ji}) & \text{if } s_{t-1}^i = 0, \exists s_{t-1}^j = 1 \text{ and } i \notin a_{t-1}^r, \\ 1 & \text{if } s_{t-1}^i = 1 \text{ and } i \notin a_{t-1}^r, \\ 0 & \text{otherwise.} \end{cases}$$

Observation. Due to limited resources, the defender can only monitor a subset of machines at each time. The states of subset of machines are referred to as observations. Formally, the observation at time step t is denoted by o_t. An observation kernel $\mathcal{Z}(\cdot|s_t, a_{t-1})$ is used to model observation generation. $\mathcal{Z}(\cdot|s_t, a_{t-1})$ presents the probability that the defender receives observation o when the system state evolves to s_t after a_{t-1} is taken. In this chapter, we simply use the states of the machines in a_{t-1}^d as observation; i.e., $o_t = (s_t^{i_1}, \cdots, s_t^{i_k})$, where $i_1, \cdots, i_k \in a_{t-1}^d$. The observation kernel $\mathcal{Z}(\cdot|s_t, a_{t-1})$ is:

$$\mathcal{Z}(o|s_t, a_{t-1}) = \begin{cases} 1 & \text{if } o = (s_t^{i_1}, \cdots, s_t^{i_k}) \text{ and } i_1, \cdots, i_k \in a_{t-1}^d, \\ 0 & \text{otherwise.} \end{cases} \tag{2}$$

Utility Function. The defender aims to find a defense policy to keep the network "secure". Utility functions are introduced to quantify security levels of the network. At time step t, after taking action a_t in state s_t, the defender receives a utility value $u_t(s_t, a_t) \triangleq r_t(s_t, a_t) - c_t(a_t)$. Here $r_t(s_t, a_t)$ represents the reward of keeping the network secure and $c_t(a_t)$ represents the cost caused by the action a_t (e.g., resources required by detection or reimage). In particular, $r_t(s_t, a_t)$ is defined according to confidentiality, integrity, and availability (CIA) triad. And we also take the dynamic environments (e.g., network traffics or machine workloads) into account. So the utility functions are also dependent on time. That is, under different environments, same pair of state-action could lead to different utility values. Note the defender does not know the utility functions because he/she is unaware of the dynamic environment.

Defender's Goal. Before introducing defender's goal, we first introduce the defender's available information at time t: $I_t \triangleq (\mathcal{Z}, P, o_1, a_1, \cdots, o_{t-1}, a_{t-1}, o_t)$. In particular, the defender knows the observation kernel, transition probability, observations and actions up to t. But he/she does not know the system state trajectory or utility functions. The defender aims to find a defense policy (with decision rules that map from $\mathcal{S}$ to $\mathcal{A}_t$) over the time horizon $T = \{0, 1, \cdots, N\}$ to maximize the aggregate utility $\sum_{t \in T} u_t(s_t, a_t)$. Essentially, the problem is an instance of POMDP problems. In what follows, we first present generic POMDP problems. Then n Sect. 5.3, we will propose the reinforcement learning algorithm to solve them.

Definition 3. A POMDP consists of a tuple $(T, \gamma, \mathcal{S}, \mathcal{A}, \mathcal{A}_t, \mathcal{O}, P_t, u_t, \mathcal{Z})$: $T \triangleq \{1, 2, \cdots, N\}$ is the time horizon with $N \leq \infty$; $\gamma \in (0, 1]$ is the discount factor; $\mathcal{S}$ is the state space; $\mathcal{A}$ is the action space and $\mathcal{A}_t \subseteq A$ is the available action set at time step t. As a result of choosing action a at state s at time step t, the agent receives a utility $u_t(s, a)$ and the state at next time step is drawn from a transition probability $P_t(\cdot|s, a)$. $\mathcal{O}$ is the observation space, and $\mathcal{Z}(o|s, a)$ is the conditional probability of observing o after the agent takes action a and the system evolves to state s.

Remark 2. The agent receives $u_t(s_t, a_t)$ after applying an action a_t at state s_t at time step t. But to emphasize that the agent does not know s_t, we denote v_t as the utility value received by the agent at time step t.

One might simply take the observation space to be the state space and treat a POMDP as an MDP. However, the transitions of the observations might not necessarily be Markovian, because one observation can represent more than one states. As a result, an optimal policy (with decision rules that map from $\mathcal{O}$ to $\mathcal{A}_t$) may not be optimal. To address the challenge, the agent maintains a probability distribution over states called *belief state*. Let $\mathcal{B}$ be the belief state space (the set of all possible probability distributions over $\mathcal{S}$) and $b(s)$ be the probability assigned to state s when the belief state is b. The belief state update law is denoted as SE, whose inputs are the last action a_{t-1}, belief state b_{t-1} and the current observation o_t and the output is the updated belief state b_t. By Bayes' rule, the updated belief state assigns the probability to state s' as follows:

$$
\begin{aligned}
SE(b_{t-1}, a_{t-1}, o_t) &= Pr(s'|a_{t-1}, o_t, b_{t-1}) \\
&= \frac{Pr(o_t|s', a_{t-1}, b_{t-1})Pr(s'|a_{t-1}, b_{t-1})}{Pr(o_t|a_{t-1}, b_{t-1})} \\
&= \frac{\mathcal{Z}(o_t|s', a_{t-1}) \sum_{s \in \mathcal{S}} P_{t-1}(s'|s, a_{t-1})b_{t-1}(s)}{\sum_{s' \in \mathcal{S}} \mathcal{Z}(o_t|s', a_{t-1}) \sum_{s \in \mathcal{S}} P_{t-1}(s'|s, a_{t-1})b_{t-1}(s)}.
\end{aligned} \tag{3}
$$

The transition probability from belief state b_{t-1} to $b_t \in \mathcal{B}$ is defined as follows:

$$
B_t(b_t|b_{t-1}, a_{t-1}) = \sum_{\{o \in \mathcal{O}|SE(b_{t-1}, a_{t-1}, o)=b_t\}} Pr(o|a_{t-1}, b_{t-1}), \tag{4}
$$

where

$$Pr(o|a_{t-1}, b_{t-1}) = \sum_{s' \in \mathcal{S}} \mathcal{Z}(o|s', a_{t-1}) \sum_{s \in \mathcal{S}} P_{t-1}(s'|s, a_{t-1}) b_{t-1}(s).$$

If there is no $o \in \mathcal{O}$ such that $SE(b_{t-1}, a_{t-1}, o) = b_t$, then $B_t(b_t|b_{t-1}, a_{t-1}) = 0$. The process of updating the belief state is proven to be Markovian [61]. That is, the current belief state only depends on the previous belief state and action.

Given the fact that the state trajectory is unknown and the belief states provide sufficient statistics of the history [69], the agent would select actions on the basis of belief states. In POMDP, $d_t : \mathcal{B} \to \mathcal{A}_t$ is the decision rule which specifies the action choice when the belief state is b_t at time step t. Therefore, a POMDP can be cast as a completely observable MDP where the state space is the belief state space $\mathcal{B}$ and the action space is $\mathcal{A}$. This MDP is formally defined as a belief-state MDP.

Definition 4. A belief-state MDP consists of a tuple $(T, \gamma, \mathcal{B}, \mathcal{A}, \mathcal{A}_t, B_t, U_t)$: $T, \gamma, \mathcal{A}, \mathcal{A}_t$ are defined in Definition 3; $\mathcal{B}$ is belief state space; $B_t(\cdot|b, a)$ is the belief transition probability at t, $U_t(b, a) = \sum_{s \in \mathcal{S}} b(s) u_t(s, a)$ is the expected immediate utility from executing action a at state s given the belief state b and U_t is referred to as the expected immediate utility function at time step t.

With the definition of belief-state MDP, the agent's problem is formulated as follows:

$$\max_{\pi \in \Pi} J_1^\pi(b) = \mathbb{E}\left[\sum_{t=1}^{N} \gamma^t U_t(b_t, a_t) \right] \tag{PA}$$

$$\text{subject to } a_t = d_t(b_t)$$
$$b_t \leftarrow B_t(\cdot|b_{t-1}, a_{t-1}),$$

where $J_1^\pi(b)$ is the expected discounted total utility from $t = 1$ to $t = N$ if policy π is used and the initial belief is b.

5.3   Reinforcement Learning for POMDP

In this subsection, we propose two reinforcement learning algorithms to solve POMDP problems with unknown utility functions. We first introduce the main idea and the informal statement of the first algorithm. Then we the computational complexity is analyzed and a Q-learning based algorithm is proposed to handle the computational complexity.

5.3.1   Main Idea

As mentioned in the previous section, a POMDP can be cast to a belief-state MDP and the goal of the agent is to solve Problem (PA). The agent wants to compute the optimal value functions $J_t^*(b) \triangleq \max_{\pi \in \Pi} J_t^\pi(b), \forall t \in T$. By the principle of optimality [70], the optimal value of belief state b is equal to the maximum sum achieved by a particular action, where the sum consists of instant utility induced by the action from b and the expected optimal value of next belief state which is discounted by γ. Therefore, the optimal value function at t can be calculated backward from the optimal value function at $t+1$ as follows:

$$J_t^*(b) = \max_{a \in \mathcal{A}_t}[U_t(b,a) + \gamma \sum_{b' \in \mathcal{B}} B_t(b'|b,a) J_{t+1}^*(b')]. \tag{5}$$

And the optimal action $d_t^*(b)$ is the one which achieves the optimal value of $J_t^*(b)$; i.e.,

$$d_t^*(b) = \arg \max_{a \in \mathcal{A}_t}[U_t(b,a) + \gamma \sum_{b' \in \mathcal{B}} B_t(b'|b,a) J_{t+1}^*(b')].$$

With the knowledge of utility functions u_t, transition probabilities P_t for all $t \in T$, and observation kernel $\mathcal{Z}$, $J_t^*(b)$ can be calculated offline based on Eq. (5) for all $t \in T$ and $b \in \mathcal{B}$.

There are two challenges to perform the backward calculation (5): 1. the utility functions are unknown; 2. the belief space $\mathcal{B}$, where J_t^* is defined, is infinite. For challenge 1, we use dynamic programming (DP) to cpmpute the estimates of the optimal value functions J_t^*. Consider the special case where U_t, B_t and $\mathcal{A}_t$ are time independent and the time horizon is infinite. The optimal value functions can be rewritten as $J^*(b) \triangleq \max_{\pi \in \Pi} \mathbb{E}\left[\sum_{t=1}^{\infty} \gamma^t U_t(b_t, d_t(b_t))\right]$, where $\pi = (d_1, d_2, \cdots)$. Start with any initial estimated value function $J^0(b)$, then the n-th estimated value function is derived from the $(n-1)$-th by the recursive equation:

$$J^n(b) = \max_{a \in \mathcal{A}_t}[U_t(b,a) + \gamma \sum_{b' \in \mathcal{B}} B_t(b'|b,a) J^{n-1}(b')]. \tag{6}$$

By applying (6) repeatedly over n, J^n will converge to the fixed point J^* [71]. In this chapter, the time horizon could be finite. Besides, the utility functions and the transition probabilities are time dependent. Therefore, given fixed U_t, B_t and $\mathcal{A}_t$, J^n may not converge to J_t^*. Further, to perform the value iteration (6), U_t is needed. However, by Definition 4, we cannot directly calculate U_t since the utility function u_t is unknown. To address above two issues, we partition the time horizon T into $M+1$ relatively long intervals $k_0, k_1, \cdots k_M$, and we use constant empirical average utility values $\hat{U}_m(b,a)$ to represent the values of $U_t(b,a)$ for all belief-action pairs $(b,a) \in \mathcal{B} \times \mathcal{A}_t$ when $t \in k_m$. Instead, we perform the following value iteration repeatedly in interval k_m:

$$J_m^n(b) = \max_{a \in \mathcal{A}_t}[\hat{U}_m(b,a) + \gamma \sum_{b' \in \mathcal{B}} B_t(b'|b,a) J_m^{n-1}(b')].$$

Then we select an action that maximizes J_m^n at the n-th time step in interval k_m.

To address the challenge of infinite belief space, we restrict the recursion on a sequence of incrementally expanded subsets of belief space $\mathcal{B}$; i.e., $\{\mathbb{B}_0, \cdots, \mathbb{B}_m, \cdots, \mathbb{B}_M\}$. In particular, we start from the initial belief state and add the new belief states which appear over time. In the rest of the chapter, the subsets of $\mathcal{B}$ are referred to as belief sets. Then one key part in the algorithm is updating $J_m^n(b)$. In particular, after receiving utility value $v_t = u_t(s_t, a_t)$, the update law is $J_m^{n+1}(b) = \max_{a \in \mathcal{A}_t}[\hat{U}_m(b, a) + \gamma \sum_{b' \in \mathbb{B}_m} B_t(b'|b, a)J_m^n(b')]$, for all $b \in \mathbb{B}_m$. To get $J_m^{n+1}(b)$ for all $b \in \mathbb{B}_m$, the agent needs to compute $B_t(b'|b, a)$ for all $b', b \in \mathbb{B}_m$ and all $a \in \mathcal{A}_t$. According to Eqs. (3) and (4), getting $B_t(b'|b, a)$ for a given b, a and b' needs $O(|\mathcal{O}||\mathcal{S}|^2)$ products. Then, getting $B_t(b'|b, a)$ for all $b', b \in \mathbb{B}_m$ and all $a \in \mathcal{A}_t$ needs $O(|\mathcal{O}||\mathcal{S}|^2|\mathbb{B}_m|^2|\mathcal{A}_t|)$ products. Even though the update only takes polynomial time, it could be very time-consuming in practice.

To address the computational complexity, we introduce an intermediate state-action value function called the Q-function:

$$Q_t(b, a) = U_t(b, a) + \gamma \sum_{b \in \mathcal{B}} B_t(b'|b, a)J_{t+1}^*(b').$$

$Q_t(b, a)$ is the aggregate utility for executing action a at time step t and following the optimal policy π^* thereafter. The relation between $Q_t(b, a)$ and $J_t^*(b)$ is given by $J_t^*(b) = \max_{a \in \mathcal{A}_t} Q_t(b, a)$. And we estimate the Q-functions Q_t instead of the optimal value functions J_t^* by applying Q-learning [72] in Algorithm 3. In particular, let $\hat{Q}_m^n$ be the n-th estimate of Q-function in interval k_m, then the $(n+1)$-th estimate is updated from $\hat{Q}_m^n$ by the recursive equation shown in Line 26 of Algorithm 3. the recursive equation makes a correction of $\hat{Q}_m^n(b_t, a_t)$; i.e., the value of belief-action pair (b_t, a_t), based on the new empirical average utility values $\hat{U}_m(b_t, a_t)$.

5.3.2 Algorithm Statement

Formally, we define the information set of the agent at time t as $I_t \triangleq (\mathcal{Z}, o_1, b_1, a_1, v_1, P_1, \cdots, o_{t-1}, b_{t-1}, a_{t-1}, v_{t-1}, P_{t-1}, o_t, b_t, P_t)$. That is, the agent knows the observation kernel, transition probabilities, belief states, and observations up to t. Additionally, the agent knows the actions and the received utility values up to $t-1$. But the visited states and utility functions are unknown. The pseudo-code of the algorithm is listed in Algorithm 3 and the details are shown as follows.

Let t_m denote the first step of the interval k_m. For the initial interval (Line 4 to 10), the agent uniformly selects an action for each time step (Line 7), records the belief states that evolve from the last belief (Line 6) and expands the belief set for interval k_1 as $\mathbb{B}_1$ (Line 11). During each interval k_m ($m \geq 1$), $\hat{U}_m$ is updated at the beginning and kept constant afterwards (Line 13), where $\mathbf{1}_{\{condition\}} = 1$ when *condition* is true and $\mathbf{1}_{\{condition\}} = 0$ otherwise. As mentioned before, it is infeasible to get the values of function $\hat{U}_m$ for all pairs of belief-action because $\mathcal{B}$ is infinite. We only consider the values of $\hat{U}_m$ for belief-action pairs $(b, a) \in \mathbb{B}_m \times \mathcal{A}_t$. The initial values of J_m^0 are set for all of belief states $b \in \mathbb{B}_m$

(Line 15). If $b_t \in \mathbb{B}_m$, then a_t is chosen to maximize the sum of $[\hat{Q}_m^n(b_t, a)$ and $\epsilon_t(a)$ if the belief state b_t is in $\mathbb{B}_m$ (Line 19). Otherwise, a_t is uniformly chosen from $\mathcal{A}_t$ (Line 21). The action is selected according to the concept of "following the perturbed leader (FPL)" [73,74], where a diminishing random perturbation for each action $\epsilon_t(a)$ is added. Then the agent receives utility value v_t (Line 23). Algorithm 3 updates $\hat{Q}_m^{n+1}$ asynchronously (Line 26). That is, it updates the value of $\hat{Q}_m^{n+1}$ for a single belief-action pair each time step. At the end of interval k_m, the belief set for next interval $\mathbb{B}_{m+1}$ is expanded by adding all new belief states which appear in interval k_m into the previous belief set $\mathbb{B}_m$ (Line 30).

5.3.3 Discussion

The agent cannot compute the values of $U_t(b, a)$ since the utility functions are unknown. In fact, the agent can only receive the utility value v_t at time step t. One key insight of Algorithm 3 is to use the empirical average utility values $\hat{U}_m(b, a)$ to estimate $U_t(b, a)$ for any belief-action pair $(b, a) \in \mathbb{B}_m \times \mathcal{A}_t$. This insight is based on the following reasons. First, the agent can only receive a collection of utility values in this problem. Second, empirical averages are most commonly used mathematical measure of central tendency [75]. By using the empirical average $\hat{U}_m(b, a)$, we can estimate the typical value of $U_t(b, a)$ in the history. Then $\hat{U}_m(b, a)$ is used in Q-learning to get the estimate of the Q-functions. Algorithm 3 chooses the most successful action which maximizes the estimate of Q-functions. It represents the exploitation phase. However, the empirical averages $\hat{U}_m(b, a)$ can deviate from the true values $U_t(b, a)$ greatly. To address this issue, Algorithm 3 introduces perturbation $\epsilon_n(a)$ using the idea of FPL. In particular, Algorithm 3 chooses arguments of the maxima of the summation of $\epsilon_n(a)$ and $\hat{Q}_m^n(b_t, a)$. It represents the exploration phase. The perturbation term $\epsilon_n(a)$ compensates the scenario where action a is optimal but the value of $\hat{Q}_m^n(b_t, a)$ is not maximum. With this perturbation, the algorithm avoids being trapped in the sub-optimal actions. The perturbation is a diminishing random vector. As time goes by, the effect of the perturbation will decrease so that Algorithm 3 will lean more on the exploitation. In order to enforce computational tractability, the value function computations are restricted to a sequence of incrementally expanded belief sets. This idea is called *point-based* [76].

5.4 Evaluation

We setup a test network similar to the one in [64]. The network, which is shown in Fig. 7, has 7 machines located in two subnets. Then we conduct numerical simulations to evaluate the performance of our algorithms. The simulations are based on real-world settings of the ACD problem on BAGs.

The Web server is located in the DMZ network while local desktops, the Gateway server, SQL server and Admin server are located in the local network. The firewall is installed to prevent remote access to the internal hosts. All communications to external parties must be passed through the Gateway server.

Algorithm 3. Point-based Q-FPL

1: Assign initial belief b_0 and arbitrarily choose a_0;
2: $\mathbb{B}_0 = \{b_0\}$;
3: Initialize the number of visits of each belief-action pair: $\mathcal{T}(b,a) = 0, \forall (b,a) \in \mathcal{B} \times \mathcal{A}$;
4: **for** $t \in k_0$ **do**
5:       Receive observation o_1;
6:       Update belief state $b_t = SE(b_{t-1}, a_{t-1}, o_t)$;
7:       Uniformly select an action $a_t \in \mathcal{A}_t$ and execute a_t;
8:       Receive utility value $v_t = u_t(s_t, a_t)$;
9:       $\mathcal{T}(b_t, a_t) = \mathcal{T}(b_t, a_t) + 1$;
10: **end for**
11: $\mathbb{B}_1 = \mathbb{B}_0 \cup \bigcup\limits_{t \in k_0} \{b_t\}$;
12: **while** $m \geq 1$ **do**
13:       $\hat{U}_m(b,a)$
$$= \begin{cases} \frac{1}{\mathcal{T}(b,a)} \sum\limits_{t=1}^{t_m-1} v_t \mathbf{1}_{\{b_t=b \text{ and } a_t=a\}}) & \text{if } \mathcal{T}(b,a) > 0 \\ -\infty & \text{if } \mathcal{T}(b,a) = 0 \end{cases}, \text{ for all } (b,a) \in \mathbb{B}_m \times \mathcal{A}_t;$$
14:       $n = 0$;
15:       Let $\hat{Q}_m^n(b,a) = \hat{U}_m(b,a)$ for all $(b,a) \in \mathbb{B}_m \times \mathcal{A}_t$;
16:       **for** $t \in k_m$ **do**
17:             Update belief state $b_t = SE(b_{t-1}, a_{t-1}, o_t)$;
18:             **if** $b_t \in \mathbb{B}_m$ **then**
19:                   $a_t \in \arg\max\limits_{a \in \mathcal{A}_t}[\hat{Q}_m^n(b_t, a) + \epsilon_t(a)]$;
20:             **else**
21:                   Uniformly select an action $a_t \in \mathcal{A}_t$ and execute a_t;
22:             **end if**
23:             Receive a utility value v_t;
24:             Observe o_{t+1};
25:             Update the subsequent belief state $b_{t+1} = SE(b_t, a_t, o_{t+1})$;
26:             $\hat{Q}_m^{n+1}(b_t, a_t) = (1 - \beta_m)\hat{Q}_m^n(b_t, a_t) + \beta_m \left[\hat{U}_m(b_t, a_t) + \max\limits_{a' \in \mathcal{A}_t} \hat{Q}_m^n(b_{t+1}, a') \right]$;
27:             $n = n + 1$;
28:             $\mathcal{T}(b_t, a_t) = \mathcal{T}(b_t, a_t) + 1$;
29:       **end for**
30:       $\mathbb{B}_{m+1} = \mathbb{B}_m \cup \bigcup\limits_{t \in k_m} \{b_t\}$;
31:       $m = m + 1$;
32: **end while**

The choices of vulnerabilities are based on [64] and listed in Table 4. These vulnerabilities can produce multiple attack scenarios. In the evaluation, we use BAG to simulate one attack scenario where the attacker starts from either the Gateway server or the Web server and tries to compromise the whole network. From either the Gateway server or the Web server, local desktop a is accessible by exploiting MS Video ActiveX buffer overflow. And from the Web server, local desktop b is accessible by exploiting LICQ buffer overflow. With local user privilege, local desktop c can be compromised by exploiting remote login. And the SQL server can be compromised through any of the three local desktops by exploiting SQL injection. Finally, with information in local desktop c and the SQL server, the Admin server can be compromised by exploiting MS SMV service Stack buffer overflow.

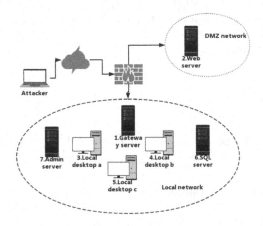

Fig. 7. Test network. (from [67], page 105)

Table 4. Vulnerabilities in the test network. (from [67], page 105)

Machine	Vulnerability	CVE#
Gateway server	Untrusted cookie in OpenSSH	2007-4752
Web server	IIS vulnerability in WebDAV service	2009-1535
Local desktop a	MS Video ActiveX stack buffer overflow	2009-0015
Local desktop b	LICQ buffer overflow	2001-0439
Local desktop c	Remote login	2008-3610
SQL server	SQL injection	2008-5416
Admin server	MS SMV service Stack buffer overflow	2008-4050

BAG. The corresponding BAG of the above attack scenario is illustrated in Fig. 8. The leaf nodes are Web server and the Gateway server which are accessible to the external attacker. The rest machines are non-leaf nodes. The edges represent the possible exploits in the network. For example, the local desktop a can be attacked after either the Gateway server or the Web server is compromised. The exploit probabilities are calculated based on the exploitability metric of CVSS scores. The exploitability metric of CVSS score consists of four

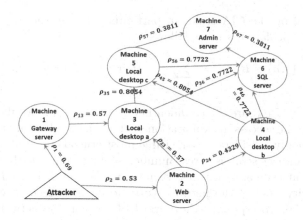

Fig. 8. BAG of the test network. (from [67], page 106)

components: Access Complexity (AC), Access Vector (AV), Privileges Required (PR) and User Interaction (UI). More details on CVSS metrics and their scoring system can be found in the CVSS document [77]. For each $(i, j) \in \mathcal{E}$, the exploit probability is calculated as follows:

$$\rho_{ij} = 2 \times AV(j) \times AC(j) \times PR(j) \times UI(j),$$

where $AV(j), AC(j), PR(j)$, and $UI(j)$ are the four exploitability components of the vulnerability on machine j.

System State. The system state space has $|S| = 2^7 = 128$ states. Each state reflects which machines are compromised.

Attacker's Knowledge and Action. The attacker wants to compromise the whole network and desires to compromise as many machines as possible. In the simulations, the attacker starts with one of the leaf nodes and he/she knows which compromised machines are recovered by the defender. And if the attacker has no available machine to exploit for next time step (e.g., if no machine is compromised, he/she will restart the attack from the leaf nodes again).

Defender's Action. Recall that the detection is implemented by manual analysis, therefore, the defender can only mornitor a subset of the machines due to limited resources. In the simulations, each defender's action is to detect 3 out of 7 machines and to reimage at most 3 out of 7 machines. Then there are total $|\mathcal{O}| = 2^3 \times \binom{7}{3} = 280$ observations and $|\mathcal{A}| = \binom{7}{3} \times (\binom{7}{0} + \binom{7}{1} + \binom{7}{2} + \binom{7}{3}) = 2240$ actions.

Utility. In the evaluation, we use time independent utility function $u(s, a) = r(s, a) - c(a)$ since we do not simulate the network dynamics. The reward part is defined based on the CIA triad. In particular, we use the impact metric of CVSS score to calculate $r(s,a)$ as follows:

$$r(s,a) = R - \sum_{\{s^i \in s | s^i = 1, i \notin a^r\}} [I_C(i) + I_I(i) + I_A(i)],$$

where $I_C(i), I_I(i), I_A(i)$ are the confidentiality, integrity, and availability impact score caused by the successful exploitation of the vulnerability on machine i. CIA triad are commonly considered as the three most crucial components of network security. Each impact score is a real number scaling from 0 to 10 and a higher score means that the network is less secure. A constant R is used to represent the base security level of the network with no compromised machine. In this network, $R = 70$. To represent the reward of keeping the network secure, we subtract the total impact score of compromised machines in the network after action a_t is taken from R. Therefore, a higher value of r represents that the network is more secure, and vice verse. And the cost caused by the defender's action is calculated as $c(a) = \sum_{\{i \in a^r | s^i = 0\}} I_A(i)$, which quantifies the cost induced by reimaging the clean machines. $c(a)$ is concerned with the availability cost. If a non-compromised machine is reimaged, some resources on the machine becomes unavailable for trusted users.

Time Line. Figure 9 describes when the main events and updates happen in the ACD problem. A time step starts with the defender receiving an observation. At the beginning of time step t, the network is in state s_t and the defender updates its belief state b_t. The attacker exploits some available vulnerabilities while the defender chooses an action based on his/her belief state. After the action is deployed, the defender receives a utility value v_t. Note that in the simulations, the received utility value is not $u(s_t, a_t)$ because the defender cannot measure the value based on the CVSS scores without knowing full state s_t. Instead, the defender uses the following measurement as the utility value $v_t = v(o_t, a_t) \triangleq R - \sum_{\{s^i \in o_t | s^i = 1, i \notin a_t^r\}} [I_C(i) + I_I(i) + I_A(i)] - \sum_{\{i \in a_t^r | s^i = 0 \text{ and } s^i \in o_t\}} I_A(i)$. That is, the defender can only measure the reward and cost based on the observation and action.

Based on the evaluation setup, we simulate the interactions among the attacker, network and defender in Python. All the simulations are conducted on an Intel(R) Core(TM) i5 machine with 8 GB memory running OS X 10.11.6. In general, $\mathcal{A}_t = \mathcal{A}$ for all t if there is no restriction on the action space (e.g., any machine can be reimaged any time in the ACD problem). In the evaluation of Algorithm 3, we let $v_t = u(o_t, a_t)$ and $\mathcal{A}_t = \mathcal{A}$ for all t. But note that the observation o_t does not include false positives. Based on this property, we propose a customized Algorithm 4 which restricts $\mathcal{A}_t$ at time step t based on o_t as follows: only the actions that reimage the compromised machines and avoid the clean machines in o_t will be considered at time step t (Lines 18 to 27).

To show the effectiveness of Algorithms 3 and 4, we compare their performances with a baseline policy. The baseline policy uniformly chooses one action

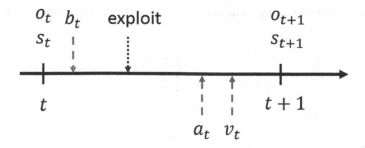

Fig. 9. Events and updates in ACD problem. (from [67], page 106)

each time step and is referred to as uniform selection policy. The duration of each simulation (from the attack begins till the attack ends) is 100 time steps and we repeat 20 identical simulations. And we assume that the defender can receive the observations without delays. The comparison results are shown in Fig. 10. We can see the aggregate utilities of Algorithm 4 are always the highest among all three policies. Besides, the aggregate utilities of Algorithm 3 are higher than those of uniform selection policy. The uniform selection policy performs worst because it does not learn anything and only randomly chooses actions, which could induce only cost at some time steps. In addition, we also compare the gaps among all three policies over time. Figure 10(b) shows that both Algorithms 3 and 4 increase their leads in general on aggregate utilities compared with uniform selection policy. We can conclude that both Algorithms 3 and 4 enable the defender to identify effective defense policies when utility functions are unknown.

In fact, to achieve observations that do not include false positives, IDS alerts should firstly be examined and correlated by security analysts. It takes some time to generate the confirmed alerts. Therefore, we consider the defender uses the observations several time steps ago. In the following simulations, the observations received by the defender are subject to delays of 4/8 time steps. In particular.

Algorithm 4. Point-based Q-FPL for ACD

...
18: $\mathcal{A}_t = \mathcal{A}$
19: **for** $a \in \mathcal{A}$ **do**
20:   **if** any $i \in a^r$ such that $s^i \in o_t$ and $s^i = 0$ **then**
21:     $\mathcal{A}_t = \mathcal{A} \setminus \{a\}$;
22:   **else**
23:     **if** any $s^i \in o_t$ and $s^i = 1$ but $i \notin a^r$ **then**
24:       $\mathcal{A}_t = \mathcal{A} \setminus \{a\}$;
25:     **end if**
26:   **end if**
27: **end for**
...

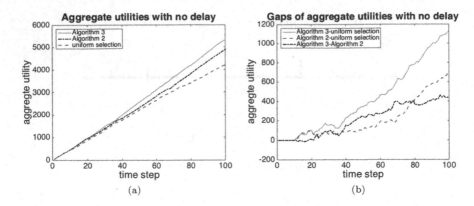

Fig. 10. (a) shows the aggregate utilities of three policies; (b) shows the gaps of aggregate utilities between Algorithms 3, 4 and uniform selection policy without observation delays. (from [67], page 107)

From the results, we can see that as delays increase, the leads of Algorithms 3 and 4 get decrease (Figs. 11 and 12).

We further discuss the differences between Algorithms 3 and 4 via evaluation results. First, Algorithm 4 outperforms Algorithm 3 in terms of aggregate utilities. Second, the gaps between Algorithms 3 and 4 enlarge but the gaps between their slops reduce as time goes by. To explain the differences, we have some conjectures based on the intuitions of the algorithms. First, the customized Algorithm 4 eliminates the actions that conflict with the current observation. Notice that the observation has no-false-positive feature. That is, the defender would not reimage clean machines. So it performs better than Algorithm 3. Second, Algorithm 4 might suffer from the false negatives induced by the observations;

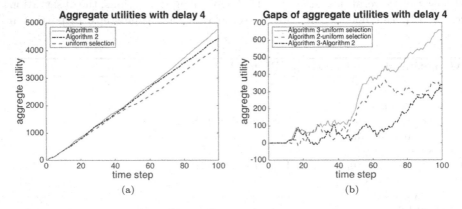

Fig. 11. (a) shows the aggregate utilities of three policies; (b) shows the gaps of aggregate utilities between Algorithms 3, 4 and uniform selection policy with 4 time steps delay.

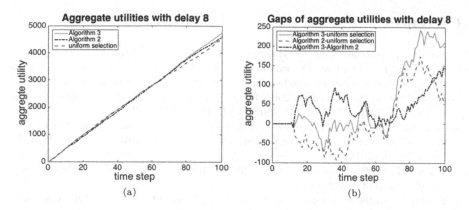

Fig. 12. (a) shows the aggregate utilities of three policies; (b) shows the gaps of aggregate utilities between Algorithms 3, 4 and uniform selection policy with 8 time steps delay.

i.e., clean machines in the current observation are already compromised when the new action is taken. Then the available action set $\mathcal{A}_t$ for Algorithm 4 might not contain optimal action at time step t while Algorithm 3 does not rule out any action at any time. Therefore, as time goes by, the advantage of Algorithm 4 might decrease.

6   Conclusion

This chapter designs, analyzes and evaluates effective ACD schemes which integrate reinforcement learning with MTD techniques. We first design an reinforcement learning algorithm against zero-day strategic attacks where the interactions between such an attacker and a defender are modeled as a non-cooperative game. The proposed algorithm allows the players to identify Nash equilibrium where each player only uses its own deployed actions and its received utility values in recent history. We next propose an adaptive defense against zero-day non-strategic random attacks where the attacker chooses its actions by following predetermined probability distribution. The proposed algorithm can guarantee that the regret is upper bounded by a logarithmic function of the number of defense cycles no matter what probability distributions the attacker follows. We finally propose reinforcement learning algorithms to defend against a kind of attacks which exploit combinations of multiple known or zero-day vulnerabilities to compromise machines in a network. The simulation results confirm that our algorithms enable the defender to identify effective defense policies when utility functions are unknown.

References

1. Shirey, R.: Internet Security Glossary, RFC 2828, RFC Editor, May 2000
2. Johansen, H., Johansen, D., van Renesse, R.: *FirePatch*: secure and time-critical dissemination of software patches. In: Venter, H., Eloff, M., Labuschagne, L., Eloff, J., von Solms, R. (eds.) SEC 2007. IIFIP, vol. 232, pp. 373–384. Springer, Boston, MA (2007). https://doi.org/10.1007/978-0-387-72367-9_32
3. Durumeric, Z., et al.: The matter of heartbleed. In: Proceedings of the 2014 Conference on Internet Measurement Conference (IMC 2014), Vancouver, BC, Canada, pp. 475–488 (2014)
4. CVE-2014-0160, Heartbleed bug (2014). https://web.nvd.nist.gov/view/vuln/detail?vulnId=CVE-2014-0160
5. Symantec, Internet security threat report (2016). https://www.symantec.com/content/dam/symantec/docs/reports/istr-21-2016-en.pdf
6. Symantec, Internet security threat report (2015). https://know.elq.symantec.com/LP=1542
7. Okhravi, H., et al.: Survey of cyber moving targets. Technical report, Massachusetts Institute of Technology Lexington Lincoln Lab (2013)
8. O'Donnell, A.J., Sethu, H.: On achieving software diversity for improved network security using distributed coloring algorithms. In: Proceedings of the 11th ACM Conference on Computer and Communications Security, CCS 2004, Washington DC, USA, pp. 121–131 (2004)
9. Fraser, T., Petkac, M., Badger, L.: Security agility for dynamic execution environments. Technical report, DTIC Document (2002)
10. Roeder, T., Schneider, F.B.: Proactive obfuscation. ACM Trans. Comput. Syst. **28**(2) (2010)
11. Larsen, P., Homescu, A., Brunthaler, S., Franz, M.: SoK: automated software diversity. In: Proceedings of the 2014 IEEE Symposium on Security and Privacy (SP 2014), San Jose, CA, USA (2014)
12. Okhravi, H., Riordan, J., Carter, K.: Quantitative evaluation of dynamic platform techniques as a defensive mechanism. In: Stavrou, A., Bos, H., Portokalidis, G. (eds.) RAID 2014. LNCS, vol. 8688, pp. 405–425. Springer, Cham (2014). https://doi.org/10.1007/978-3-319-11379-1_20
13. Bigelow, D., Hobson, T., Rudd, R., Streilein, W., Okhravi, H.: Timely rerandomization for mitigating memory disclosures. In: ACM SIGSAC Conference on Computer and Communications Security (CCS 205), Denver, Colorado, USA, pp. 268–279 (2015)
14. Sidiroglou, S., Giovanidis, G., Keromytis, A.D.: A dynamic mechanism for recovering from buffer overflow attacks. In: Zhou, J., Lopez, J., Deng, R.H., Bao, F. (eds.) ISC 2005. LNCS, vol. 3650, pp. 1–15. Springer, Heidelberg (2005). https://doi.org/10.1007/11556992_1
15. Giuffrida, C., Kuijsten, A., Tanenbaum, A.S.: Enhanced operating system security through efficient and fine-grained address space randomization. In: USENIX Conference on Security Symposium (Security 2012), Bellevue, WA, pp. 475–490, August 2012
16. Xin, Z., Chen, H., Han, H., Mao, B., Xie, L.: Misleading malware similarities analysis by automatic data structure obfuscation. In: Burmester, M., Tsudik, G., Magliveras, S., Ilić, I. (eds.) ISC 2010. LNCS, vol. 6531, pp. 181–195. Springer, Heidelberg (2011). https://doi.org/10.1007/978-3-642-18178-8_16

17. Chen, P., Xu, J., Lin, Z., Xu, D., Mao, B., Liu, P.: A practical approach for adaptive data structure layout randomization. In: Pernul, G., Ryan, P.Y.A., Weippl, E. (eds.) ESORICS 2015. LNCS, vol. 9326, pp. 69–89. Springer, Cham (2015). https://doi.org/10.1007/978-3-319-24174-6_4
18. Cybenko, G., Jajodia, S., Wellman, M.P., Liu, P.: Adversarial and uncertain reasoning for adaptive cyber defense: building the scientific foundation. In: Prakash, A., Shyamasundar, R. (eds.) ICISS 2014. LNCS, vol. 8880, pp. 1–8. Springer, Cham (2014). https://doi.org/10.1007/978-3-319-13841-1_1
19. Bertsekas, D., Tsitsiklis, J.: Neuro-Dynamic Programming. Athena Scientific, Belmont (1996)
20. Barto, A.G.: Reinforcement Learning: An Introduction. MIT Press, Cambridge (1998)
21. mprotect(2) - Linux man page. http://linux.die.net/man/2/mprotect
22. Silberman, P., Johnson, R.: A comparison of buffer overflow prevention implementations and weaknesses. IDEFENSE, August 2004
23. Nash, J.: Non-cooperative games. Ann. Math. **54**, 286–295 (1951)
24. Alpcan, T., Başar, T.: Network Security: A Decision and Game-Theoretic Approach. Cambridge University Press, Cambridge (2010)
25. Manshaei, M., Zhu, Q., Alpcan, T., Basar, T., Hubaux, J.: Game theory meets network security and privacy. ACM Comput. Surv. **45**(3), 25–39 (2013)
26. Roy, S., Ellis, C., Shiva, S., Dasgupta, D., Shandilya, V., Wu, Q.: A survey of game theory as applied to network security, Hawaii, USA, pp. 1–10 (2010)
27. Bohacek, S., Hespanha, J., Lee, J., Lim, C., Obraczka, K.: Game theoretic stochastic routing for fault tolerance and security in computer networks. IEEE Trans. Parallel Distrib. Syst. **18**(9), 1227–1240 (2007)
28. Zhu, Q., Clark, A., Poovendran, R., Başar, T.: Deceptive routing games. In: 2012 IEEE 51st IEEE Conference on Decision and Control (CDC), Maui, Hawaii, USA, pp. 2704–2711 (2012)
29. Clark, A., Sun, K., Bushnell, L., Poovendran, R.: A game-theoretic approach to IP address randomization in decoy-based cyber defense. In: Khouzani, M.H.R., Panaousis, E., Theodorakopoulos, G. (eds.) GameSec 2015. LNCS, vol. 9406, pp. 3 21. Springer, Cham (2015). https://doi.org/10.1007/978-3-319-25594-1_1
30. Basar, T., Olsder, G.: Dynamic Noncooperative Game Theory. SIAM Classics in Applied Mathematics, Philadelphia (1999)
31. Fudenberg, D., Levine, D.K.: The Theory of Learning in Games, vol. 2. MIT Press, Cambridge (1998)
32. Sandholm, W.H.: Population Games and Evolutionary Dynamics. MIT Press, Cambridge (2010)
33. Young, H.P.: Individual Strategy and Social Structure: An Evolutionary Theory of Institutions. Princeton University Press, Princeton (2001)
34. Arrow, K., Debreu, G.: Existence of an equilibrium for a competitive economy. Econometrica **22**, 265–290 (1954)
35. Facchinei, F., Kanzow, C.: Generalized Nash equilibrium problems. 4OR **5**(3), 173–210 (2007)
36. Rosen, J.: Existence and uniqueness of equilibrium points for concave N-person games. Econometrica **33**(3), 520–534 (1965)
37. Pang, J.-S., Scutari, G., Facchinei, F., Wang, C.: Distributed power allocation with rate constraints in Gaussian parallel interference channels. IEEE Trans. Inf. Theory **54**(8), 3471–3489 (2008)

38. Yin, H., Shanbhag, U., Mehta, P.: Nash equilibrium problems with scaled congestion costs and shared constraints. IEEE Trans. Autom. Control **56**(7), 1702–1708 (2011)
39. Palomar, D., Eldar, Y.: Convex Optimization in Signal Processing and Communications. Cambridge University Press, Cambridge (2010)
40. Zhu, M., Martínez, S.: Distributed coverage games for energy-aware mobile sensor networks. SIAM J. Control Optim. **51**(1), 1–27 (2013)
41. Koshal, J., Nedić, A., Shanbhag, U.V.: Regularized iterative stochastic approximation methods for stochastic variational inequality problems. IEEE Trans. Autom. Control **58**, 594–609 (2013)
42. Yousefian, F., Nedić, A., Shanbhag, U.V.: A distributed adaptive steplength stochastic approximation method for monotone stochastic Nash games. In: 2013 American Control Conference, pp. 4765–4770, June 2013
43. Young, H.P.: The evolution of conventions. Econ.: J. Econom. Soc. **61**, 57–84 (1993)
44. Hu, Z., Zhu, M., Chen, P., Liu, P.: On convergence rates of game theoretic reinforcement learning algorithms. Automatica **104**(6), 90–101 (2019)
45. Okhravi, H., Comella, A., Robinson, E., Haines, J.: Creating a cyber moving target for critical infrastructure applications using platform diversity. Int. J. Crit. Infrastruct. Prot. **5**(1), 30–39 (2012)
46. Takahashi, S., Yamamori, T.: The pure Nash equilibrium property and the quasi-acyclic condition. Econ. Bull. **3**(22), 1–6 (2002)
47. Lai, T.L., Robbins, H.: Asymptotically efficient adaptive allocation rules. Adv. Appl. Math. **6**(1), 4–22 (1985)
48. Auer, P., Cesa-Bianchi, N., Fischer, P.: Finite-time analysis of the multiarmed bandit problem. Mach. Learn. **47**(2–3), 235–256 (2002)
49. Zurschmeide, J.B.: IRIX Advanced Site and Server Administration Guide. Silicon Graphics (1994)
50. Kuleshov, V., Precup, D.: Algorithms for multi-armed bandit problems. CoRR, abs/1402.6028 (2014)
51. SPEC CPU benchmark suite (2000). http://www.spec.org/cpu2000/
52. R. LLC, World's most used penetration testing software (2016). http://www.metasploit.com/
53. Kullback, S., Leibler, R.A.: On information and sufficiency. Ann. Math. Stat. **22**, 79–86 (1951)
54. CVE-2002-0656, Apache openSSL heap overflow exploit (2002). http://www.phreedom.org/research/exploits/apache-openssl/
55. Lin, Z., Riley, R.D., Xu, D.: Polymorphing software by randomizing data structure layout. In: Flegel, U., Bruschi, D. (eds.) DIMVA 2009. LNCS, vol. 5587, pp. 107–126. Springer, Heidelberg (2009). https://doi.org/10.1007/978-3-642-02918-9_7
56. Crispin, C., et al.: Stackguard: automatic adaptive detection and prevention of buffer-overflow attacks. In: Proceedings of the 7th Conference on USENIX Security Symposium (SSYM 1998), San Antonio, Texas, pp. 63–78, January 1998
57. CVE-2001-0144, SSH CRC-32 compensation attack detector (2001). http://www.securityfocus.com/bid/2347/discuss
58. CVE-2015-0235, Ghost: glibc gethostbyname buffer overflow (2015). https://www.qualys.com/2015/01/27/cve-2015-0235/GHOST-CVE-2015-0235.txt
59. CVE-1999-0071, Apache-cookie bug (1999). http://seclab.cs.ucdavis.edu/projects/testing/vulner/39.html
60. Chen, P., Hu, Z., Xu, J., Zhu, M., Liu, P.: Feedback control can make data structure layout randomization more cost-effective under zero-day attacks. Cybersecurity **1**, 3 (2018)

61. Åström, K.J.: Optimal control of Markov processes with incomplete state information. J. Math. Anal. Appl. **10**(1), 174–205 (1965)
62. Miehling, E., Rasouli, M., Teneketzis, D.: Optimal defense policies for partially observable spreading processes on Bayesian attack graphs. In: Proceedings of the Second ACM Workshop on Moving Target Defense, MTD 2015, Denver, Colorado, USA, pp. 67–76. ACM (2015)
63. Liu, Y., Man, H.: Network vulnerability assessment using Bayesian networks. In: Data Mining, Intrusion Detection, Information Assurance, and Data Networks Security 2005, pp. 61–71 (2005)
64. Poolsappasit, N., Dewri, R., Ray, I.: Dynamic security risk management using Bayesian attack graphs. IEEE Trans. Dependable Secur. Comput. **9**, 61–74 (2012)
65. Nguyen, T.H., Wright, M., Wellman, M.P., Baveja, S.: Multi-stage attack graph security games: heuristic strategies, with empirical game-theoretic analysis. In: Proceedings of the 2017 Workshop on Moving Target Defense, MTD 2017, Dallas, Texas, USA, pp. 87–97. ACM (2017)
66. Schiffman, M.: Common vulnerability scoring system (CVSS) (2017). http://www.first.org/cvss
67. Hu, Z., Zhu, M., Liu, P.: Online algorithms for adaptive cyber defense on Bayesian attack graphs. In: Fourth ACM Workshop on Moving Target Defense in Association with 2017 ACM Conference on Computer and Communications Security, Dallas, pp. 99–109, October 2017
68. Zambon, E., Bolzoni, D.: Network intrusion detection systems: false positive reduction through anomaly detection (2006). http://www.blackhat.com/presentations/bh-usa-06/BH-US-06-Zambon.pdf
69. Sondik, E.J.: The optimal control of partially observable Markov processes over the infinite horizon: discounted costs. Oper. Res. **26**(2), 282–304 (1978)
70. Bellman, R.E.: Dynamic Programming. Dover Publications, Mineola (2003)
71. Bellman, R.E., Dreyfus, S.E.: Applied Dynamic Programming. Princeton University Press, Princeton (1962)
72. Watkins, C.J.C.H., Dayan, P.: Q-learning. Mach. Learn. **8**, 279–292 (1992)
73. Hutter, M., Poland, J.: Adaptive online prediction by following the perturbed leader. J. Mach. Learn. Res. **6**, 639–660 (2005)
74. Kalai, A., Vempala, S.: Efficient algorithms for online decision problems. J. Comput. Syst. Sci. **71**, 291–307 (2005)
75. Weisberg, H.: Central tendency and variability. No. 83, Sage University Paper Series on Quantitative Applications in the Social Sciences (1992)
76. Shani, G., Pineau, J., Kaplow, R.: A survey of point-based POMDP solvers. Auton. Agent. Multi-Agent Syst. **27**, 1–51 (2013)
77. Schiffman, M.: Common vulnerability scoring system v3.0: specification document (2017). https://www.first.org/cvss/cvss-v30-specification-v1.7.pdf

Moving Target Defense Quantification

Massimiliano Albanese[1]([⊠]), Warren Connell[1], Sridhar Venkatesan[2],
and George Cybenko[3]

[1] George Mason University, Fairfax, VA, USA
{malbanes,wconnel2}@gmu.edu
[2] Perspecta Labs, Basking Ridge, NJ, USA
svenkatesan@perspectalabs.com
[3] Dartmouth College, Hanover, NH, USA
george.cybenko@dartmouth.edu

Abstract. Moving Target Defense (MTD) has the potential to increase
the cost and complexity for threat actors by creating asymmetric uncertainty in the cyber security landscape. The tactical advantages that MTD
can provide to the defender have led to the development of a vast array
of diverse techniques, which are designed to operate under different constraints and against different classes of threats. Due to the diverse nature
of these various techniques and the lack of shared metrics to assess their
benefits and cost, comparing multiple techniques is not a trivial task. We
addressed this gap by designing a framework to enable a uniform approach to the analysis and quantification of MTD techniques. This framework looks at each MTD technique in terms of the attacker's knowledge
it is capable of compromising, thus enabling direct comparison of any
two techniques or set of techniques.

1 Introduction

Moving Target Defense (MTD) offers a great potential in turning the typical
asymmetry of the cyber security landscape in favor of the defender [13], and
many different techniques have been developed since the term first surfaced in
the literature. However, each of these techniques only addresses a narrow subset of potential attack vectors and different techniques tend to measure their
effectiveness in different and often incompatible ways. Additionally, in order to
provide a comprehensive security solution, multiple MTD mechanisms should be
jointly deployed, but solving this problem requires the development of methods
for selecting an optimal subset of available techniques. Although several surveys
have identified scenarios where certain MTDs might not work well together [22],
or give a qualitative estimate of their effectiveness and cost [10], a quantitative
framework that can accommodate any existing or future MTDs is still needed for
this area of research to progress beyond specialized, isolated solutions. To address

The work presented in this chapter was supported by the Army Research Office under
grant W911NF-13-1-0421.

S. Jajodia et al. (Eds.): Adaptive Cyber Defense, LNCS 11830, pp. 94–111, 2019.
https://doi.org/10.1007/978-3-030-30719-6_5

this pressing need, we have developed a novel framework that captures, through probabilistic measures, the relationships between available MTDs and the information such MTDs are intended to protect from attackers [7]. Our model also captures the relationships between services, their weaknesses, and the knowledge required to exploit such weaknesses. Using this model, we can probabilistically estimate the effectiveness of any given technique or set of techniques, regardless of how they operate. Indeed, the capability to quantify MTD techniques is critical to effectively support the control and game theoretic approaches discusses in Chaps. 1, 2, 3, and 6 of this books.

Our framework presents the following desirable attributes: (i) **generality** – the relationships between MTDs and the knowledge they protect defines an interface that enables to plug any MTD into the framework; (ii) **extensibility** – the model can be extended to accommodate future MTDs by introducing new elements, such as additional knowledge blocks or classes of weaknesses; (iii) **resilience** – as the framework addresses generic classes of weaknesses rather than specific vulnerabilities, the model can address both known and unknown (zero-day) attacks; (iv) **usability** – the framework is simple and intuitive, can be used to compute utility estimates at different levels of granularity, and can incorporate cost in the estimation of utility.

The remainder of the chapter is organized as follows. Section 2 discusses related work. Section 3 briefly discusses the threat model and our assumptions. The proposed framework is presented in detail in Sect. 4 using a simple running example, whereas a more complex case study is considered in Sect. 4.3. Then, Sect. 5 discusses two practical applications of the proposed model, and Sect. 6 discusses two other important aspects of the problem, namely the probability of attack success and the optimal reconfiguration rate of an MTD. Finally, Sect. 7 gives some concluding remarks and discusses potential future work.

2 Related Work

Many different metrics have been proposed to measure the effectiveness of MTDs, such as attacker's success rate [4], or metrics for deception, deterrence, and detectability [12]. Still others utilize multiple metrics (productivity, success, confidentiality, and integrity) for both the attacker and the defender [23], leading to confusion over the multiple dimensions. However, the majority of these techniques are designed to protect systems against a very narrow set of attack vectors such as SQL injection [2], data exfiltration [20], and distributed DoS attacks [15,16]. One expert survey provides a thorough assessment of the effectiveness and cost of many techniques across the spectrum of existing MTDs [10], but the survey is qualitative in nature and potentially subject to reviewer's bias.

Cai *et al.* [3] developed a performance evaluation and comparison model for existing MTDs based on stochastic Petri Nets. Although more general than most existing approaches, this model still has some limitations, as the authors focus on MTD techniques that can be deployed on a web server, whereas the model we present in this chapter is agnostic of the specific nature of the hosts being defended or the MTDs being deployed.

Our work leverages existing work on attack graphs [14], particularly those approaches that evaluate security by looking at how the probability of a successful attack propagates over an attack graph [21]. The TREsPASS project[1] provides a holistic view of an organization's information security risk. It provides a visualization framework that combines the impact of vulnerability exploitation, physical security breach and social engineering on the target organization. This framework can be used to analyze several properties of multi-step attacks such as the required effort or time, and likelihood of success. However, attack graphs cannot be readily used with every MTD, as they are often tied to specific vulnerabilities. In fact, several MTDs can drastically alter a system's attack surface, requiring to generate an entirely new attack graph every time the MTD changes the system's configuration, which is not feasible in practice. To address this limitation, our framework operates at a higher level of abstraction, and models general classes of weaknesses rather than implementation-specific vulnerabilities.

3 Threat Model and Assumptions

The general nature of our model lets us make very broad, worst-case assumptions about the cyber threats we are trying to protect against. In particular, we assume that *attackers can exploit any known attack vector and can potentially discover zero-day vulnerabilities*. Most techniques described in the literature only protect against a narrow subset of possible attacks and no single MTD can protect against all possible attack vectors. This is handled in our model by offering the capability of combining multiple MTDs in a defense-in-depth approach.

We also make the worst-case assumption that *no static defense can prevent an attack*, as the attacker has virtually unlimited time to plan and execute an attack and zero-day exploits can always evade static defenses. Only the deployment of MTDs may have an effect on the attacker's success rate, and even when MTDs are deployed, there is a residual probability of attacker's success, as no MTD is perfect and capable of completely preventing an attack.

We assume that *attackers can be stopped or at least delayed by preventing them from acquiring accurate knowledge* about the target system. Our primary focus here is on the reconnaissance phase, when that knowledge is gathered prior to planning and executing attacks. Our goal can be achieved by either preventing attackers from accessing that knowledge or delaying them until that knowledge is no longer useful.

Finally, we make several additional simplifying assumptions throughout the chapter that we summarize here. Future work will allow us to revise many of our assumptions in order to further generalize our approach. We assume that services and weaknesses are time-invariant, thus no services are added or removed over time. We also assume that both services and knowledge blocks are independent of one another, but our framework could be easily extended to handle multiple interdependent services. We currently assume that each MTD has a predefined optimal configuration of its parameters, and that, if multiple MTDs

---

[1] http://www.trespass-project.eu.

affect a knowledge block, they do not interact and only the most effective one is considered.

4   MTD Quantification Framework

In this section, we present the proposed quantification framework, which, as shown for the motivating example of Fig. 1, consists of four layers: (i) a time-invariant service layer representing the set $\mathscr{S}$ of services to be protected; (ii) a weakness layer representing the set $\mathscr{W}$ of general classes of weaknesses that may be exploited; (iii) a knowledge layer representing the set $\mathscr{K}$ of all possible knowledge blocks required to exploit those weaknesses; and (iv) an MTD layer representing the set $\mathscr{M}$ of available MTD techniques.

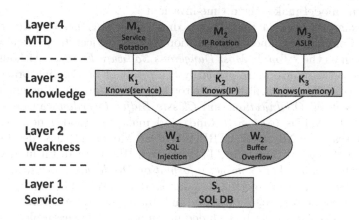

Fig. 1. Quantification framework layers

4.1   Mathematical Model

The proposed MTD quantification framework can be formally defined as a 7-tuple $(\mathscr{S}, \mathscr{R}_{SW}, \mathscr{W}, \mathscr{R}_{WK}, \mathscr{K}, \mathscr{R}_{KM}, \mathscr{M})$, where: (i) $\mathscr{S}$, $\mathscr{W}$, $\mathscr{K}$, $\mathscr{M}$ are the sets of services, weaknesses, knowledge blocks, and MTD techniques, respectively; (ii) $\mathscr{R}_{SW} \subseteq \mathscr{S} \times \mathscr{W}$ represents relationships between services and the common weaknesses they are vulnerable to; (iii) $\mathscr{R}_{WK} \subseteq \mathscr{W} \times \mathscr{K}$ represents relationships between weaknesses and the knowledge blocks required for an attacker to exploit them; and (iv) $\mathscr{R}_{KM} \subseteq \mathscr{K} \times \mathscr{M}$ represents relationships between knowledge blocks and the MTD techniques that affect them. The proposed model induces a k-partite graph, with $k = 4$, $G = (\mathscr{S} \cup \mathscr{W} \cup \mathscr{K} \cup \mathscr{M}, \mathscr{R}_{SW} \cup \mathscr{R}_{WK} \cup \mathscr{R}_{KM})$.

4.1.1   Layer 1: Service Layer
The first layer represents the set $\mathscr{S}$ of services we wish to protect against attacks. We assume that the services are time-invariant, i.e., the functionality of the

services does not change over time, and services cannot be taken down to prevent attacks, as this action would result in a denial-of-service condition. We only consider one service in the case studies presented in this chapter, but the model can be easily extended to consider multiple interdependent services, similarly to how a chain of interdependent vulnerability exploits is modeled in an attack graph [14, 21].

4.1.2   Layer 2: Weakness Layer

The second layer represents the set of weaknesses $\mathscr{W}$ that services are vulnerable to. We choose general classes of weaknesses rather than specific vulnerabilities because there are too many vulnerabilities to enumerate, some vulnerabilities are unknown, and, depending on the MTD used (e.g., OS rotation), specific vulnerabilities may change over time. Using general classes of weaknesses when building the model makes them time-invariant.

The examples used in this chapter draw these weaknesses primarily from MITRE's Common Weakness Enumeration (CWE) project [6], particularly from those known as the *"Top 25 Most Dangerous Software Errors."* Although many of the top software errors are primarily the result of bad coding practices and better solved at development time, the top software errors enabling exploits such as *SQL Injection*, *OS Injection*, and *Classic Buffer Overflow* can be addressed at runtime by MTDs (e.g., SQLRand) and make for good general categories of weaknesses. The Microsoft STRIDE Threat Model [11] has also been used as a source of general threats in MTD research [19] and can fill in areas where CWE may be lacking. For example, *Information Disclosure* (eavesdropping) and *Denial of Service* are not specifically addressed by CWE. The example of Fig. 1 shows two weaknesses, *SQL Injection* and *Buffer Overflow*. More weaknesses, such as *OS Injection*, might be included in a more complex example, while other weaknesses, such as *Cross-Site Scripting*, would not be applicable to this service.

4.1.3   Layer 3: Knowledge Layer

The third layer represents the knowledge blocks $\mathscr{K}$ required to effectively exploit weaknesses. An attacker may need some of this knowledge (such as a victim's IP address) to plan an attack even when no MTD is deployed, whereas other knowledge blocks may be specifically required to attempt circumventing a deployed MTD. For example, SQLRand [2] adds a keyword to SQL commands, which must be known for a malicious user to perform SQL injection. We assume that knowledge blocks are independent and must be acquired using different methods. For instance, IP address and port number should not be modeled as separate knowledge blocks because a method to determine one would also reveal the other.

The relationship between the knowledge and weakness layers is many-to-many. A weakness may require several pieces of knowledge to be exploited, and a knowledge block may be key to exploiting several weaknesses. This layer may also be extended as new MTDs – disrupting new and different aspects of an attacker's knowledge – are developed.

In our example, we assume that, in order to execute a SQL Injection attack, the attacker must gather information about the service (e.g., name and version of the specific DBMS) and the network configuration (e.g., IP address). In order to execute a Buffer Overflow attack, an attacker must know the IP address and some information about the vulnerable memory locations. A higher-fidelity version of this model may take a knowledge block and break it down into smaller, more specific items that are specifically targeted by available MTDs.

4.1.4 Layer 4: MTD Layer

The fourth layer of the model represents the set $\mathcal{M}$ of available MTDs. As MTD techniques provide probabilistic security, we model the impact of an MTD M_i on the attacker's effort to acquire knowledge K_j by associating a probability $P_{i,j}$ – representing the attacker's success rate – with the relation (K_j, M_i). As mentioned in Sect. 3, when only static defenses are deployed (i.e., no MTD), an attacker will acquire the necessary knowledge without significant effort, which we model by associating a probability of 1.

For example, if technique M_1 in Fig. 1 (*Service Rotation*) reduces an attacker's likelihood of acquiring knowledge block K_1 (i.e., correct version of the service) by 60%, we would label that edge with $P_{1,1} = 0.4$. If an MTD delays an attacker by some factor, we can also express that as a probability that the attacker will not gather the correct information in a timely manner. For example, an MTD that expands addressable memory by a factor of 10 might reduce the attacker's probability of success to 0.1, so $P_{i,j} = 0.1$. The exact methodology for determining the value of $P_{i,j}$ may vary from MTD to MTD, and we are investigating this problem as a separate line of research. Although this aspect of the problem goes beyond the scope of this chapter, we provide a brief discussion and some details in Sect. 6.1. Specifically, we are developing a general approach to model the tradeoff between cost and effectiveness of MTD techniques, as we vary the values of a technique's tunable parameters and other aspects of the attacker/defender interaction. Ultimately, this approach will enable us to identify the optimal configuration for each technique. Therefore, in this chapter, we assume that such optimal configuration has already been identified for each available MTD technique, along with the corresponding value of $P_{i,j}$ and the corresponding cost.

Expressing MTD effectiveness in terms of the probability an attacker will succeed in acquiring required knowledge enables us to analyze multiple techniques using a uniform approach, with a theoretically perfect MTD yielding $P_{i,j} = 0$, and a completely ineffective MTD yielding $P_{i,j} = 1$. In our example, we use *service rotation* to disrupt knowledge about the version of the service, and assume that rotating between 4 services reduces the attacker's probability of gathering the correct information to $P_{1,1} = 0.25$. We apply an *IP address rotation* scheme to mask the victim's IP address. It has been shown that perfect shuffling reduces the attacker's likelihood of guessing the correct IP address by 37% [4]. Using a conservative estimate, we assume $P_{2,2} = 0.75$. Finally, to protect knowledge of the memory layout, we use a dynamic ASLR scheme. Although dynamic ASLR only adds a single bit of entropy compared to typical ASLR [18], this further delays the attacker, resulting in a probability $P_{3,3} = 0.5$ of gathering the correct information.

4.2 Computing MTD Effectiveness

We compute an MTD's effectiveness starting from layer 4 of the model and working our way down the layered model to find the overall probability of attacker's success. First, we define $P(K_j)$ as the probability that the attacker has the correct information about knowledge block K_j, and compute $P(K_j)$ for each K_j in layer 3, based on the active MTDs deployed to protect K_j. If there is no active MTD, we assume that the attacker is guaranteed to obtain that information, i.e., $P(K_j) = 1$.

In our example, each knowledge block is affected by one MTD only. When multiple MTDs affect the same knowledge block, we make the simplifying assumption that the resulting effect is driven by the best-performing MTD. Thus:

$$P(K_j) = \begin{cases} 1, & \text{if } \nexists M_i \in \mathcal{M} \text{ s.t. } (K_j, M_i) \in \mathcal{R}_{KM} \wedge active(M_i) \\ \min_{M_i \in \mathcal{M} \text{ s.t. } (K_j, M_i) \in \mathcal{R}_{KM}} P_{i,j} \wedge active(M_i), & \text{otherwise} \end{cases} \quad (1)$$

A possible improvement to the model would be to capture the effect of multiple MTDs acting on the same knowledge block by using a function modeling either diminishing returns or some other interaction between MTDs.

Next, we determine the probability $P(W_k)$ that an attacker has gained all the knowledge required to exploit a given weakness W_k. Since each knowledge block is independent, this is simply the product of the probabilities associated with all knowledge blocks enabling $P(W_k)$, as shown by Eq. 2.

$$P(W_k) = \prod_{K_j \in \mathcal{K} \text{ s.t. } (W_k, K_j) \in \mathcal{R}_{WK}} P(K_j) \quad (2)$$

In our example, when calculating $P(W_1)$ and $P(W_2)$ for *SQL Injection* and *Buffer Overflow*, respectively, we obtain $P(W_1) = 0.25 \cdot 0.75 = 0.1875$ and $P(W_2) = 0.75 \cdot 0.50 = 0.375$.

Finally, we determine the defender's utility U gained by deploying MTD techniques based on the reduced probability of exploit for each class of weaknesses. In this work, the utility is defined as a function of the probability $P(S_l)$ that an attacker can compromise a service S_l by exploiting any of the weaknesses leading to it. $P(S_l)$ can be computed as the probability of the union of non-mutually exclusive events, using the *Inclusion-Exclusion Principle* [5]. With respect to our running example, $P(S_1)$ can be computed as follows:

$$P(S_1) = P(W_1 \cup W_2) = P(W_1) + P(W_2) - P(W_1 \cap W_2) \quad (3)$$

As W_1 and W_2 are not necessarily independent (as shown in this example), we cannot assume $P(W_1 \cap W_2) = P(W_1) \cdot P(W_2)$. Instead, we must express each $P(W)$ in terms of its corresponding independent knowledge blocks K_j, that is $P(W_1) = P(K_1) \cdot P(K_2)$, $P(W_2) = P(K_2) \cdot P(K_3)$, and $P(W_1 \cap W_2) = P(K_1) \cdot P(K_2) \cdot P(K_3)$, and then express $P(S_1)$ as a function of probabilities $P(K_j)$:

$$P(S_1) = P(K_1) \cdot P(K_2) + P(K_2) \cdot P(K_3) - P(K_1) \cdot P(K_2) \cdot P(K_3)$$

which results in

$$P(S_1) = 0.25 \cdot 0.75 + 0.75 \cdot 0.5 - 0.25 \cdot 0.75 \cdot 0.5 = 0.469$$

For graphs with 3 or more weaknesses $\mathscr{W}^* \subseteq \mathscr{W}$, we can expand Eq. 3 to the generalized form of the *Inclusion-Exclusion Principle* [5]:

$$P \left(\bigcup_{W_k \in \mathscr{W}^*} W_k \right) = \sum_{i=1}^{|\mathscr{W}^*|} \left((-1)^{i-1} \cdot \sum_{\mathscr{W}' \in 2^{\mathscr{W}} \text{ s.t. } |\mathscr{W}'|=i} P \left(\bigcap_{W_j \in \mathscr{W}'} W_j \right) \right)$$

Computing the probability of the union of multiple events is an NP-hard problem that cannot be solved in better than $O(2^n)$ time [5]. However, the general nature of the weaknesses in layer 2 of the model limits their number – as opposed to vulnerabilities which may number in the thousands – keeping the computing time manageable.

After computing $P(S_l)$, we can easily compute the defender's utility as $U = 1 - P(S_l)$. Besides this simple approach, the utility could be a sigmoid function of $P(S_l)$ with an inflection point centered around a desired effectiveness. Such functions are commonly used in autonomic computing [1]. The complete computation for each of the values in our example is shown in Fig. 2. Note that this choice of utility function relies upon the expectation that at least some measure of protection will be guaranteed for at least one knowledge block for each weakness, otherwise the attacker will be guaranteed to exploit that weakness and reduce the utility to 0. To handle this issue, utility can be defined as a function of the probabilities to exploit each weakness.

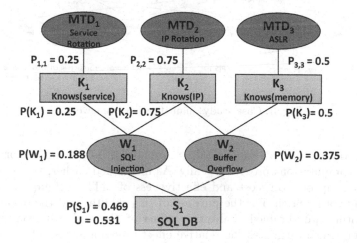

Fig. 2. Computing MTD effectiveness

4.3   Evaluation

We now present a more complex example to demonstrate the capabilities of our model. As seen in Fig. 3, we consider the same basic service but protect against two additional classes of weaknesses, OS Injection [6] and Eavesdropping (related to Information Disclosure from the STRIDE model [11]).

In this case study, more fine-grained knowledge blocks have been considered in order to provide more detail or to fit the specific MTDs selected for the case study. For example, knowledge block *Knows(memory)* has been broken down into separate blocks related to system call mapping, memory address, and stack direction. Similarly, SQL Injection now explicitly requires knowledge of keywords appended to SQL commands and some knowledge of the database schema, both of which are protected by SQLRand. Most importantly, we can now observe the many-to-many relationships between weaknesses, knowledge blocks, and MTDs, and conclude that finding the optimal solution is no longer trivial. However, using approximate yet reasonable values of $P_{i,j}$ for each MTD and cost constraints, we can determine the final utility as a function of selected MTDs using the steps previously shown and find an optimal solution using a problem solving method, such as stochastic hill climbing or evolutionary methods.

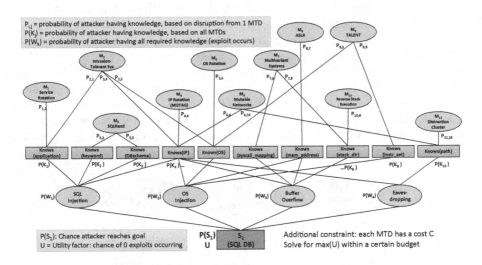

Fig. 3. Case study quantification framework

As a proof of concept, we can take the model in Fig. 3 and perform all the necessary computations programatically. As mentioned earlier, we are studying the relationship between cost and effectiveness of MTD techniques as part of another line of research. For the purpose of this chapter and the evaluation we are presenting, we obtained qualitative values of $P_{i,j}$ and cost from an expert survey [10], which estimates the relative effectiveness and cost of several MTD techniques by grouping them into coarse-grained categories of Low, Medium, or High. Whether or not an MTD is active can be treated as a Boolean variable, with inactive MTDs implying an attacker's probability of success of 1 and a cost of 0. The values from a sample MTD setup are shown in Table 1. The interim calculations for the probabilities of each knowledge block being acquired and each weakness being potentially exploited are also shown.

5   Applications

In these section, we discuss two different applications of our framework, showing how it can be used to compare different MTDs and to select optimal sets of MTDs to be jointly deployed.

Table 1. Sample case study evaluation

MTD	$P_{i,j}$		Cost	Active?	$P_{i,j}$ (effective)	Cost (effective)
M_1 (Service Rotation)	$P_{1,1}$	0.500	15	No	1.000	0
M_2 (Intrusion Tolerant Systems)	$P_{2,1}$	0.900	25	No	1.000	0
	$P_{2,4}$	0.900			1.000	
	$P_{2,5}$	0.900			1.000	
M_3 (SQLRand)	$P_{3,2}$	0.300	20	No	1.000	0
	$P_{3,3}$	0.300			1.000	
M_4 (IP Rotation/MOTAG)	$P_{4,4}$	0.900	25	No	1.000	0
M_5 (OS Rotation)	$P_{5,5}$	0.700	15	No	1.000	0
M_6 (Mutable Networks)	$P_{6,4}$	0.500	20	Yes	0.500	20
	$P_{6,10}$	0.500			0.500	
M_7 (Multivariant Systems)	$P_{7,6}$	0.500	20	No	1.000	0
	$P_{7,8}$	0.500			1.000	
M_8 (ASLR)	$P_{8,7}$	0.500	10	Yes	0.500	10
M_9 (TALENT)	$P_{9,5}$	0.500	20	No	1.000	0
	$P_{9,9}$	0.500			1.000	
M_{10} (Reverse Stack Execution)	$P_{10,8}$	0.500	20	No	1.000	0
M_{11} (Distraction Cluster)	$P_{11,10}$	0.500	20	No	1.000	0

Knowledge:

Knows(application)	1.000
Knows(keyword)	1.000
Knows(DBschema)	1.000
Knows(IP)	0.500
Knows(OS)	1.000
Knows(syscall_mapping)	1.000
Knows(mem_address)	0.500
Knows(stack_dir)	1.000
Knows(instr_set)	1.000
Knows(path)	0.500

Total Cost	30
Total Budget	120

Cost:	
High	25
Medium	15
Low	5

Effectiveness:	
High	0.3
Medium	0.5
Low	0.9

Chance of attack success:

SQL Injection	0.500
OS Injection	0.250
Buffer Overflow	0.250
Easvesdropping	0.250

Chance of attacker success:	0.500
Utility	**0.500**

5.1   Comparing MTDs

Given a set $\mathcal{M}$ of MTD techniques, we want to identify the technique that provides the highest overall utility. With respect to the example of Fig. 3, we start from the baseline deployment, shown earlier in Table 1, including M_6 (Mutable Networks) and M_8 (ASLR) to ensure we have a utility value to compare with. We then measure the updated utility value after individually adding each of the other MTDs to our baseline deployment. From the results reported in Table 2, we find that M_3 (SQLRand) offers the greatest increase in utility, with M_1, M_2, and M_3 being the only ones offering any increase at all. To explain these results, we observe that there is a lower bound on $P(S_1)$ that translates into an upper bound on U, defined by $\max(P(W_1), P(W_2), P(W_3), P(W_4))$.

In other words, the overall defense can only be as strong as the protection against exploitation of its most vulnerable weakness, which in turn benefits from the deployment of multiple MTDs. Therefore, given the baseline conditions, only an MTD that affects the most vulnerable weakness will yield any improvement in our utility value. This procedure could be used iteratively in an attempt to find an optimal solution in a greedy manner, but there would have to be some way to handle cases where no MTD adds any utility (such as random selection).

Table 2. Improvement from adding MTDs

MTD	M_1 (service rotation)	M_2 (intrusion tolerant systems)	M_3 (SQLRand)	All others
Utility	0.5625	0.513	0.614	0.5
Delta	0.0625	0.013	0.114	0.0

5.2   Selecting Optimal Defenses

Given a set $\mathcal{M}$ of MTDs and a budget, we would like to select the optimal set of MTDs that yield the highest utility with a total cost within the budget. As we now have a tool to evaluate the utility of any MTD deployment, we can also solve for the optimal selection of MTDs, given the constraints that the deployment of each MTD is a Boolean variable (either active or not) and that the sum of the costs of selected MTDs be under our budget. For the purpose of evaluating our framework and making the problem interesting, we selected a value of the budget (120) halfway between 0 and the total cost of deploying all available MTDs (i.e., 210) in the example of Fig. 3. This choice ensured that a solution with utility greater than 0 would be found and that approximately half the MTDs would be chosen as part of the optimal solution. We solved using the *generalized reduced gradient non-linear algorithm* [17] with random restarts to avoid finding local maxima. After solving, we obtained the solution shown in Fig. 4, with the selected MTDs highlighted with a thicker red outline. Detailed results, including margins of error for our estimates of effectiveness, are shown in Table 3.

We can observe that our choice of a utility function forces the selection of a variety of MTDs such that each weakness has at least one MTD protecting one of its knowledge blocks and that protection is evenly distributed over the 4 weaknesses. Visually, we can also observe that an MTD with the ability to protect multiple knowledge blocks is inherently more powerful than one that only protects one. However, if its cost is too high or effectiveness too low, it will still not be chosen as part of an optimal solution. Similarly, an MTD that only protects one knowledge block may be chosen if it is effective, low-cost, or affects a knowledge block that still receives relatively weak protection from other MTDs.

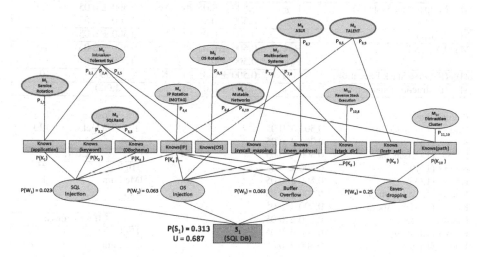

Fig. 4. Case study optimal configuration

6   Discussion

6.1   Analysis of Attack Success Probability

Estimating the time required for an attacker to gather sufficient knowledge during the reconnaissance phase is critical to assess the attacker's ability to successfully compromise a system. As our focus is on disrupting an attacker's reconnaissance effort, we can – without loss of generality – define the probability that an attacker succeeds as the probability to gather sufficient information to plan and execute an attack, which in turn is a function of the time available to complete the reconnaissance phase. In other words, we are implicitly assuming that, once accurate information is available to the attacker, the attack will always be successful. The probability $P_s(t)$ that an attacker succeeds in t time units is important in determining the required *reconfiguration rate*, i.e., the rate at which resources needs to be reconfigured by an MTD mechanism.

Table 3. Case study optimal configuration

MTD	$P_{i,j}$		C	Active?	Pi,j (effective)	C (effective)
M_1 (Service Rotation)	$P_{1,1}$	0.500 ± 0.05	15	Yes	0.500 ± 0.05	15
M_2 (Intrusion Tolerant Systems)	$P_{2,1}$	0.900 ± 0.05	25	No	1.000	0
	$P_{2,4}$	0.900 ± 0.05			1.000	
	$P_{2,5}$	0.900 ± 0.05			1.000	
M_3 (SQLRand)	$P_{3,2}$	0.300 ± 0.05	20	Yes	0.300 ± 0.05	20
	$P_{3,3}$	0.300 ± 0.05			0.300 ± 0.05	
M_4 (IP Rotation/MOTAG)	$P_{4,4}$	0.900 ± 0.05	25	No	1.000	0
M_5 (OS Rotation)	$P_{5,5}$	0.700 ± 0.05	15	No	1.000	0
M_6 (Mutable Networks)	$P_{6,4}$	0.500 ± 0.05	20	Yes	0.500 ± 0.05	20
	$P_{6,10}$	0.500 ± 0.05			0.500 ± 0.05	
M_7 (Multivariant Systems)	$P_{7,6}$	0.500 ± 0.05	20	Yes	0.500 ± 0.05	20
	$P_{7,8}$	0.500 ± 0.05			0.500 ± 0.05	
M_8 (ASLR)	$P_{8,7}$	0.500 ± 0.05	10	Yes	0.500 ± 0.05	10
M_9 (TALENT)	$P_{9,5}$	0.500 ± 0.05	20	Yes	0.500 ± 0.05	20
	$P_{9,9}$	0.500 ± 0.05			0.500 ± 0.05	
M_{10} (Reverse Stack Execution)	$P_{10,8}$	0.500 ± 0.05	20	No	1.000	0
M_{11} (Distraction Cluster)	$P_{11,9}$	0.500 ± 0.05	20	No	1.000	0

Knowledge:

Knows (1,application)	0.500 ± 0.05
Knows (1,keyword)	0.300 ± 0.05
Knows (1,DBschema)	0.300 ± 0.05
Knows (1,IP)	0.500 ± 0.05
Knows (1,OS)	0.500 ± 0.05
Knows (1, syscall_mapping)	0.500 ± 0.05
Knows (1, Mem_Address)	0.500 ± 0.05
Knows (1,stack_dir)	0.500 ± 0.05
Knows (1,instr_set)	0.500 ± 0.05
Knows (1,path)	0.500 ± 0.05

Total Cost	105
Total Budget	120

Cost:	
High	25
Medium	15
Low	5

Effectiveness:	
High	0.3 ± 0.05
Medium	0.5 ± 0.05
Low	0.9 ± 0.05

Chance of attack success:

SQL Injection	0.023 ± 0.006
OS Injection	0.063 ± 0.013
Buffer Overflow	0.063 ± 0.013
Easvesdropping	0.250 ± 0.035

Chance of attacker success:	0.313 ± 0.043
Utility	**0.687 ± 0.043**

Figure 5 shows two examples of $P_s(t)$, namely, linear and exponential functions. The linear function, $P_s(t) = t/T_s$, indicates that the probability of attack success increases linearly with time and reaches 1 (i.e., success) at time T_s. The exponential function (see for instance Eq. 4) indicates a situation in which the attacker initially accumulates knowledge at a low rate and then becomes exponentially more knowledgeable over time and succeeds at time T_s.

$$P_s(t) = 1 - \frac{1 - e^{(t-T_s)}}{1 - e^{-T_s}}. \tag{4}$$

As an example, consider an IP sweep combined with a port scan, where the attacker's goal is to discover the IP address of the machine running a specific service within the target network. The attack consists in sequentially scanning all IP addresses in a given range. Assuming an IP space of n addresses and that t^* time units are required to scan a single IP, we obtain $T_s = n \cdot t^*$ and $P_s(t) = \frac{t}{T_s} = \frac{t}{n \cdot t^*}$. As another example, consider the following DoS attack. The attacker initially compromises n hosts, which takes t^* time units. Then, each of the newly compromised hosts compromises additional n hosts, which takes additional t^* time units. At any given time t, the total number of compromised hosts, including the attacker's machine, is $N(t) = 1 + n + n^2 + \ldots + n^k = \frac{1 - n^{k+1}}{1-n}$, where $k = \lfloor t/t^* \rfloor$. We can assume that the attacker's success probability is proportional to the aggregate amount of flood traffic that compromised hosts can send to the victim, compared to the victim's capacity to handle incoming traffic. Let V denote the volume of traffic the victim can handle per time unit and let v denote the amount of traffic each compromised node can send per time unit. Then,

$$P_s(t) = \min\left\{1, \frac{N(t) \cdot v}{V}\right\} = \min\left\{1, \frac{v}{V} \cdot \frac{1 - n^{\lfloor t/t^* \rfloor + 1}}{1 - n}\right\}$$

6.2 Optimal Reconfiguration Rate

As mentioned earlier, we assume that each MTD is deployed with its own predefined optimal configuration. However, we could relax this assumption and further generalize the framework in order to enable network administrators to not only find an optimal set of MTDs but also configure them optimally.

The most critical configuration parameter for any MTD is the *reconfiguration frequency* α, that is the frequency at which the value of a given parameter (e.g., IP address in IP hopping) is updated. In [8], we presented a quantitative analytic model for assessing the performance of MTDs in terms of availability of the resources being periodically reconfigured. In fact, while a resource is being reconfigured, it is not available to handle service requests, thus impacting overall system performance. Based on this model, we also developed a method to

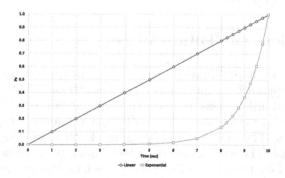

Fig. 5. Probability of success P_s vs. time for $T_s = 10$

determine the reconfiguration rate that minimizes the attacker's probability of success while meeting performance and stability constraints. Choosing the optimal value of α is a critical problem: in fact, as shown in Fig. 6, a larger value of α may cause peaks in the system's response time, whereas a lower value may not provide enough security.

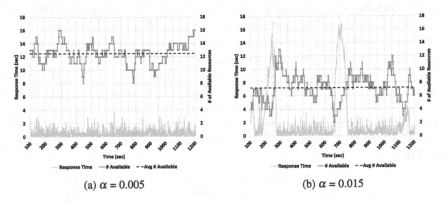

(a) $\alpha = 0.005$ (b) $\alpha = 0.015$

Fig. 6. Number of available resources and response time

6.3   Extending the Framework

Our framework can accommodate any existing MTD as long as we can identify the knowledge blocks it affects, the extent to which it disrupts that knowledge, and how it relates to the weaknesses we plan to protect against. Another important feature of our framework is the ability to be extended to accommodate any future MTD that may be developed. A new MTD that affects existing knowledge blocks may be simply added to the MTD layer of the model, while an MTD that works in ways we have not yet considered might also require the addition of new knowledge blocks. Even a new class of weaknesses could be added to the model if the situation warrants it, making our model "future-proof" against new developments in cyber threats.

7   Conclusions and Future Work

In this chapter, we have discussed a framework for quantifying moving target defenses. Our approach to quantifying the benefits of MTDs yields a single, probability-based utility measure that can accommodate any existing or future MTD, regardless of their nature. Our multi-layered approach captures the relationship between MTDs and the knowledge blocks they are designed to protect, and the relationship between knowledge blocks and generic classes of weaknesses that can be exploited using that knowledge. We have shown through case studies that we can compute the joint effectiveness of multiple MTDs as a function of their individual effectiveness and, by doing so, we can make informed decisions about which MTD or set of MTDs provide better protection based on the security requirements or cost constraints.

Although the work presented in this chapter represents a significant step towards effective MTD quantification, several limitations still exist and will be addressed as part of our planned future work. Specifically, limitations exist in the following areas: (i) **probability computation** – our methods for computing the probability $P_{i,j}$ provide rough estimates, so a procedure needs to be developed to accurately assess the effectiveness of any MTD; (ii) **cost modeling** – currently, we adopt a very simple notion of cost, and use cost just as an additional constraint, whereas a more sophisticated notion of cost could be introduced and taken into account in the computation of utility values; and (iii) **choice of utility function** – the proposed utility function is based on the assumption that all weaknesses need to be at least partially protected by MTDs to prevent the utility from dropping to 0, therefore, if the risk of leaving a specific weakness unprotected can be accepted, other classes of utility functions could be explored. To address these limitations and further refine our model, we plan to work on several aspects of the framework, as briefly described below.

Implementation and Validation. To validate the model, we plan to deploy multiple MTDs on our computing infrastructure and then study their effectiveness both in isolation – in order to determine the value of $P_{i,j}$ for each MTD – and jointly – in order to accurately study the combined cost and performance. Preliminary experiments along this line of research were presented in [9].

Application to Multiple Attack Phases. Our model aims at disrupting an attacker's knowledge in the reconnaissance phase of the cyber kill chain. While this may be the most cost-effective way to approach cyber security, no defense is perfect, and we need to ensure multiple layers of defense. Some MTDs can disrupt an attacker's ability to maintain a foothold in the system, so we plan to extend our framework to model this additional class of MTDs.

Application to Dependent Services. Our framework currently models only independent services. Similar to attack graphs, an attacker may need to execute a sequence of exploits to reach a specific goal. Thus, we plan to extend our framework by introducing a meta-model that captures the relationships between services and the MTDs that can protect them from multi-step attacks.

Heuristics. Because of the $O(2^n)$ runtime to evaluate utility with the current model, it may be necessary to develop heuristics to speed up the evaluation in the case that the number of weaknesses grows to the point where using the model becomes infeasible.

Confidence Intervals. Because of the level of uncertainty of our probabilistic values, we may not have a completely accurate utility value. With enough experimental samples, we could introduce confidence intervals into our assertion that a certain MTD or set of MTDs has a higher utility.

References

1. Alomari, F., Menascé, D.A.: An autonomic framework for integrating security and quality of service support in databases. In: Proceedings of the 6th International Conference on Software Security and Reliability (SERE 2012), pp. 51–60. IEEE, Gaithersburg, June 2012
2. Boyd, S.W., Keromytis, A.D.: SQLrand: preventing SQL injection attacks. In: Jakobsson, M., Yung, M., Zhou, J. (eds.) ACNS 2004. LNCS, vol. 3089, pp. 292–302. Springer, Heidelberg (2004). https://doi.org/10.1007/978-3-540-24852-1_21
3. Cai, G., Wang, B., Luo, Y., Hu, W.: A model for evaluating and comparing moving target defense techniques based on generalized stochastic Petri Net. In: Wu, J., Li, L. (eds.) ACA 2016. CCIS, vol. 626, pp. 184–197. Springer, Singapore (2016). https://doi.org/10.1007/978-981-10-2209-8_16
4. Carroll, T.E., Crouse, M., Fulp, E.W., Berenhaut, K.S.: Analysis of network address shuffling as a moving target defense. In: IEEE International Conference on Communications (ICC 2014), pp. 701–706. IEEE, Sydney, June 2014
5. Chen, S.G.: Reduced recursive inclusion-exclusion principle for the probability of union events. In: Proceedings of the IEEE International Conference on Industrial Engineering and Engineering Management (IEEM 2014), pp. 11–13. IEEE, Bandar Sunway, December 2014
6. Christey, S.: 2011 CWE/SANS top 25 most dangerous software errors (2011). http://cwe.mitre.org/top25/
7. Connell, W., Albanese, M., Venkatesan, S.: A framework for moving target defense quantification. In: De Capitani di Vimercati, S., Martinelli, F. (eds.) SEC 2017. IAICT, vol. 502, pp. 124–138. Springer, Cham (2017). https://doi.org/10.1007/978-3-319-58469-0_9
8. Connell, W., Menascé, D.A., Albanese, M.: Performance modeling of moving target defenses. In: Proceedings of the 4th ACM Workshop on Moving Target Defense (MTD 2017), pp. 53–63. ACM, Dallas, October 2017
9. Connell, W., Pham, L.H., Philip, S.: Analysis of concurrent moving target defenses. In: Proceedings of the 5th ACM Workshop on Moving Target Defense (MTD 2018), pp. 21–30. ACM, Toronto, October 2018
10. Farris, K.A., Cybenko, G.: Quantification of moving target cyber defenses. In: Proceedings of SPIE Defense + Security 2015, Baltimore, MD, USA, April 2015
11. Howard, M., LeBlanc, D.: Writing Secure Code. Developer Best Practices Series, 2nd edn. Microsoft Press, Redmond (2002)
12. Jafarian, J.H., Al-Shaer, E., Duan, Q.: Spatio-temporal address mutation for proactive cyber agility against sophisticated attackers. In: Proceedings of the 1st ACM Workshop on Moving Target Defense (MTD 2014), pp. 69–78. ACM, Scottsdale, November 2014
13. Jajodia, S., Ghosh, A.K., Swarup, V., Wang, C., Wang, X.S. (eds.): Moving Target Defense: Creating Asymmetric Uncertainty for Cyber Threats. Advances in Information Security, vol. 54. Springer, New York (2011). https://doi.org/10.1007/978-1-4614-0977-9
14. Jajodia, S., Noel, S., O'Berry, B.: Topological analysis of network attack vulnerability. In: Kumar, V., Srivastava, J., Lazarevic, A. (eds.) Managing Cyber Threats: Issues, Approaches, and Challenges, Massive Computing, vol. 5, pp. 247–266. Springer, Boston (2005). https://doi.org/10.1007/0-387-24230-9_9

15. Jia, Q., Sun, K., Stavrou, A.: MOTAG: moving target defense against internet denial of service attacks. In: Proceedings of the 22nd International Conference on Computer Communications and Networks (ICCCN 2013). IEEE, Nassau, August 2013

16. Jia, Q., Wang, H., Fleck, D., Li, F., Stavrou, A., Powell, W.: Catch me if you can: a cloud-enabled DDoS defense. In: Proceedings of the 44th Annual IEEE/IFIP International Conference on Dependable Systems and Networks (DSN 2014), pp. 264–275. IEEE, Atlanta, June 2014

17. Lasdon, L.S., Fox, R.L., Ratner, M.W.: Nonlinear optimization using the generalized reduced gradient method. RAIRO Recherche opérationnelle **8**(V3), 73–103 (1974)

18. Shacham, H., Page, M., Pfaff, B., Go, E.J., Modadugu, N., Boneh, D.: On the effectiveness of address-space randomization. In: Proceedings of the 11th ACM Conference on Computer and Communications Security (CCS 2004), pp. 298–307. ACM, Washington DC, October 2004

19. Soule, N., et al.: Quantifying & minimizing attack surfaces containing moving target defenses. In: Proceedings of the Resilience Week (RWS 2015), pp. 220–225. IEEE, Philadelphia, August 2015

20. Venkatesan, S., Albanese, M., Cybenko, G., Jajodia, S.: A moving target defense approach to disrupting stealthy botnets. In: Proceedings of the 3rd ACM Workshop on Moving Target Defense (MTD 2016), pp. 37–46. ACM, Vienna, October 2016

21. Wang, L., Islam, T., Long, T., Singhal, A., Jajodia, S.: An attack graph-based probabilistic security metric. In: Atluri, V. (ed.) DBSec 2008. LNCS, vol. 5094, pp. 283–296. Springer, Heidelberg (2008). https://doi.org/10.1007/978-3-540-70567-3_22

22. Ward, B.C., et al.: Survey of cyber moving targets. Technical report 1228. MIT Lincoln Laboratory, Lexington, MA, USA, January 2018

23. Zaffarano, K., Taylor, J., Hamilton, S.: A quantitative framework for moving target defense effectiveness evaluation. In: Proceedings of the 2nd ACM Workshop on Moving Target Defense (MTD 2015), pp. 3–10. ACM, Denver, October 2015

Empirical Game-Theoretic Methods for Adaptive Cyber-Defense

Michael P. Wellman[✉], Thanh H. Nguyen, and Mason Wright

University of Michigan, Ann Arbor, USA
wellman@umich.edu

Abstract. Game-theoretic applications in cyber-security are often restricted by the need to simplify complex domains to render them amenable to analysis. In the empirical game-theoretic analysis approach, games are modeled by simulation, thus significantly increasing the level of complexity that can be addressed. We survey applications of this approach to scenarios of adaptive cyber-defense, illustrating how the method operates, and assessing its strengths and limitations.

1 Introduction

Strategic analysis of a cyber-security situation starts with the understanding that attacker and defender are engaged in an adversarial interaction, driven by (largely) opposing objectives, and armed with distinct tools for assessing and shaping the cyber environment. Formalizing these elements almost inevitably leads the analyst to describe the situation in *game-theoretic* terms: available actions and observations of the respective actors (players), and utility functions representing objectives. Thus, it is not surprising to observe a large expansion of the literature on game theory applied to cyber-security,[1] and an associated increase in development of tools and applications (Manshaei et al. 2013; Roy et al. 2010; Sinha et al. 2018).

Many game-theoretic treatments of cyber-security domains start with major simplifications, due to the analytic complexity of high-fidelity representations of realistic environments. Analysis of such stylized models can often shed valuable light on a strategic situation. For example, Edwards et al. (2017) employ a coarse-grained "blame game" model to identify qualitative considerations for deciding how and when to attribute responsibility for suspected state-sponsored cyber-attacks. Simplicity in modeling facilitates reasoning and allows a given model to cover a broad class of relevant scenarios. Choosing the right abstractions to isolate exactly the strategic issues of interest is central to the game theorist's art, and when done well, it can provide deep insight for decision makers.

There are two significant drawbacks to the stylized approach, however. First, the models analyzed tend to be generic, and so do not necessarily help for

---

[1] Including dedicated annual conferences, such as GameSec (Bushnell et al. 2018; Rass et al. 2017a).

© Springer Nature Switzerland AG 2019
S. Jajodia et al. (Eds.): Adaptive Cyber Defense, LNCS 11830, pp. 112–128, 2019.
https://doi.org/10.1007/978-3-030-30719-6_6

determining particular solutions to specific situations. Work in the framework of Stackelberg security games (Tambe 2011) has effectively addressed this issue, by supporting decisions for specified problem instances rather than generic scenarios. Second, for complex scenarios there is danger that the abstractions applied may discard essential detail, and thus the resulting guidance is incomplete, or worse—potentially misleading. Cyber-systems are inherently complex environments, typically involving numerous computationally interacting entities, with considerable state and complicated patterns of communication and observation. Experts familiar with the intricacies of such systems are likely to view stylized game models as toy versions of reality, and thus take a skeptical stance to conclusions from such models.

Since any modeling approach will entail some abstraction of the real world, there is no way for an analysis method to completely avoid this second drawback. Simplification is a matter of degree, so extending game-theoretic reasoning to accommodate greater complexity will enable the models to capture more of the richness of realistic cyber-security situations. This is particularly important for treatments of *adaptive* cyber-defense, since the dynamic evolution of configuration and information is the essence of adaptation. To be considered adaptive, a defense policy must take into account the attack state of the system, in consideration of how successful attacks require a succession of actions to gain knowledge about and eventually compromise targeted resources (Evans et al. 2011). Incorporating dynamics in the game model is therefore an absolute requirement for this domain. Dynamic information in turn poses significant technical challenges for game-theoretic methods (Tavafoghi et al. 2019).

One interesting effort to capture complex security dynamics in an abstract game model is the *FlipIt* framework introduced by Dijk et al. (2013). In FlipIt, two players vie for control of a single resource. Each has a single action, which takes control of the resource at some cost. Neither player can observe when the other has acted, and so is uncertain about the state of control except at the instant it performs its own action. Though the FlipIt model is quite abstract, it captures key elements of system security not well-supported by previous models (Bowers et al. 2012). Analysis of FlipIt has led to interesting insights about the interplay of various strategy classes, the value of aggressive play, and the significance of information advantages. As a stylized model, however, the generic version of FlipIt misses many relevant features of adaptive cyber-defense and is not suitable for decision making in a particular situation. Extensions of FlipIt have covered additional relevant scenario features (Farhang and Grossklags 2016; Jones et al. 2015; Laszka et al. 2013, 2014; Pham and Cid 2012). These add to practical realism, but seriously complicate analysis of the FlipIt game, which to date has eluded complete analytic solution, even in its basic version.

Which brings us finally to the approach described in this chapter: *empirical game-theoretic analysis* (EGTA) (Wellman 2016). Rather than build an analytic model that may be amenable to direct game-theoretic solution, EGTA starts with a detailed environment model described in procedural form, that is, by a simulation. We then introduce a set of specific dynamic strategies, and systematically run simulations over combinations of these strategies. The simulation data form the basis for estimating a game model, which can be solved using standard techniques.

The advantage of simulation is its ability to handle complex, stochastic, and temporally extended scenarios. This allows us to include realistic features of adaptive cyber-defense domains, going beyond generic and toy models. In its iterative form, EGTA also supports exploration, allowing us to address a rich set of strategic questions without premature simplification, such as isolating all the key strategic variables in advance. There are also limitations, particularly regarding the difficulty of generalizing game-theoretic conclusions beyond the specific environments and strategy instances studied. Overall, we regard EGTA as a complement to traditional game-theoretic treatments, which sacrifice complexity for generality (within the simplified model).

2   Empirical Game-Theoretic Analysis

The general idea of EGTA is to apply game-theoretic reasoning to models derived from agent-based simulation. The approach is designed to combine the advantage of simulation models in accommodating complexity with principles of strategic analysis expressible in the framework of game theory.

2.1   Basic Steps

The basic steps of EGTA are as follows, illustrated in Fig. 1.

1. Define a space of strategies for each player.
2. Simulate various combinations, or *profiles*, of agent strategies.
3. Induce or estimate an *empirical game model* from the accumulated simulated payoff data.
4. Analyze the resulting empirical game model, for example to identify Nash equilibria or otherwise characterize solutions of the game.

We elaborate on each step in turn.

Define Strategy Space. In EGTA, we typically constrain attention to a strict subset of the strategies that could be implemented in principle, for example imposing a parameterized form for strategies or adopting a particular agent

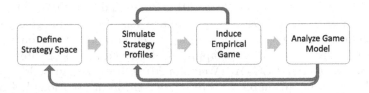

Fig. 1. Basic steps of empirical game-theoretic analysis. Feedback arrows show common patterns for iteration.

architecture. For this reason the available options are sometimes referred to as *heuristic strategies*. Though in general, a game may have any number of players, cyber-security games commonly focus on two: *attacker* (A) and *defender* (D). Let S_A and S_D denote their respective strategy sets.

Simulate Strategy Profiles. In a cyber-security game, we would simulate profiles (s_A, s_D) for various choices of $s_A \in S_A$ and $s_D \in S_D$. Each simulation yields a sample *payoff vector*, giving a numeric representation of the value of the outcome received by each player from one play of the given profile. Given stochastic factors in the simulation, we would typically require many samples of a profile to produce a reliable estimate of the expected payoffs to A and D.

Induce Empirical Game. In the most straightforward implementation of this step, we estimate a normal-form game model by sampling every profile $s \in S_A \times S_D$ a sufficient number of times. The payoff to player A in s, $u_A(s)$ is simply the sample average of A's payoffs in these simulations (and similarly $u_D(s)$ for player D). If the strategy spaces are very large, machine learning methods may be employed to generalize over the data to estimate payoffs for profiles not explicitly simulated (Vorobeychik et al. 2007).[2]

Analyze Game Model. The goal of analysis is to calculate Nash equilibria or another chosen solution concept, typically using off-the-shelf techniques. In the cyber-security context, let us define a *mixed profile* (σ_A, σ_D), with $\sigma_A \in \Delta(S_A)$ a probability distribution over A's strategy set (and similarly for σ_D) to be a joint strategy where each player independently chooses a strategy according to these distributions. Then (σ_A, σ_D) is a *Nash equilibrium* iff $\mathbb{E}[u_A(\sigma_A, \sigma_D)] \geq \mathbb{E}[u_A(s_A, \sigma_D)]$ for all $s_A \in S_A$, and similarly $\mathbb{E}[u_D(\sigma_A, \sigma_D)] \geq \mathbb{E}[u_D(\sigma_A, s_D)]$ for all $s_D \in S_D$.

Game analysis may also include reasoning about strategic relationships, such as dominance or ranking responses to particular opponents. Sensitivity analysis or statistical reasoning about candidate solutions would also be included in the game analysis step.

2.2   Iterative EGTA

It would be unusual for an EGTA study to proceed linearly according to steps 1-2-3-4 and complete. In practice, preliminary results at one step may inform reconsideration or elaboration of work at previous steps, and so the procedure would be iterative in nature. The key feedback links are shown in Fig. 1.

The simulation of strategy profiles (step 2) generates a collection of payoff samples. The number of samples required may not be straightforward to determine in advance. Feedback arrows from the game induction and analysis suggest that the results of these computations may be relevant in determining whether the collection is adequate, and if not, where additional simulation-based sampling is required. Such determination can be made on a principled basis through

---

[2] Such generalization is also often needed for the more general case of games where there are many players (Sokota et al. 2019; Wiedenbeck et al. 2018).

statistical analysis (Wiedenbeck et al. 2014), considering properties of the data collected and goals of the game analysis.

The longer feedback arrow from game analysis to strategy space represents *strategy exploration* (Jordan et al. 2010). Analysis of an accurate game model in step 4 gives us solutions to the game defined by the strategy space (S_A, S_D) defined in step 1. Since S_A and S_D are strict subsets of the true strategy sets available to players A and D, it is quite likely that the solutions found are not actually equilibria of the true game. We can bolster our confidence by considering additional strategies, thus defining augmented strategy sets $S'_A \supset S_A$ and $S'_D \supset S_D$. Solutions to the game over strategy space (S'_A, S'_D) are not actually guaranteed to be better approximations with respect to the full game (except in the limit when all strategies are included), but all else equal we expect improvement as more strategies are considered.

Of course, the interesting question in strategy exploration is which strategy or strategies to add at each iteration. A natural approach is to try to improve on the current equilibrium, by computing a *best response* to the other-player strategy. It turns out that the best response is generally not the optimal strategy to add in an iterative EGTA procedure (Jordan et al. 2010), as it does not consider opponent strategies outside the equilibrium, and it may not diversify the strategy set enough. Nevertheless, it is often a good heuristic, particularly if some stochastic exploration is conducted as well.

3   Example: A Moving Target Defense Game

We illustrate the EGTA approach to cyber-security by sketching the study of Prakash and Wellman (2015), which addressed an abstract scenario in moving-target defense (MTD). MTD covers a broad class of adaptive defenses where the main object is to defeat the attacker's ability to gain sufficient knowledge to compromise or take over a system (Jajodia et al. 2011). There are many MTD techniques, which accomplish this objective in various ways, generally involving some adaptation of the system to confuse the attacker or render its existing knowledge obsolete. We sought an abstract model that could fit the MTD approach broadly, without committing to a particular technology or system context. We thus adopted an extended version of the FlipIt model (van Dijk et al. 2013) discussed above in Sect. 1. The extension adds some complicating features present in prior work, such as multiple servers (Laszka et al. 2014) and asymmetric stealth (Laszka et al. 2013). It also incorporates a progressive concept of attack, in that unsuccessful attempts to compromise a server yield information that make subsequent attempts more likely to succeed (absent defender adaptation). This last feature is essential for capturing the primary dynamic of MTD (Albanese et al. 2019; Evans et al. 2011).

3.1   Game Description

In the specific MTD game studied, an attacker and defender compete for the control of 10 servers. (We could scale to many more servers with linear growth in the simulation time). Servers start out in control of the defender. The key actions are *probe* for the attacker, and *reimage* for the defender. A probe is essentially an attempt to compromise a server. If it succeeds, the attacker gains control, and if not, the attacker gains some information (not modeled explicitly) that increases its chance of succeeding on the next attempt. A reimage action by the defender takes a server down and resets its state. That is, any progress the attacker may have made on that server through probing is erased, such that probe success probability is reduced to its initial value.

The simulation proceeds for $T = 1000$ time steps. At each time step, the attacker may decide to probe any subset of the servers, and similarly the defender may choose some servers to reimage. Each faces a tradeoff, in that their actions help them achieve their goal of gaining or maintaining control of servers—but at a cost. For attackers, the probe actions bear an explicit cost, and for defenders the cost of reimaging is implicit in the downtime (7 time units in our setting) incurred for performing that action.

The state of the system at any point can be described by which player controls each server, and if the defender controls: whether it is down or up, and how many probes the attacker has attempted since the last reimage.

3.1.1   Observation Model

As argued above, cyber-security games are generally characterized by complex dynamics of state and observations, and this game is no exception. Technically, when agents cannot reliably observe each other's actions, the game is said to exhibit *imperfect information*. In this game, neither agent can perfectly observe the other. Precisely characterizing the model of what is and is not observed by each player is crucial for capturing the strategic interaction in an imperfect information game.

In the example MTD game, the defender has a partial ability to detect probes executed on any server. Specifically, if the server is up, the defender detects the probe with a specified probability, which varies across environment settings. However, the defender cannot tell whether a detected probe succeeded in compromising its target. The defender does of course know when it performs a reimage, and it is only at that point (and for the following downtime) that it can be sure it controls the server.

The attacker, on the other hand, does become aware when a probe succeeds. It also finds out when a server it controls is retaken by the defender through reimaging. Therefore, the attacker always knows the state of control of every server. However, it can only imperfectly track its progress in increasing success probability through probes, because it cannot tell when a defender reimages a server not in its control.

3.1.2   Utility

The primary objective of each player is to control servers. This is reflected in their utility functions, which quantify the value they attribute to any trajectory of states and actions. In the MTD game, players accrue utility each time period, based on the fraction of servers up and in their control, and also the fraction of servers *not* in the *other* player's control (i.e., either up and in the player's control, or not up).

This functional form of the utility function is designed to accommodate a variety of preference patterns, including objectives from the classic "CIA" (confidentiality, availability, integrity) triad (Pfleeger and Pfleeger 2012). For example, the *confidentiality* objective can be expressed through parameters encoding the defender's strong aversion to allowing the attacker to control servers. *Availability* from the defender's perspective can be expressed as requiring that a sufficient fraction of servers are in the defender's control and not down. We can categorize attacker utility in an analogous way. An attacker that accrues utility only by having servers in its control is termed a *control* attacker, whereas an attacker that accrues utility by having servers in its control or down is termed a *disrupting* attacker.

The utility function also includes threshold parameters governing the level of contention for servers in the associated environment. For example, by setting the threshold to 1/2 we impose the constraint that significant utility is accrued only if at least a majority are in control.

Finally, the utility model accounts for the cost of actions. The attacker pays a specified cost in utility per probe. The cost of the defender action is expressed implicitly in the utility function as the difference in utility accrued by servers being down as opposed to in the defender's control.

In the best case, a player accrues one utility unit per time period for keeping servers in their desired state, at no cost. The maximum overall utility for a game run is therefore T. The minimum is unbounded, as players may take unlimited costly actions without achieving their objective.

3.1.3   Strategies

In the EGTA approach, we focus on parameterized families of heuristic strategies, characterized by regular structures and patterns of behavior over time. Defining this strategy space is the first key step of EGTA (Fig. 1). The heuristic strategies defined for the MTD game generate actions based on the passage of time, or observed events in the system. If the actions are triggered by passage of time (in either a deterministic or probabilistic manner), we call the strategy *periodic*. The remaining strategies are triggered by observed events. They may apply actions to servers based on observations of that server, or based on combinations of observations across servers.

Specific families of heuristic strategies are defined for both attacker and defender. Within each family, there may be parametric options, so a large or even infinite number of possible instances. Overall, we considered 12 distinct attacker strategies and 20 defender strategies (i.e., $|S_A| = 12, |S_D| = 20$). These include for each player the *No-Op* strategy, in which the agent never takes any action.

Attacker Strategies. We consider two forms of periodic attacker strategy:

- *Uniform-Uncompromised.* Selects uniformly among those servers under the defender's control.
- *MaxProbe-Uncompromised.* Selects the server that has been probed the most since last reimage (that the attacker knows about), among those servers under the defender's control, breaking ties uniformly.

We also include one non-periodic attacker strategy that generates probe actions based on the number of servers that an attacker controls.

- *Control-Threshold.* If the attacker controls less than a threshold fraction of the servers, it chooses to probe the server that has been probed the most since last reimage (as far as it is aware) among those it does not currently control. Ties are broken uniformly among all eligible servers. A minimum waiting period of one time unit separates any two consecutive actions.

Defender Strategies. We consider periodic defender strategies employing two different criteria for server selection:

- *Uniform.* Selects uniformly among all up servers.
- *MaxProbe.* Selects the server that has been probed most since its last reimage, breaking ties uniformly.

The non-periodic defender strategies trigger a reimage operation based on probe activity or inactivity.

- *ProbeCount-or-Period* (PCP). Reimages a server whenever it detects that a threshold number of probes since the last reimage, *or* if it has been probed at least once but not within the specified period. The rationale for reimaging a server that is not being probed is that this could be an indication that the attacker has already compromised it and thus ceased attack.
- *Control-Threshold.* Analogous to the attacker's strategy by the same name, we include a defender strategy that performs a reimage action when the fraction of servers controlled falls below a threshold. Unlike the attacker, however, the defender cannot directly observe control state. Instead, the defender estimates the number of servers compromised based on the probes it has observed since reimaging on each server.
- *Control-Target.* Like Control-Threshold, except based on a target rather than a threshold.

3.2 Simulation and Analysis

We performed EGTA over a variety of environment and agent utility settings. The experiments covered a variety of environment settings, with systematic analysis over the possible combinations. Specifically, we varied games over the following features:

- Defender objective: confidentiality or availability.
- Attacker objective: disruption or control.

- Threshold on server control: low, majority, or high.
- Attacker probe cost: low, medium, or high.
- Defender probe detection: perfect, or imperfect at levels low, medium, or high.

Altogether, these settings define 144 distinct game instances. For 43 of these instances we conducted a full empirical game analysis: steps 2 through 4 of Fig. 1. These include all 18 instances with availability defender objective and perfect probe detection (Fig. 2), and another 18 with availability defender objective, imperfect probe detection, and medium p;robe cost (Fig. 3). We also ran seven with confidentiality defenders; as discussed below the confidentiality objective generally leads to an obvious equilibrium. For steps 2 and 3, we estimated through repeated simulation the joint payoffs for all $12 \times 20 = 240$ strategy profiles (s_A, s_D). Finally (step 4) we computed Nash equilibria for each game instance, using the Gambit software package (McKelvey et al. 2014). Most of the games had multiple equilibria—often similar, but sometimes quite diverse.

Our goal for this analysis was to gain strategic insight into a generic MTD scenario. As such, we were interested not so much in specifics of individual equilibria, but rather understanding at a qualitative level the kinds of equilibria observed. We found that equilibria could be classified into four qualitatively distinct groups.

1. *MaxDef.* In a maximal defense equilibrium, the defender reimages so aggressively that it is futile for the attacker to even try to compromise the servers. Aggressive defense means a frequent periodic reimaging strategy or one that reimages based on a low-threshold probe trigger. Faced with such an aggressive defense, the attacker plays No-Op. As a result, the attacker gets no utility and the defender may get near maximum.
2. *MaxAtt.* We classify a profile as maximal attack if the attacker probes aggressively and in response the defender either plays No-Op or reimages only infrequently or ineffectively. This category is the dual of MaxDef, and corresponds to outcomes that are poor for the defender.
3. *Share.* We classify an equilibrium profile as a sharing if attack and defense levels are moderate, and both players are able to achieve their objectives.
4. *Fight.* Fight equilibria are characterized by robust attack and defense activity, resulting in persistent contention such that neither player achieves its objective to a satisfactory degree.

First, we observe that games with confidentiality defenders always have MaxDef equilibria. Such defenders care only that the attackers do not control their servers, and they can trivially achieve this objective by frequently reimaging—essentially keeping the servers down and unavailable. This result actually shows that a focus purely on confidentiality is not very realistic, so we devote the main part of our attention to games where defenders have an availability objective.

Figure 2 presents results for the 18 games where the defender has the availability objective and can perfectly detect attacker probes. The games cover all combinations of settings for attacker objective, probe cost, and utility thresholds for control of servers. As we see, the various game instances lead to different qualitative categories of equilibria. For the disrupting attacker (table on left), we have MaxDef equilibria for cases of high threshold or high probe cost. Those settings are particularly challenging for the attacker, enabling the defender to effectively deter attack through aggressive reimaging. Since the threshold setting applies to both players, the high threshold games also have MaxAtt equilibria, where an aggressive attack can cause the defender to give up. With low thresholds, both players need only achieve their objective with a minority of servers, so sharing equilibria are possible. Some of the intermediate settings support fight equilibria, where both players accrue some utility, but neither can keep the majority of servers in their preferred state on a consistent basis.

For the control attacker (table on right), the objective is more challenging than disruption. As a result, the defender always has the possibility of deterring attacks through sufficient aggression in a MaxDef equilibrium. MaxAtt can be sustained under the high threshold, or with majority threshold and low probe cost. Sharing equilibria appear for a couple of the low threshold environments, and fight equilibria in all the high threshold environments.

Results for 18 environments with imperfect probe detection are presented in Fig. 3. By comparing the two figures, it is obvious that maintaining a MaxDef equilibrium is much harder when the defender may miss some probes. On the other hand, degraded detection opens the door for aggressive attack, as MaxAtt and Fight are the only equilibria found in the majority or high threshold environments. With low threshold, sharing remains possible, and indeed this equilibrium is most prevalent.

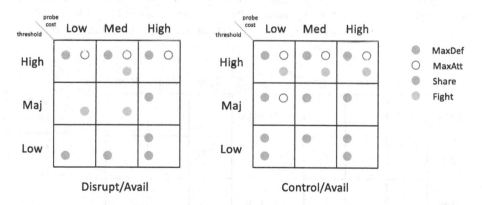

Fig. 2. Qualitative categorization of equilibria across 18 game settings, with availability defender and perfect probe detection. The left table is for a disruptive attacker, and the right for a control attacker. In each cell, colored circles indicate which categories of equilibria are found. (Color figure online)

3.3 Discussion

This example is meant to illustrate general several features of EGTA for cyber-security domains. First, that the method can address a strategically complex scenario, and evaluate a variety of heuristic strategies. Second, that through systematic exploration, we can uncover regular qualitative patterns of strategic behavior. Once identified, these patterns can deepen our understanding of the strategic tradeoffs in the domain. In this case, the findings can all be rationalized straightforwardly. Cases where multiple behaviors are possible (e.g., instances with both MaxDef and MaxAtt equilibria) are natural candidates for further study, toward characterizing refinements that would support one or the other.

4 Survey of Literature

The first application of EGTA to a security domain was the study of privacy attacks by Duong et al. (2010). This work started from the well-understood fact that an attacker's ability to compromise the privacy of a target depends on the background knowledge it already has about the target. In a scenario with multiple attackers, a coalition can increase their collective prospects of privacy breach by sharing background knowledge. There is a tradeoff, however, in that the value of a successful attack may decrease if it is non-exclusive. The study employed EGTA to characterize rational sharing in a variety of settings. The ability to predict sharing is relevant in particular to a database publisher, who must decide how much to degrade the published information (at a cost) in order to protect privacy.

A second security domain studied using EGTA by some of the same authors addressed incentives for compliance with a network security protocol (Wellman et al. 2013). Compliance is an important strategic problem for security,

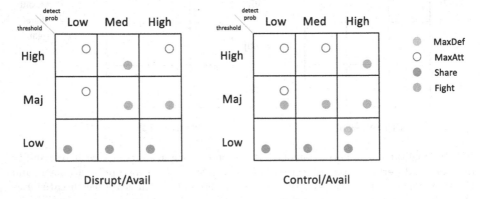

Fig. 3. Qualitative categorization of equilibria for imperfect probe detection. Columns represent three levels where the probability the defender detects a given probe action is 0.2 (low), 0.5 (med), or 0.7 (high).

as often participants will have an incentive to free-ride on the security contributions of others (Čagalj et al. 2005; Naghizadeh and Liu 2016). This study included several methodological innovations, including a systematic procedure to extend the strategy space through local search, and scaling the number of agents by exploiting symmetry across multiple roles. Specifically, the work modeled the introduction-based routing protocol (Frazier et al. 2011) on a network with four kinds of nodes: clients, ISPs, roots, and servers. The game is *role-symmetric*, meaning that players corresponding to a given role (in this context, node type) had the same strategy sets and utility functions, but these generally varied between roles. This enabled use of an aggregation technique called *player reduction* (Wellman et al. 2005), in which a many-player game is approximated by an empirical game with much fewer players. For example, one reported analysis simulated a 4956-node network to estimate a game with six players. Results for that instance are shown in Fig. 4. As we can see, tendency toward compliance varies by role, and there are qualitatively distinct equilibria. Overall, we found over several game settings that compliance was not universal, but typically at a sufficient level to deter attacks.

More recently, we have conducted several EGTA studies within a broader project on adaptive cyber-defense. The first was the MTD study illustrated in Sect. 3 (Prakash and Wellman 2015; Wellman and Prakash 2014). The second employed EGTA to evaluate a moving-target defense against distributed denial of service (DDoS) attacks (Wright et al. 2016). The defense, called MOTAG, had originally been designed and modeled in non-game-theoretic terms (Jia et al. 2013; Venkatesan et al. 2016). Like the MTD game study, the MOTAG

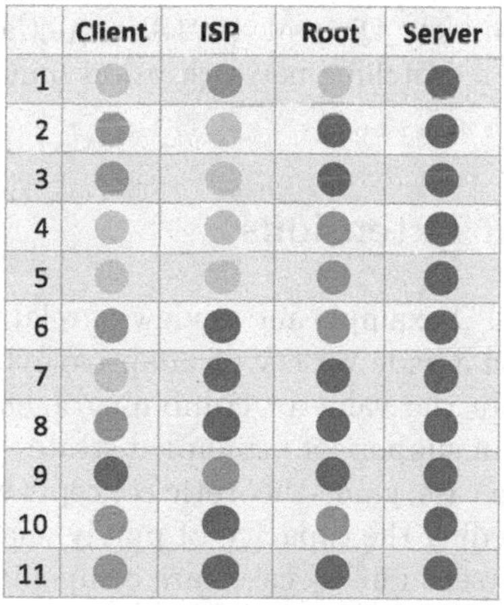

	Client	ISP	Root	Server
1		●		●
2	●		●	●
3	●		●	●
4		●	●	●
5			●	●
6	●	●	●	●
7		●	●	●
8	●	●	●	●
9	●	●	●	●
10	●	●		●
11	●	●	●	●

Fig. 4. Top 11 approximate symmetric mixed equilibria for a 4956-node instance of the introduction-based routing compliance game. Strategies are classified as compliant or non-compliant. Each row represents a mixed profile, indicating whether the role plays strategies that are compliant (green), non-compliant (red), or a mixture of these (yellow). (Color figure online)

investigation covered a two-player game with 10–20 strategies per player, and systematically evaluated a set of parametric variations on the game environment (41 game instances overall). We found that strategy ideas proposed in prior literature for this setting can be effective under certain conditions, but the ideal strategies varied considerably across these conditions. The study was helpful for making these conditions precise, and generally illuminating the strategic landscape for DDoS mitigation in the MOTAG framework.

The third EGTA study in this broader project addressed strategic behavior in domains that can be modeled by attack graphs (Nguyen et al. 2017). The basic idea of an attack graph model is to represent the progress of an attack in terms of following paths in a graph of security conditions (Kordy et al. 2014; Phillips and Swiler 1998). The work in this project specifically builds on a Bayesian framework for attack graphs developed by Miehling et al. (2015). The EGTA study extended the framework to a game, where at each time the attacker chooses edges representing available exploits, and a defender chooses nodes to defend. The strategy sets for both attacker and defender were populated by sophisticated heuristics developed as approximate solution of corresponding optimization problems. The study found that these heuristics successfully beat several baselines, and were robust to variation in the environment settings.

In work outside of this project, Chapman (2016) developed an abstract cybersecurity game based on an extension of hide-and-seek game models. The extensions were motivated by adaptive attack behavior in network security, and render the model infeasible for analytic solution. Chapman therefore adopted a simulation-based approach, and appealed to EGTA methods for game-theoretic treatment. Rass et al. (2017b) likewise appeal to EGTA for a game involving mitigation of advanced persistent threats, citing uncertainty as a complicating factor requiring this approach. Qi et al. (2018) model a scenario similar to the MTD game of Sect. 3 on a switching network using simulation to estimate game payoffs.

5   Conclusion and Extensions

As established by the MTD example and review of related literature, EGTA has by now been employed in a wide variety of adaptive cyber-defense applications. These works demonstrate the value of combining agent-based simulation and game-theoretic analysis in support of principled strategic reasoning for complex security domains. In each case, game-theoretic concepts were applied to scenarios of a complexity far exceeding the capacity of purely analytic methods to tackle.

Results of these analyses in many cases are compelling, though not necessarily definitive. Since by definition an EGTA study restricts attention to chosen strategies, conclusions are always subject to refutation based on refined analysis. Moreover, as for any modeling approach, assumptions incorporated in simulation or approximation methods are open for debate, or relaxation in subsequent studies. Indeed, there remain many areas where improvement in technique could significantly increase the power and scope of EGTA methodology. Here we briefly catalog some of the open issues and opportunities for extensions of EGTA in service of cyber-security analysis.

Covering Large Strategy Spaces. For a two-player game, profile space grows quadratically with strategy sets. This often allows consideration of a rich variety of attack and defense strategy candidates, albeit far from the full space of strategies available (typically highly dimensional or even infinite). Moreover, it is often possible to identify equilibria without evaluating all strategy combinations (Fearnley et al. 2013), which can sometimes dampen even quadratic growth. Limitations on strategy space become more acute when there are greater than two players. Though the standard setup in cyber-security domains is attacker versus defender, some scenarios naturally feature a broader set of strategic actors.

Automating Strategy Search. An effective approach to dealing with limitations on strategy space is to incrementally extend coverage, based on an iterative exploration using feedback from analysis of games of progressively increased size (Jordan et al. 2010). Given some formal description of the strategy space, strategy exploration can be automated in terms of a search in that space. Previous work has employed automated strategy generation for EGTA using local search (Wellman et al. 2013) or reinforcement learning (Lanctot et al. 2017; Schvartzman and Wellman 2009; Wright and Wellman 2018). Recent advances in deep learning have demonstrated breakthrough performance on two-player board games (Silver et al. 2017), and are demonstrating promise in cyber-security games as well (Wang et al. 2019; Wright et al. 2019).

Statistical Reasoning About Results. In the EGTA approach, the game model is estimated or induced from simulation data. The simulations are generally samples of a stochastic system, which means that results are subject to sampling error. This error may be mitigated by devoting more resources to sampling, though naturally that presents tradeoffs regarding alternative uses of that computation (e.g., to exploring more strategies or profiles). There has been some progress on developing principled methods for statistical reasoning in EGTA (Vorobeychik 2010; Wiedenbeck et al. 2014), but further work in this area is needed.

Generalizing Over Environments. The results produced from EGTA studies apply directly to the game instance modeled by the given simulator. Often in security settings, guidance about action in a specific instance is exactly what we care about. However, deriving broad insights about strategic issues in cyber-securities entails lifting results from specific instances to broad categories of game scenarios. The current state of art in EGTA is to systematically explore a range of environments, and attempt to identify patterns in the mapping to solutions. This approach is illustrated well by the qualitative characterization of equilibrium patterns in Figs. 2 and 3. Further work should attempt to codify and automate this systematic search and generalization process.

Acknowledgment. This work was partially supported by the Army Research Office under grant W911NF-13-1-0421.

References

Albanese, M., Connell, W., Venkatesan, S., Cybenko, G.: Moving target defense quantification. In: Jajodia et al. (2019)

Bowers, K.D., van Dijk, M., Griffin, R., Juels, A., Oprea, A., Rivest, R.L., Triandopoulos, N.: Defending against the unknown enemy: applying FlipIt to system security. In: Grossklags, J., Walrand, J. (eds.) GameSec 2012. LNCS, vol. 7638, pp. 248–263. Springer, Heidelberg (2012). https://doi.org/10.1007/978-3-642-34266-0_15

Bushnell, L., Poovendran, R., Başar, T. (eds.): GameSec 2018. LNCS, vol. 11199. Springer, Cham (2018). https://doi.org/10.1007/978-3-030-01554-1

Čagalj, M., Ganeriwal, S., Aad, I., Hubaux, J.-P.: On selfish behavior in CSMA/CA networks. In: 24th IEEE International Conference on Computer Communications, pp. 2513–2524 (2005)

Chapman, M.: Cyber Hide-and-Seek. Ph.D. thesis, King's College London (2016)

Duong, Q., LeFevre, K., Wellman, M.P.: Strategic modeling of information sharing among data privacy attackers. Informatica 34, 151–158 (2010)

Edwards, B., Furnas, A., Forrest, S., Axelrod, R.: Strategic aspects of cyberattack, attribution, and blame. Proc. Natl. Acad. Sci. 114, 2825–2830 (2017)

Evans, D., Nguyen-Tuong, A., Knight, J.: Effectiveness of moving target defenses. In: Jajodia et al. (2011)

Farhang, S., Grossklags, J.: FlipLeakage: a game-theoretic approach to protect against stealthy attackers in the presence of information leakage. In: Zhu, Q., Alpcan, T., Panaousis, E., Tambe, M., Casey, W. (eds.) GameSec 2016. LNCS, vol. 9996, pp. 195–214. Springer, Cham (2016). https://doi.org/10.1007/978-3-319-47413-7_12

Fearnley, J., Gairing, M., Goldberg, P., Savani, R.: Learning equilibria of games via payoff queries. In: 14th ACM Conference on Electronic Commerce (2013)

Frazier, G., Duong, Q., Wellman, M.P., Petersen, E.: Incentivizing responsible networking via introduction-based routing. In: McCune, J.M., Balacheff, B., Perrig, A., Sadeghi, A.-R., Sasse, A., Beres, Y. (eds.) Trust 2011. LNCS, vol. 6740, pp. 277–293. Springer, Heidelberg (2011). https://doi.org/10.1007/978-3-642-21599-5_21

Jajodia, S., Ghosh, A.K., Swarup, V., Wang, C., Sean Wang, X. (eds.): Moving Target Defense: Creating Asymmetric Uncertainty for Cyber Threats. Springer, New York (2011). https://doi.org/10.1007/978-1-4614-0977-9

Jajodia, S., Cybenko, G., Liu, P., Wang, C., Wellman, M.P. (eds.): Adversarial and Uncertain Reasoning for Adaptive Cyber Defense. Springer, Champ (2019). https://doi.org/10.1007/978-3-030-30719-6

Jia, Q., Sun, K., Stavrou, A.: MOTAG: moving target defense against internet denial of service attacks. In: 22nd International Conference on Computer Communications and Networks (2013)

Jones, S., et al.: Evaluating moving target defense with PLADD. Technical report 8432R, Sandia National Lab (2015)

Jordan, P.R., Schvartzman, L.J., Wellman, M.P.: Strategy exploration in empirical games. In: 9th International Conference on Autonomous Agents and Multi-Agent Systems, pp. 1131–1138 (2010)

Kordy, B., Piètre-Cambacédès, L., Schweitzer, P.: DAG-based attack and defense modeling: don't miss the forest for the attack trees. Comput. Sci. Rev. 13, 1–38 (2014)

Lanctot, M., et al.: A unified game-theoretic approach to multiagent reinforcement learning. In: 31st Annual Conference on Neural Information Processing Systems (2017)

Laszka, A., Johnson, B., Grossklags, J.: Mitigating covert compromises. In: Chen, Y., Immorlica, N. (eds.) WINE 2013. LNCS, vol. 8289, pp. 319–332. Springer, Heidelberg (2013). https://doi.org/10.1007/978-3-642-45046-4_26

Laszka, A., Horvath, G., Felegyhazi, M., Buttyán, L.: FlipThem: modeling targeted attacks with FlipIt for multiple resources. In: Poovendran, R., Saad, W. (eds.) GameSec 2014. LNCS, vol. 8840, pp. 175–194. Springer, Cham (2014). https://doi.org/10.1007/978-3-319-12601-2_10

Manshaei, M.H., Zhu, Q., Alpcan, T., Başar, T., Hubaux, J.-P.: Game theory meets network security and privacy. ACM Comput. Surv. 45(25), 1–39 (2013)

McKelvey, R.D., McLennan, A.M., Turocy, T.L.: Gambit: software tools for game theory, Version 13.1.2 (2014). www.gambit-project.org

Miehling, E., Rasouli, M., Teneketzis, D.: Optimal defense policies for partially observable spreading processes on Bayesian attack graphs. In: Second ACM Workshop on Moving Target Defense, pp. 67–76 (2015)

Naghizadeh, P., Liu, M.: Opting out of incentive mechanisms: a study of security as a non-excludable public good. IEEE Trans. Inf. Forensics Secur. 11, 2790–2803 (2016)

Nguyen, T.H., Wright, M., Wellman, M.P., Singh, S.: Multi-stage attack graph security games: heuristic strategies, with empirical game-theoretic analysis. In: Fourth ACM Workshop on Moving Target Defense, pp. 87–97 (2017)

Pfleeger, C.P., Pfleeger, S.L.: Analyzing Computer Security: A Threat/Vulnerability/Countermeasure Approach. Prentice Hall, Upper Saddle River (2012)

Pham, V., Cid, C.: Are we compromised? Modelling security assessment games. In: Grossklags, J., Walrand, J. (eds.) GameSec 2012. LNCS, vol. 7638, pp. 234–247. Springer, Heidelberg (2012). https://doi.org/10.1007/978-3-642-34266-0_14

Phillips, C., Swiler, L.P.: A graph-based system for network-vulnerability analysis. In: Workshop on New Security Paradigms, pp. 71–79 (1998)

Prakash, A., Wellman, M.P.: Empirical game-theoretic analysis for moving target defense. In: Second ACM Workshop on Moving Target Defense, pp. 57–65 (2015)

Qi, C., Jiangxing, W., Cheng, G., Ai, J., Zhao, S.: Security analysis of dynamic SDN architectures based on game theory. Secur. Commun. Netw. 4123736, 2018 (2018)

Rass, S., An, B., Kiekintveld, C., Fang, F., Schauer, S. (eds.): Decision and Game Theory for Security. LNCS, vol. 10575. Springer, Cham (2017a). https://doi.org/10.1007/978-3-319-68711-7

Rass, S., König, S., Schauer, S.: Defending against advanced persistent threats using game-theory. PLoS ONE 12, e0168675 (2017b)

Roy, S., Ellis, C., Shiva, S.G., Dasgupta, D., Shandilya, V., Wu, Q.: A survey of game theory as applied to network security. In: 43rd Hawaii International Conference on System Sciences (2010)

Schvartzman, L.J., Wellman, M.P.: Stronger CDA strategies through empirical game-theoretic analysis and reinforcement learning. In: 8th International Conference on Autonomous Agents and Multi-Agent Systems, pp. 249–256, Budapest (2009)

Silver, D.: Mastering chess and shogi by self-play with a general reinforcement learning algorithm. Technical report, arXiv 1712.01815 (2017)

Sinha, A., Fang, F., An, B., Kiekintveld, C., Tambe, M.: Stackelberg security games: looking beyond a decade of success. In: 27th International Joint Conference on Artificial Intelligence, pp. 5494–5501 (2018)

Sokota, S., Ho, C., Wiedenbeck, B.: Learning deviation payoffs in simulation-based games. In: 33rd AAAI Conference on Artificial Intelligence, pp. 1266–1273 (2019)

Tambe, M.: Security and Game Theory: Algorithms, Deployed Systems, Lessons Learned. Cambridge University Press, Cambridge (2011)

Tavafoghi, H., Yi, O., Teneketzis, D., Wellman, M.P.: Game theoretic approaches to cyber security: issues, results and challenges. In: Jajodia et al. (2019)

van Dijk, M., Juels, A., Oprea, A., Rivest, R.L.: FlipIt: the game of "stealthy takeover". J. Cryptol. **26**, 655–713 (2013)

Venkatesan, S., Albanese, M., Amin, K., Jajodia, S., Wright, M.: A moving target defense approach to mitigate DDoS attacks against proxy-based architectures. In: IEEE Conference on Communications and Network Security (2016)

Vorobeychik, Y.: Probabilistic analysis of simulation-based games. ACM Trans. Model. Comput. Simul. **20**(3), 16:1–16:25 (2010)

Vorobeychik, Y., Wellman, M.P., Singh, S.: Learning payoff functions in infinite games. Mach. Learn. **67**, 145–168 (2007)

Wang, Y.: Deep reinforcement learning for green security games with real-time information. In: 33rd AAAI Conference on Artificial Intelligence (2019)

Wellman, M.P.: Putting the agent in agent-based modeling. Auton. Agents Multi-Agent Syst. **30**, 1175–1189 (2016)

Wellman, M.P., Prakash, A.: Empirical game-theoretic analysis of an adaptive cyber-defense scenario (preliminary report). In: Poovendran, R., Saad, W. (eds.) GameSec 2014. LNCS, vol. 8840, pp. 43–58. Springer, Cham (2014). https://doi.org/10.1007/978-3-319-12601-2_3

Wellman, M.P., Reeves, D.M., Lochner, K.M., Cheng, S.-F., Suri, R.: Approximate strategic reasoning through hierarchical reduction of large symmetric games. In: 20th National Conference on Artificial Intelligence, pp. 502–508 (2005)

Wellman, M.P., Kim, T.H., Duong, Q.: Analyzing incentives for protocol compliance in complex domains: a case study of introduction-based routing. In: Twelfth Workshop on the Economics of Information Security (2013)

Wiedenbeck, B., Cassell, B.-A., Wellman, M.P.: Bootstrap techniques for empirical games. In: 13th International Conference on Autonomous Agents and Multi-Agent Systems, pp. 597–604 (2014)

Wiedenbeck, B., Yang, F., Wellman, M.P.: A regression approach for modeling games with many symmetric players. In: 32nd AAAI Conference on Artificial Intelligence, pp. 1266–1273 (2018)

Wright, M., Wellman, M.P.: Evaluating the stability of non-adaptive trading in continuous double auctions. In: 17th International Conference on Autonomous Agents and Multi-Agent Systems, pp. 614–622 (2018)

Wright, M., Venkatesan, S., Albanese, M., Wellman, M.P.: Moving target defense against DDoS attacks: an empirical game-theoretic analysis. In: Third ACM Workshop on Moving Target Defense (2016)

Wright, M., Wang, Y., Wellman, M.P.: Iterated deep reinforcement learning in games: history-aware training for improved stability. In: 20th ACM Conference on Economics and Computation (2019)

MTD Techniques for Memory Protection Against Zero-Day Attacks

Ping Chen[1](✉), Zhisheng Hu[1], Jun Xu[2](✉), Minghui Zhu[1], Rob Erbacher[3],
Sushil Jajodia[4], and Peng Liu[1](✉)

[1] The Pennsylvania State University, State College, USA
Ping.Chen@jd.com, hzsxiaoyi@gmail.com, muz16@psu.edu, pliu@ist.psu.edu
[2] Stevens Institute of Technology, Hoboken, USA
jxu69@stevens.edu
[3] U.S. Army Research Office, Research Triangle Park, Durham, USA
robert.f.erbacher.civ@mail.mil
[4] George Mason University, Fairfax, USA
jajodia@gmu.edu

Abstract. During the past 25 years, the arms race between attacks exploiting memory corruption and memory protection techniques has drawn tremendous attention. This book chapter seeks to give an in-depth review of the newest research progress made on applying the MTD methodology to protect memory corruption exploits. The new research progress also represents the current phase of the arms race in the MTD perspective. In particular, on one hand, at the frontier of defending against control-hijacking attacks, we will give an in-depth review on the shift of defense strategy from static ASLR to dynamic ASLR. On the other hand, at the frontier of defending against data-oriented attacks, we will give an in-depth review on the shift of defense strategy from static DSLR to dynamic DSLR.

1 Introduction

During the past 25 years, the arms race between cyber attacks exploiting memory corruption vulnerabilities and memory protection techniques has drawn tremendous attention. It started with control-hijacking attacks exploiting memory corruption vulnerabilities (e.g., buffer overflow, format string and integer overflow). In a control-hijacking attack, the adversary manipulates the control flow objects and shifts the execution to malicious logic. The earliest attacks hijack the control flow to execute injected code. To defend against those code-injection attacks, Data Execution Prevention (DEP) [32,43] techniques were proposed. DEP ensures that a memory page is either writable or executable, but not both. It should be noted that across this phase of the arms race, Moving Target Defense (MTD) was not really considered by the defender.

During the next phase of the arms race, as a counteraction against DEP, adversaries switched from code-injection attacks to code-reuse attacks such as return-to-libc and Return-Oriented-Programming (ROP). These code-reuse

S. Jajodia et al. (Eds.): Adaptive Cyber Defense, LNCS 11830, pp. 129–155, 2019.
https://doi.org/10.1007/978-3-030-30719-6_7

attacks have motivated a very large amount of research on defense. As a result, two fundamentally important defense technologies have been developed:

- Address Space Layout Randomization (ASLR) randomizes the base addresses of data and code segments in the memory [5, 6, 29, 33, 42].
- Control Flow Integrity (CFI) disables deviations from the target program's legit control-flow graph [25].

We note that in this phase, MTD techniques have been playing an essential role. To date, ASLR has been the most widely-recognized and widely-deployed defense against code-reuse attacks. The fact that ASLR is nowadays a standard component of most mainstream operating systems indicates: (a) before MTD was adopted, the static nature of memory protection had provided the attacker with an incredible advantage, which had led to ineffective defenses in many ways; (b) MTD techniques can significantly increase the resiliency of a computer system when defending against control-hijacking attacks that exploit memory corruption vulnerabilities.

Although ASLR and CFI are effective when they were invented, the arms race did not end. In fact, the arms race has been continuing along two branches: the counterattacks against ASLR as well as CFI and data-oriented attacks. In the wake of defending control-hijacking attacks, data-oriented attacks [16, 27] have emerged. Data-oriented attacks do not modify control flow objects. Instead, they read/write security-sensitive data objects for malicious goals [16, 27]. Recently, it has been shown that data-oriented attacks are Turing-complete and can result in arbitrary behaviors.

Among data-oriented attacks, data structure manipulation attack (DSMA) is a major category. DSMA exploits memory corruption vulnerabilities to manipulate security sensitive fields in encapsulated data objects (e.g., struct and class). For example, the attack against openssh (CVE-2001-0144) overwrites a particular instance of data structure passwd to achieve privilege escalation. Pioneering research was conducted and shows that DSMA is able to circumvent the most effective defenses against control-hijacking attacks—DEP, ASLR and CFI.

To date, only two defense techniques have demonstrated their effectiveness[1]: Data Flow Integrity (DFI) [11, 38] and Data Structure Layout Randomization (DSLR) [30, 39, 45]. DFI maintains the definition-use relationship from the Data Flow Graph, and checks whether the definition of each data object is legit at run-time. By theory, DFI can defend against DSMA. However, DFI introduces performance overhead as high as 103% [11], making it impractical for deployment.

Comparing with DFI, DSLR has similar defense effectiveness but substantially less cost. Several years ago, researchers proposed static DSLR [30, 39, 45]. At the time of compilation, static DSLR randomly reorders the fields or adds dummy fields in encapsulated data objects. Static DSLR can prevent DSMA

---

[1] Fine-grained ASLR also provides certain level of effectiveness but it can be easily circumvented by attacks such as rootkits.

from correctly locating target fields and further manipulating them. However, its randomization is fixed at run-time and vulnerable to brute force attacks. Further, static DSLR requires manual efforts to determine which data structures can be randomized.

The goal of this book chapter is to give an in-depth review of the newest research progress made on applying the MTD methodology to protect memory against remote exploits. These new research progress also represents the current phase of the arms race in the MTD perspective. In particular, on one hand, at the frontier of defending against control-hijacking attacks, we will give an in-depth review on the shift of defense strategy from static ASLR to dynamic ASLR. On the other hand, at the frontier of defending against data-oriented attacks, we will give an in-depth review on the shift of defense strategy from static DSLR to dynamic DSLR.

The rest of the paper is organized as follows. In Sect. 3, we will give an in-depth review of the defense side shift from static ASLR to adaptive ASLR. In Sect. 3, we will give an in-depth review of the defense side shift from static DSLR to adaptive DSLR. In Sect. 4, we conclude the paper.

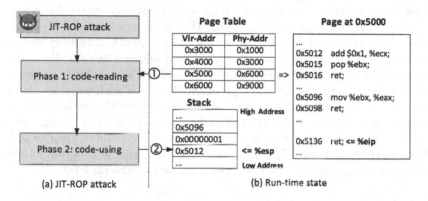

Fig. 1. A demonstrative JIT-ROP attack. It proceeds with two phases: ①exploiting a memory disclosure to read code pages and ② constructing an ROP payload on stack (from [14], page 3).

2   From Static ASLR to Adaptive ASLR

2.1   The History

Return-Oriented Programming. The era of code injection essentially ends since the availability of Data Execution Prevention (DEP). In its place comes Return-Oriented Programming (ROP) [10,12,34,36] attacks. An ROP attack hijacks the control flow of a program to elaborately chained instruction sequences (i.e., "gadgets"). It has been proved that ROP attacks are Turing-complete, which can result in arbitrary operations.

Static ASLR. Against the threats of ROP attacks, address space layout randomization (ASLR) has been invented. It is a technique that randomizes the locations of code segments in the address space. The rationale behind ASLR is as follows. ROP attacks leverages the practice that code segments are loaded at identical addresses (in different machines). ASLR changes this practice by introducing non-deterministic to the code layout. Research in the past develops various techniques to enforce randomization at different granularity, including segment offset level [42], code page level [6], function level [5,29], and even basic block or instruction level [33,44]. In the early stage, ASLR techniques enforce one-time randomization during either code compilation or program loading [5–7,26,29,33,42,44,46]. As such, they are referred to as static ASLR. Despite of its effectiveness against conventional ROP, static ASLR is vulnerable to memory-leakage assisted attacks, in particular Just-In-Time ROP (JIT-ROP) [37].

JIT-ROP Attacks. A JIT-ROP attack consists of two phases: *code reading* and *code using*. In the first phase, it exploits memory disclosure [13,40] to obtain code from memory and therefore breaks static ASLR. Then the adversary analyzes the obtained code and identifies valid code gadgets. In the second phase, using information gained through the first phase, the adversary crafts ROP payloads and sets up the stack as required by malicious intentions. In the following, we re-use the figure from [14] to illustrate an JIT-ROP attack.

As is presented in Fig. 1, the adversary first exploits a memory-disclosure vulnerability to obtain the code page at virtual address 0x5000. He then disassembles this page and finds two gadgets at 0x5012 and 0x5096, respectively. In the following step, the attacker prepares the stack with payload [0x5012, 0x00000001, 0x5096]. In this payload, 0x5012 and 0x5096 are the addresses of the above two gadgets and 0x00000001 is the data to be used by the first gadget. To launch the exploit, the attacker will redirect stack pointer %esp to 0x5012 and PC %eip to 0x5136. This will finally compromise the program execution and convert the control flow to the gadgets chosen by the attacker.

2.2 The New Age: Dynamic ASLR

As a counteractive movement against JIT-ROP, static ASLR gets further developed and evolves to dynamic ASLR. Differing from static ASLR, dynamic ASLR performs adaptive randomization at run-time [8,17,21]. By doing this re-randomization frequently, dynamic ASLR instantly obsoletes information gathered through memory disclosure, essentially making JIT-ROP in-feasible. To date, plenty of dynamic ASLR techniques have been designed and implemented. They share similar threat model and encounter similar challenges. In the following, we first summarize the mainstream dynamic ASLR techniques and then summarize the threat model as well as the challenges behind.

Mainstream Techniques. Across the literature in the past few years, dynamic ASLR techniques have been proposed with different design emphases [8,21,24]. TASR [8] applies re-randomization to the memory layout whenever the execution produces an output. TASR works on C source code and instruments the program with information that is required to track pointer locations for dynamic re-randomization. Isomeron [21] is a hybrid approach. It combines code randomization and run-time execution-path randomization to mitigate JIT-ROP. Isomeron requires neither source code nor static analysis. Alternatively, it relies on dynamic instrumentation to keep track of randomized information and maintain correct execution semantics. RuntimeASLR [31] prevents clone-probing based ROP attacks via a semantics-preserving and runtime-based approach. This approach works by re-randomizing the address space of every child after fork but keeping the parent's state. To maintain the semantic correctness after re-randomization, RuntimeASLR devises a pointer tracing mechanism to identify pointers inside the address space.

A Common Threat Model. Dynamic ASLR techniques share a similar threat model. In this model, the adversary can construct JIT-ROP exploits. As such, it assumes that the software has memory disclosure vulnerabilities which can enable the adversary to leak code memory and also hijack the control flow.

As DEP [32] and ALSR [42] have been applied in most of the modern systems, the threat model typically assumes these two defense are deployed. In addition, code layouts are randomized at a fine-grained granularity, making the registers [33] used and instruction locations within a function [29] or a basic block [44] different. The reason behind this assumption is that an adversary may be able to find code pointers (e.g., virtual table entries) in non-executable memory (e.g., stack and heap), infers the code layout of the rest of the memory without directly reading them, and finally mounts an indirect JIT-ROP attack [18].

Finally, the threat model usually excludes data only attacks, since data only attacks do not rely on knowledge about code layout, which are out of the defense scope of dynamic ASLR. It also should be noted that binary-compatible techniques usually assume the code is not obfuscated.

Design Challenges. Dynamic ASLR usually encounters two types of challenge. From the perspective of correctness, adaptive randomization techniques can easily break the program semantics especially when only binary code is available. For instance, Remix [17] and TASR [8] require to accurately pinpoint code pointers. However, locating code pointers in binary is still an open challenge [31]. From the perspective of performance, adaptive randomization techniques could introduce significant run-time overhead. In particular, techniques such as Isomeron [21] and RuntimeASLR [31] perform dynamic code instrumentation, incurring substantial slowdown.

2.3 A Representative Adaptive ASLR Technique: Chameleon

In this chapter, we elaborate on a representative technique, Chameleon. Our goal is to demonstrate the principles behind the design and implementation of dynamic ASLR. We select Chameleon because of two-fold considerations. One one side, Chameleon well addresses the aforementioned two challenges. This makes Chameleon an appropriate example for principle explanation. One the other side, `Chameleon` is fully implemented as a software solution, which has no requirement on hardware or architecture. This makes `Chameleon` have broader applicability than other solutions (e.g., [23,41]).

2.3.1 `Chameleon` Overview

Recall that a JIT-ROP attack typically leverages memory disclosure vulnerabilities to read contents in code pages. Facilitated by information in those code pages, the attacker then manipulates the stack with an ROP payload which can convert the execution to perform malicious activities. `Chameleon` works on obfuscating the addresses of code pages. As such, we make it in-feasible for the attacker to obtain the actual addresses of code pages, and thus interrupt reading on code pages. Differing from previous techniques that aim to stop the attacks after memory disclosure, `Chameleon` serves as a first-line defense that counteracts attacks by preventing code page disclosure directly.

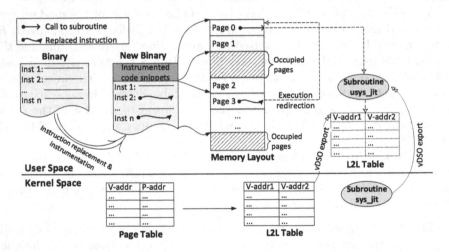

Fig. 2. The high level work-flow of `Chameleon` (from [14], page 4).

Figure 2 presents the overall design of `Chameleon`. Technically speaking, `Chameleon` introduces a layer between a user space process its page table in the kernel space. This layer is implemented as a mapping table (i.e., L2L Table), which maps the virtual address of each code page to a random one that is yet available. In the original page table, `Chameleon` replaces the original virtual address with its mapped target in the L2L Table. When an adversary attempts on reading a code page using the original virtual address, the memory management unit (MMU) will not locate it in the page table and hence, issue a page fault. Apparently, this interrupts the code reading process and makes JIT-ROP in-feasible.

Despite its effectiveness, the above scheme breaks the normal execution in the sense that control flow transfers cannot find the target either. To remedy this issue, Chameleon opt to instrument the target binary. Specifically, Chameleon rewrites the instructions that may break the control flow (i.e., instructions that transfer program control flow and the last instruction of each page) and guides them to find the correct targets.

2.3.2 Design and Implementation of Chameleon

As is shown in Fig. 2, Chameleon is designed with two major component. The first component performs randomization over the page table and thwarts JIT-ROP attacks. The second component fixes the control flow that is interrupted by the randomization. For a better understanding of how Chameleon works, we present the two components with more details.

Page Table Transformation. After the birth of a process but right before its execution, the OS kernel will already set up the page table and load all the code segments. In this page table, SALADS searches for all code pages and couples the virtual address of each code page with a virtual address that is still available. In the meanwhile, Chameleon also remembers this replacement in the L2L table. At the high level, the above idea is straightforward. It, however, involves many details that must be taken into considerations for assurance of randomization.

First, simply mapping a virtual page address to a random unoccupied address provide insufficient security. In particular, in the case that Chameleon randomizes a virtual page address to its adjacent page address, an adversary may still succeed to leak the code page by over-reading a large continuous memory. To address this problem, Chameleon regulates that a code address is not permitted to map to neighboring addresses.

Second, the MMU can make the above protection invalid. To be specific, Chameleon replaces code page addresses with random addresses from an unoccupied space. Once the adversary exploits to read a code page using the original page address, the MMU will find that this page is file-backed and yet has no corresponding page table entry. As a result, the MMU will allocate a physical page and inserts into the page table with an entry—mapping the attacker-specified virtual address and the new physical page. It will also load into this physical page with code from the file. Hence, the reading will succeed regardless of the page table randomization. Chameleon fixes this flaw by inserting into the page table with dummy entries. These entries maps each of original code pages to the NULL physical address. This prevents the MMU to re-allocate page table entries as above mentioned.

Third, one-time randomization can be bypassed simply by brute-force. The design of Chameleon mitigates this problem by making the randomization repetitive at run-time. In each round of randomization, Chameleon repeats the above process. In terms of the time to do re-randomization, Chameleon provides two schemes. On the one hand, it supports re-transformation on the basis of a predefined interval. On the other hand, it can also perform re-randomization on the occurrence of a system call.

Last but not least, it's dangerous to re-randomize the code page that is under execution. Otherwise the synchronization between re-randomization and the execution has to be enforced at instruction level, which is practically unacceptable. As a counter-movement against this problem, Chameleon pauses re-randomization on the code page that is under execution.

Address Translation. In principle, the idea of page table transformation obfuscates the addresses of code pages, which effectively prevents the adversaries from reading contents in code pages. Despite of the security strength, this randomization inevitably breaks the normal execution. When the execution needs to transfer to a new code page, it will neither find the correct address. To handle this critical issue, Chameleon instruments the target binary to correct address translation for control flow transfers. We describe more details in the following.

Basically, Chameleon first disassembles the target binary with a customized algorithm. Then, it rewrites the binary code to replace each of the error-leading instructions with a translation routine. This routine would keep the execution correct at run-time. Coordinating with an instrumented binary, the page table transformation can work smoothly without interruptions. Despite this idea is intuitive at the high level, there are many hurdles that must be crossed.

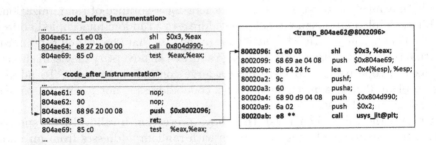

Fig. 3. A demonstrative example of code replacement and instrumentation (from [14] page 7).

First and foremost, disassembling a stripped binary is usually considered unreliable (in particular in the x86 architecture). To reduce the uncertainty and enhance reliability, Chameleon reuses the state-of-art disassembling algorithms [22,47]. Going beyond that, Chameleon enforces a strategy towards higher conservativeness. More specifically, Chameleon skips a code page once it encounters uncertainty in the disassembling process. To ensure the correctness of execution, Chameleon prevents randomization over the pages that are not disassembled. This conservative strategy does not significantly reduce the effectiveness of Chameleon, and more details are explained shortly.

The disassembling provides the basis of instrumentation. However, it's still challenging to determine the instrumentation spots. Missing a single error-leading instruction would introduce exceptions while too much instrumentation incurs unacceptable cost. Chameleon addresses this issue by leveraging a design aforementioned. Recall that Chameleon pauses re-randomization on the code page that is currently under execution. Therefore, the execution will not be interrupted before it leaves a page that has been correctly entered. This indicates that there is only necessity to remedy instructions that may incur cross-page visits. These instructions only include direct jmp, indirect jmp (including indirect jmp, indirect call and return) and the last instruction in each code page. To be specific, the design of Chameleon replace each of these instructions with a pair of push and ret.

In the above design, the push inserts to the stack an address which points to a routine. This address is then taken by the ret instruction as the destination. At run-time, this routine corrects the interrupted execution and makes the program to follow the expected logic.

Implementation of Address Translation. Following the aforementioned design, Chameleon implements the scheme of address translation on Linux systems. Here, we explain the important details that are not covered yet.

On the basis of disassembled results, Chameleon replaces each cross-page instruction with the pair of push and ret. This, which seems to be intuitive, is however difficult. In a system running the X86 architecture, instructions vary in size. Our replacement may enlarge the space occupied by the original instructions. This will manipulate the layout of entire code segment and result in program exceptions. Chameleon handles this issue through layout offsetting: ① For a cross-page instruction that has equal length with the pair of push and ret, Chameleon simply performs the replacement; ② For a case where the cross-instruction has larger size than the pair of push and ret, Chameleon rewrites it with the instruction pair and following padding instructions (nop); ③ Lastly, in the case where the cross-page instruction is short in size, Chameleon groups the neighboring instructions prior to this cross-page instruction and replaces this group to the pair of push and ret. To make up the execution of those neighboring instructions, Chameleon moves them to the beginning of the aforementioned routine.

For a better understanding, we demonstrate such a case corresponding to ③ in Fig. 3. In this case, the cross-page instruction is a call instruction, which consists of 5 bytes and falls short for the pair of push and ret. To resolve this case, Chameleon relocates [shl $0x3,%eax] together with [call 0x804d990] out and fills in [push $0x8002096] and ret. In the trampoline that locates at [push $0x8002096], Chameleon reallocates the additionally removed instruction [shl $0x3,%eax].

By enforcing the above implementation, Chameleon makes the execution transfer from cross-page instructions to the trampoline. To fully mitigate the side effects of the page table randomization, Chameleon adds into the trampoline two extra sets of instructions. One set of instructions are used to store and restore the execution status (in particular the registers), while the other set of instructions prepare a subroutine which remedies the address translation. For the ease of interpretation, we use the example shown in Fig. 3 for illustration again. In the original code, the execution invokes a function at 0x804d990 via instruction [call 0x804d990]. This functions later returns back to 0x804ae69.

Through the aforementioned instrumentation, the execution will be redirected to the trampoline at \$0x8002096. In this trampoline, Chameleon first maintains the original return address 0x804ae69 on the stack. This is to ensure that the execution will be correctly returned back and more details will be shortly discussed. Then, this trampoline saves all the registers via pushf and pusha. Going beyond that, this trampoline invokes a call to a subroutine usys_jit through a set of three instructions. The beginning two instructions prepare usys_jit with two arguments 0x804d990 and 0x2. In this case, 0x804d990 is the target address of the original call instruction and 0x02 indicates this translation serves a call instruction. In addition, this trampoline also invokes a lea instruction, which pre-allocate space on the stack. This space will later be used to store the translated target address.

As shown in the above example, the trampoline is placed at a deterministic location. This requires Chameleon to preserve the locations of all the trampolines. To this end, Chameleon inserts those trampolines to head of the code segment in the target binary. Since the code segment is loaded into fixed memory location, the addresses of those trampolines are also determined before execution. Note that enforcing the aforementioned randomization will invalidate this scheme, Chameleon ensures the trampolines and the code segment are loaded into different pages and it disables randomization over the trampolines.

In the above case, Chameleon invokes the subroutine usys_jit with an argument carrying the expected target address. This subroutine then accomplishes the expected address translation. To be more specific, usys_jit first verifies the necessity of translation. For an replaced instruction that does not actually transfer the control flow to another page (e.g., a conditional jump with a false condition and an indirect jump that transfers the execution in the same page), usys_jit simply restores the execution status and redirect the execution to the original address. With regard to an instruction that indeed leads the execution to another page, usys_jit looks up the L2L table with the target address. By doing so, usys_jit learns the after-randomization value of the target address and then sets this value as the new target for the execution to follow.

For a better view of the above process, we present a demonstrative example in Fig. 4. In this case, usys_jit examines its arguments on the stack and determines the type of the replaced instruction. Corresponding to this example, the

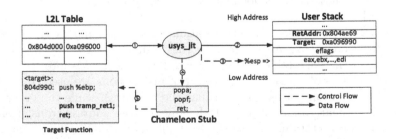

Fig. 4. The work flow of usys_jit. From [14], page

instruction is a `call` and it truly has a target in a different page. As a result, `usys_jit` searches the L2L table, using the original address 0x804d990 as the index (①). The search result 0xa096990 is stored on stack (②). Following that, `usys_jit` moves the stack pointer to the location where the registers are stored and then delivers the execution to a stub (④). This stub recovers all the registers by retrieving their values from the stack and then transfers the execution to the translated address. As such, the code page containing the original logic will be correctly visited and executed.

In the implementation of `Chameleon`, `usys_jit` needs to interact with the L2L table in the kernel space. An intuitive reaction is to issue a system call and perform the interaction. This, however, requires to cross the privilege boundary between user space and kernel space. To avoid this time consuming process, `Chameleon` uses the scheme of virtual Dynamic Shared Object (vDSO) and maps the L2L table to user space. In this way, `usys_jit`) can access it for address translation without switch between different privilege levels.

2.4   Evaluation of Adaptive ASLR

2.5   Evaluation Design

To comprehensively evaluate a randomization-based defense, the literature has developed a de facto evaluation scheme. It typically consists of three requirements.

First and foremost, the strength of the defense has to be thoroughly reasoned under the assumption that memory disclosure occurs, considering that a single channel of such disclosure could be sufficient to break the entire defense. Note that this evaluation cannot be accomplished via quantification or experiments, simply because it's practically impossible to gain awareness of all the memory disclosure vulnerabilities.

Second, an adversary may attempt to brute force the randomization and therefore escape from the defense. This is usually quantified by the number of attempts required to succeed through brute force. This number essentially represents the resources that the adversary needs to invest. The requirement of resources beyond a threshold would make the attack unlikely practical.

Last but not least, the defense needs to be measured from the perspective of practicality. This is usually achieved by deploying the defense against a set of standard benchmarks and observing the overhead that the defense brings to the computation resources.

2.6   Evaluation on Adaptive ASLR

In this chapter, we demonstrate the application of the above measurement in the case of `Chameleon`.

Reasoning over `Chameleon`. As is explained in the threat model, `Chameleon` is designed to maintain its security under disclosure of data in the memory.

Its robustness against memory disclosure is enforced by it's re-randomization scheme in an isolated privilege level. All related details are presented in the sections for design and implementation of `Chameleon`. We avoid to repeat them in order to reduce space.

Defense Strength Under Brute Force. To bypass the defense of `Chameleon`, an adversary may thrive to brute force the address of a code page and dump the contents inside. This brute force is evaluated via a theoretical model presented in [14]. To be more specific, this model divides the entire address space into N code pages. Out of them, y pages are pertaining to the target program p and need to be randomized by `Chameleon`. To construct a successful ROP chain, the adversary has to collect gadgets from x code pages.

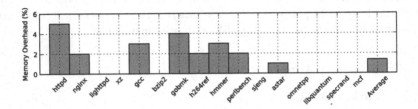

Fig. 5. Memory overhead of `Chameleon` (from [14] page 10).

Following the above model, the probability for the attacker to capture a code page in one attempt is $\frac{y}{N}$. Therefore, $\frac{N}{y}$ attempts are required to dump the first code page. Continuing that, the attacker needs to issue $\frac{N-1}{y-1}$ attempts before he can obtain the second code page and $\frac{N-(x-1)}{y-(x-1)}$ attempts for the xth code page. Accumulatively, the adversary needs in total $\frac{N}{y} + \frac{N-1}{y-1} + ... + \frac{N-(x-1)}{y-(x-1)}$ attempts. This number is larger than $x \cdot \frac{N}{y}$.

To better understand the above model, we apply it on an example from real-world settings. In this example, we assume the operating system is a 64-bit version of Linux, in which half of the address space is allocated to serve the user space. According to the statistics with real-world programs, a large program and its linked libraries usually needs less than 500 code pages. To correctly capture a code page via brute force, the attacker has to perform at least 2^{42} attempts, i.e., $f(x) \geq x \cdot \frac{r \cdot 2^n}{s} = \frac{x \cdot r \cdot 2^n}{y \cdot s}$ where $x = 1$. Making this tremendous volumes of attempts, which may trigger almost the same number of software errors, would be impossible to be successful without being detected by security facilities such as Introduction Detection Systems.

Overhead Evaluation. `Chameleon` introduces extra operations to the target program, both in the user space and the kernel space. By intuition, this would bring additional overhead to both of the computation and memory. To verify the level of overhead that `Chameleon` introduces, the work [14] presented a thorough evaluation on `Chameleon` from the perspective of performance overhead.

It collected a group of programs that are widely used as benchmarks. All the evaluation experiments are performed on a machine running with Intel(R) Core2 Duo CPU E8400 3.00 GHz and 2 GB RAM.

Chameleon instruments the machine code to correct runtime execution and it also consistently keeps all the code pages in memory. These two schemes both increase consumption of memory. To this end, Chameleon is evaluated in terms of memory overhead. In this experiment, the target programs are first executed without Chameleon. Then, all these target programs run with Chameleon deployed and all their code pages locked in memory.

Figure 5 demonstrates the memory overhead Chameleon introduces to the target programs. On average, the memory overhead is around 1.38%. This, in essence, illustrates that the page table randomization in Chameleon barely reduces the practicality of the target programs from the perspective of memory consumption.

Table 1. The frequencies at which a program needs to interact with a trampoline. Note that the figure is at the scale of 10^3 (from [14], page 9).

Program	httpd	nginx	light-httpd	xz	gcc	bzip2	gobmk	h264ref
	1,400	4,205	2,129	2	5,792	46,996	2,754	3,949
Program	hmmer	perlbench	sjeng	astar	omnetpp	libquantum	specrand	mcf
	1,010	2,500	3,099	4,278	7,830	20	0.1	897

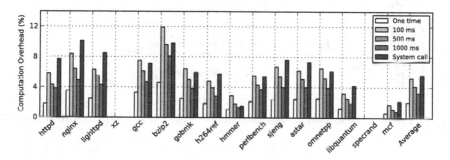

Fig. 6. Computation overhead of Chameleon in different randomization strategies (from [14] page 10).

Since Chameleon introduces re-randomization over the page table and address translation to maintain the correctness of execution, it inevitably slows down the target programs. Therefore, the work [14] also designs experiments on the computation overhead that Chameleon incurs. To more closely inspect how different randomization strategies affect the performance overhead, this experiment runs SPEC CPU2006 benchmark with Chameleon performing code page randomization at different frequencies as well as randomization at each system call invoked. To be more specific, it runs the SPEC benchmarks with recommended setting, web service applications by using ApachBench with command `ab -k -n50000 -c100 HOST/index.html`, and compression program xz by performing compression on a 16MB binary file for 1,000 times.

The results of the above evaluation is shown as Fig. 6. It can be easily observed that the performance overhead varies as the randomization strategy changes. Not surprisingly, Chameleon brings higher overhead while the frequency of re-randomization increases. Zooming into the evaluation results, we can find that Chameleon incurs negligible effects on performance in the case of xz and specrand, even the re-randomization is performed at a very high frequency. The experiment also counts the number of executions on those trampolines. The results are presented as Table 1. By associating Fig. 6 and Table 1, we can learn that Chameleon has reduced intrusiveness to the program when the program issues less frequent address translation. In particular, the tests on xz and specrand visited the trampolines for very few times and correspondingly, they imposed nearly zero overhead. When the volumes of address translation is increased, the computation overhead will also grow.

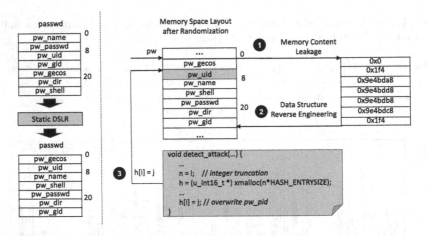

Fig. 7. Privilege escalation in openssh under ASLR and static DSLR (from [15] page 4).

3   From Static DSLR to Adaptive DSLR

3.1   The History

Control Flow Hijacking. Control flow hijacking has been enabling a wide spectrum of popular attacks in the past decades, such as code injection attacks and ROP attacks. At the high level, control flow hijacking manipulates the targets of indirect control flow transfers (e.g., return addresses) and converts the execution to run malicious code that is either injected or chained from the original code. Note that hijacking the control flow to injected code is less feasible nowadays due to the deployment of DEP. Therefore, the only option is hijacking the control flow to code already in the program, which depends on two conditions. On the one hand, it requires to exploit a vulnerability such as buffer overflow to manipulate control flow targets. On the other hand, it needs knowledge about the location of code that encodes malicious logic.

Code Layout Randomization. To mitigate attacks based on control flow hijacking, plenty of solutions have been developed to randomize the layout of code space. The intuition behind is to prevent adversaries from understanding the location of code in need. These techniques covers a wide spectrum of randomization granularity, including segment offset level [42], code page level [6], function level [5,29], and even instruction level blocks [33,44]. In addition, techniques developed in the old days are mostly static, which enforce one-time randomization during either code compilation or program loading [5–7,26,29,33,42,44,46]. These static techniques are later proven to be vulnerable against memory leakage. As a counteractive movement against leakage-assisted control flow hijacking, static ASLR gets further developed and evolves to dynamic ASLR. Differing from static ASLR, dynamic ASLR performs adaptive randomization at runtime [8,17,21]. These technique instantly obsolete information leaked and works effectively against attacks with memory leakages.

Data-Only Attacks. The development of ASLR, in particular adaptive ASLR, largely reduces the space of control flow hijacking. This pushes attacks switch to data-only attacks [16]. Such attacks exploit vulnerabilities to manipulate data objects other than targets of indirect control flow. In the early stage, the belief is that data-only attacks can only cause issues such as information leakage and privilege escalation. Recent research demonstrates that data-only attacks can achieve arbitrary execution like ROP attacks [28]. As data only attacks require zero knowledge about code layout, it cannot be thwarted by ASLR techniques.

For a better understanding, we present an example of data-only attack in Fig. 7. In this example, the adversary exploits a heap overflow in openssh-2.1.1 (CVE-2001-0144) [19] to achieve privilege escalation. To be more specific, he overflows the buffer h on heap and overwrites the neighboring data structure pw. By carefully managing the overflow, the attack is able to control two special members of pw: pw_uid and pw_gid. As these two fields marks the privilege level of the session user (i.e., the attacker), this attack enables the attacker to arbitrarily change his privileges. As is indicated by the attack process, it requires zero knowledge about code layout and hence, can fully work under deployment of fine-grained adaptive ASLR.

Static Data Space Layout Randomization. Data-only attacks require no knowledge about code layout. Alternatively, it needs to understand the layout of data objects. Take the attack against openssh-2.1.1 in Fig. 7 as an example. To correctly manipulate the privilege fields and avoid corrupt other sensitive members (e.g., pointers), the attacker has to determine the members inside pw and their spatial distribution. This characteristic inspires the development of data space layout randomization (DSLR) [15,30]. The high level idea of DSLR is to randomizes the space arrangement of data objects as well as their internal layout. As DSLR interrupts the knowledge that adversaries have about the data layout, it will significantly raise the bar for data structure manipulation attacks.

Extant research of DSLR has been mainly focusing on randomizing internal layout of data structures. The initial development proposes static DSLR [30]. Static DSLR re-orders the fields or inserts dummy fields in the definition of data structures. The left part in Fig. 7 shows an example of randomizing data structure `passwd` (`pw` is an instance of `passwd`).

Despite the effectiveness of static DSLR in mitigating data-only attacks, it can be bypassed when memory leakage occurs. Here we explain how to launch the aforementioned privilege escalation (Fig. 7) under static DSLR. Note that as ASLR has been a standard in mainstream operating systems, we also assume ASLR is deployed. To begin with, the adversary exploits a memory leakage vulnerability (e.g., buffer over-read [37] and side channel [9,35,48]), which enables him to locate the memory page containing `pw` and dump the contents in this page (①). Thereafter, the attacker searches the "signature" of `struct passwd`. By signature, we mean the characteristics pertaining to fields in the data structure. Take `pw` for an example. The fields `pw_uid` and `pw_gid` identifies privilege levels which usually keep small values; `pw_passwd`, `pw_name`, `pw_shell`, and `pw_dir` are pointers and their values usually have difference of 16; `pw_gecos` is 0. This signature matching enables the attacker to locate `pw` in the leaked page and then infer the layout after randomization (②). As such, the attacker bypasses the static DSLR and finally launches the attack (③).

3.2   The New Era: Dynamic DSLR

On account of weakness of static DSLR, recent research advancement moves towards dynamic DSLR [30]. It permutates the internal layout of data objects in a similar manner as static DSLR. However, instead of performing the randomization only once, dynamic DSLR repeats the randomizing process on demand at run-time. The goal is to promptly make leaked information outdated.

Despite the promise of dynamic DSLR in battling data-only attacks, there are not many solutions. Presumably, this is mainly because dynamic DSLR has to address a strong threat model and overcome plenty of challenges. In the following, we give more details about the threat model and those challenges.

A Common Threat Model. Solutions of dynamic DSLR are expected to work under a strong threat model. In this model, the adversaries are able to exploit memory-related vulnerabilities to manipulate arbitrary data objects. In addition, the adversaries have channels to leak both code and data pages. Lastly, the adversaries are further assumed to only perform data-only attacks.

On the defender side, DEP and ASLR are assumed to be deployed by default. Going beyond that, it requires that code layout is randomized at a fine-grained granularity, making control flow hijacking less possible. As existing techniques can barely infer data object from binary code in a reliable manner, dynamic DSLR typically assumes the availability of source code.

Design Challenges. Design of dynamic DSLR faces similar challenges as that of dynamic ASLR. On the one hand, it requires fine-grained layout randomization as well as frequent re-randomization to enforce security. On the other hand, it has to maintain both the correctness and performance of the program. Correctness wise, dynamic DSLR necessarily needs to ensure all permutated data objects are correctly re-organized when they are accessed. Performance wise, frequent re-randomization may easily incur high overhead.

3.3 SALADS: A Practical Approach for Adaptive Data Structure Layout Randomization

To date, only the technique **SALADS** presented in [15] achieves dynamic DSLR, following the aforementioned threat model and addressing the above challenges. As such, we elaborate on the insights and technique details of **SALADS**.

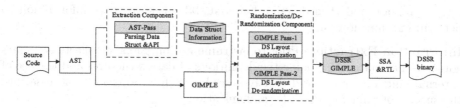

Fig. 8. System Overview of **SALADS**. The grey boxes are components introduced by **SALADS** on the basis of GNU GCC (from [15] page 5).

3.3.1 SALADS Overview.

The design of **SALADS** is to randomize the arrangement of fields in encapsulated data objects of C/C++ programs, including `struct` and `class`. In the following we first explain how this defeats data-only attacks with an example and then given an overview of the system.

Recall that in the attack as shown in Fig. 7, the adversary learns about the layout of data objects at Step ② and leverages this information at Step ③. **SALADS** aims at disconnecting Step ② and Step ③ by frequently re-randomizing the data objects. Once the target data object gets re-randomized between the two steps, the attack will have low chance to succeed. To enforce the above randomization scheme, **SALADS** is built as a plug-in on the top of GNU GCC to rewrite the target program and instrument the logic in need.

At the high level, **SALADS** follows the architecture as presented in Fig. 8. It consists of two main components, including a component to extract required information and another component to do the instrumentation. Given a program, **SALADS** follows several steps to make it a Data Structure Self-Randomizing (DSSR) program. **SALADS** starts with parsing the source code to an Abstract Syntax Tree (AST). Taking this AST as input, **SALADS** runs the extracting component to gather information which is used later. Thereafter, the AST is transferred into GIMPLE representation—the intermediate representation (IR) of GCC. On the top of the GIMPLE IR, **SALADS** replaces each of the statements that access data structures with a Data Structure Self-Randomizing (DSSR) routine (GIMPLE Pass-1).

In the course of execution, this routine will adjust the access to correct locations and perform re-randomization when necessary. Considering that randomization may interrupt certain accesses that SALADS cannot rectify at run-time, SALADS also inserts special statements before each of such accesses to prevent randomization. In the rest of this chapter, those accesses are referred to as *dangerous accesses*. Once the instrumentation is accomplished, SALADS further compiles the GIMPLE IR into the final binary (DSSR binary). A DSSR binary is able to achieve re-randomization/de-randomization by itself at the time of execution.

3.3.2   SALADS Design and Implementation.

In this section, we include the important details of design and implementation in SALADS. We first explain the types of information SALADS extracts from the target program and then describe how SALADS relies on those information to perform instrumentation.

Information Extraction. Prior to instrumentation, SALADS collects two categories of information, including the definitions of data structures and the use of external functions. As we will shortly describe, such information is necessary to the success of randomization/de-randomization.

While SALADS iterates the AST, it records each of the encountered data structures. Specifically, pertaining to a data structure, SALADS records its type and its name. Also SALADS maintains information about all the members inside the data structure, including their names, sizes and offsets in the data structure. In particular, there are two challenges in determining the offsets. On the one hand, compiler may follow special rules to align different fields (e.g.,#param pack(n)). When such rules are specified, SALADS uses them to calculate the offsets. Otherwise it determines the offsets based on standard alignment rules (e.g., architecture optimization rules). On the other hand, some arrays inside data structures have flexible sizes [1] that can only be determined at run-time. To handle such cases, SALADS places such arrays at the end of the corresponding data structures and excludes them from randomization.

SALADS also collects the uses of external functions that are not defined by the target program (e.g., library functions). Those invocations are marked as dangerous statements, since the external functions may take randomized data structures as parameters while they are not instrumented to achieve correct accesses. SALADS determines a function as external function if its definition is not included in the target program. Going beyond, SALADS identifies functions that are defined in the target program but exported to other programs. SALADS labels such functions as dangerous functions, considering that data structures randomized inside might interrupt the execution of programs that import these functions. Exported functions can usually be identified with the export table.

Instrumentation for Randomization. Using the information as extracted above, SALADS performs instrumentation on the GIMPLE IR. At the high level, this instrumentation replaces each of the data structure accesses with a DSSR routine. This routine takes charge of correcting the access and makes re-randomization on demand.

The instrumentation to achieve randomization takes several steps. At the first place, SALADS traverses the GIMPLE IR and identify accesses to data structures. Given a GIMPLE statement, SALADS disassembles it into expressions following the order from right to left. Note that compound expressions are simplified as atomic ones. For instance, the expression var1 op var2 is split into var1 and var2. On encountering an expression which represents an data structure access, SALADS keeps track of (1) the type and address of the data structure and (2) the identify of the accessed field. In the case where the access is nested, SALADS records both of the parent and child access. Take DA->DB.mem for an example. SALADS sequentially parse the outer access A->B and inner access B.x.

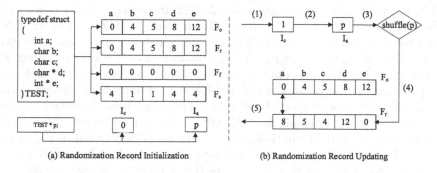

(a) Randomization Record Initialization   (b) Randomization Record Updating

Fig. 9. Initialization and updating of a randomization record (from [15] page 7).

1 typedef struct {	1 main()	1 main()	16	p = (struct TEST *) D.2052;
2 int a;	2 {	2 {	17	p.0 = (int)p;
3 char b;	3 void * D.1962;	3 void * D.2052;	18	Initialize_Record(0,p.0);
4 char c;	4 struct TEST * p;	4 int p.0;	19	Update_Record(0,p.0);
5 char * d;	5 extern void * malloc	5 int * D.2054;	20	D.2057 = Offset_Diff(0,1);
6 int * e;	(unsigned int);	6 int q.1;	21	D.2054 = &p->a;
7 } TEST;	6 struct TEST q;	7 char * D.2056;	22	D.2058 = D.2054 + D.2057;
8	7 D.1962 = malloc(16);	8 struct TEST * p;	23	*D.2058 = 1;
9 void main()	8 p = (struct TEST *)	9 extern void * malloc	24	q.1 = (int) &q;
10 {	D.1962;	(unsigned int);	25	Initialize_Record(0,q.1);
11 TEST * p;	9 p->a = 1;	10 struct TEST q;	26	Update_Record(0,q.1);
12 p = (TEST *)malloc	10 q.c = 97B;	11 int D.2057;	27	D.2059 = Offset_Diff(0,3);
(sizeof(TEST));	11 }	12 int * D.2058;	28	D.2056 = (char *)&q.c;
13 TEST q;		13 int D.2059;	29	D.2060 = D.2056 + D.2059;
14 p->a = 1;		14 char * D.2060;	30	*D.2060 = 97B;
15 q.c = 'a';		15 D.2052 = malloc(16);	31 }	
16 }				

(a) Source Code   (b) GIMPLE Output by GCC-4.5.0   (c) DSSR GIMPLE Output by SALADS

Fig. 10. An example showing how DSSR program generated by SALADS works (from [15] page 8).

Prior to an access to a data structure, SALADS plants a DSSR routine which consists of a group of DSSR statements. These statements first conducts a round of randomization over the involved data structure(s) (if necessary) and then corrects the access based on the accessing type. If the access is a read, the DSSR statements store the original value in a temporary variable. Otherwise, these statements maintains a pointer which points to the randomized location of the field. Lastly, SALADS replaces the statement of access with another DSSR statement which completes the access using the temporary variable or the new pointer.

For a better understanding about the above instrumentation, we present a demonstrative example in Fig. 10. To be specific, the source code of the program, the native GIMPLE IR, and the instrumented GIMPLE IR are presented in Fig. 10(a), (b) and (c), respectively. It's worth noting that GIMPLE is a IR in static single assignment form [4] and it uses temporary variables to store intermediate values in complex expressions.

In this example, the program allocates a data structure of type TEST on the heap (line 12 Fig. 10(a)). It corresponds to the GIMPLE statements at line 15 of Fig. 10(c). We next demonstrate how SALADS instruments the access to this data structure at line 14 in Fig. 10(a). Given that the definition of TEST has been extracted, SALADS inserts a GIMPLE statement which invokes the Initialize_Record procedure (line 18 Fig. 10(c)). Initialize_Record takes two arguments with 0 specifying that this data structure is being accessed for the first time and p.0 carrying the address of this data structure. Internally, this procedure records this new data structure by its address and initializes meta-data pertaining to each filed (F_o: original offset; F_r: offset after randomization; F_s: size; F_f: randomizable or not). The initialization results are demonstrated in Fig. 9(a).

The above initialization is followed by a call to Update_Record at line 19 in Fig. 9. Update_Record first increments the number of accesses to the data structure by 1. If the number of accumulated accesses exceeds W_m, SALADS then uses a routine Shuffle(p) to shuffle the layout of the target data structure and clear the access counter. This process is shown as step-(1) to step-(3) in Fig. 9. The offsets of fields after randomization are shown in Fig. 9 following step-(4). Thereafter, SALADS makes a GIMPLE statement which calls Offset_Diff (line 20 Fig. 10(c)). As is depicted as step-(5) in Fig. 9, this function calculates the difference between the randomized offset and the original offset pertaining to the target field. In this example, the difference on a is 8. According to that offset difference, SALADS calculates a temporary pointer D.2058 to hold the correct location (line 22 Fig. 10(c)) and issues the write to the location referred to by this pointer (line 23 Fig. 10(c)).

Instrumentation for De-randomization. Randomization on certain data structures may incur exceptions, even we enable the instrumentation as aforementioned. For instance, passing a randomized data structure to a library function will likely cause errors, because the library is usually un-instrumented and hence cannot access the data correctly. In general, there are two categories of accesses which might be interrupted by SALADS. In the following, we present their details and explain how SALADS address them.

In program developed in C/C++ language, a pointer is often cast into a different type. This causes problems to SALADS. For example, a program may cast a pointer P of type struct X to point Q with type struct Y (struct X and struct Y are different). As struct X and struct Y usually have different layout, accessing Y based on the randomization record on X will likely cause errors. To handle such cases, SALADS simply discards the randomization over the data structure pointed to by P. In addition, a C/C++ program may also assign the address of a field to a pointer (e.g.,int *p=&z.a). Once SALADS randomizes the location of the field (&z.a), accesses to this field via dereference on the pointer (p) will be interrupted. In such a case, SALADS simply marks the de-referenced field (&z.a) as un-randomizable.

Table 2. Defense results of DSSR applications (from [15] page 11).

Programs	CVE #	Bugs	Data structure	ASLR& DSLR	SALADS
openssl-0.9.6d	CVE-2002-0656	KEY ARG [20]	ssl_session_st	×	√
openssh-2.1.1	CVE-2001-0144	CRC-32 [19]	passwd	×	√

Recall that randomization on data structures passed to external functions or involved in exported functions is also dangerous. SALADS also handles such cases. To be more specific, it de-randomizes data structures that are passed into external functions. Likewise, it prevents randomization over data structures that are involved in exported functions.

SALADS implements GIMPLE Pass-2 to achieve the above de-randomization. Before each of the statements that may involve dangerous randomization, it inserts GIMPLE statements to complete de-randomization or prevent randomization.

3.3.3   Evaluation on SALADS

Similar to evaluation on dynamic ASLR techniques, a dynamic DSLR technique should also be measured following the set of principles we discussed in Sect. 2.4 For the consideration of space, we omit the details. In the following, we present the evaluation on SALADS, following the guide of those principles.

Experimental Settings. To measure the idea behind SALADS from the perspective from both security and practicality, the developers implement SALADS on the top of gcc-4.5.0 with around 11K lines of extra C code. To serve the evaluation, they measure SALADS against a set of benchmark vulnerabilities and programs on a machine running Fedora Core Release 8 with Intel(R) Core(TM) i5 and 4 GB RAM. In the following, we describe how these benchmarks are prepared.

Preparation of Benchmarks. For the goal of security evaluation, the developers configured a group of two real attacks against the programs under protection of static DSLR and SALADS. To better mimic the real world, both ASLR and DEP are enabled. The first attack is the one presented in Fig. 7. The second attack involves an exploit against a buffer overflow on the array key_arg in a data structure session (of type ssl_session_st) in openssl [20]. In this attack, the adversary overflows key_arg and injects malicious code. Without randomization, the shell code has a constant distance from another field ciphers in session. This enables the adversary to correctly locate the shell code and jump to it for malicious intentions.

For the goal of performance evaluation, the developers collect a set of bench-mark programs, including SPECInt2000 [3], `httpd-2.0.6`, `openssh-2.1`, and `openssl-0.9.6d`. SALADS compiles each of them to a DSSR binary. Note that instead of randomizing all the data structures, their instrumentation only randomly covers 20% of them. They also additionally cover the set of security sensitive data structures that are manually selected.

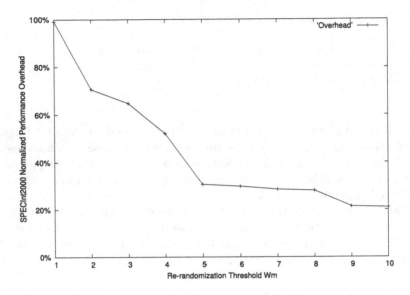

Fig. 11. Effects of W_m on performance (from [15] page 13).

Effectiveness of DSSR. The results of attacking experiments are shown in Table 2. As indicated by this table, the combination of ASLR and static DSLR can be bypassed under the condition of memory leakage. To the contrary, SALADS successfully defeats both of the attacks. In the attack against `openssh`, the adversary relies on the location of `pw_uid` to succeed. However, the request from the attack triggers more than 5^2 accesses to `pw` before it finally launch the attacks. As such, the layout of `pw` gets re-randomized prior to the final step and prevents the attacks from succeeding. The attack against `openssl` is prevented in a similar manner—the attacking request triggers over 17 accesses to the target data structure `session` and triggers re-randomization before the target data object is used for exploitation.

Performance Overhead. Recall that SALADS performs a round of re-randomization after a data structure has been accessed for W_m times. Selection of W_m is a balance between security and performance. A smaller W_m makes more frequent re-randomization while introduces higher overhead. The design of

---

[2] SALADS uses 5 as the threshold of accesses to trigger de-randomization. Reasons are explained in the evaluation on performance of SALADS.

SALADS relies on empirical tests to determine W_m. To be specific, they run the SPECInt2000 benchmark under SALADS and observe the changes of performance overhead as W_m ranges from 1 to 10. The results of averaged performance overhead are shown in Fig. 11. Not surprisingly, the extend of overhead increases as W_m becomes larger. In particular, the overhead reduces dramatically while W_m switches from 4 to 5 and the overhead does not significantly decrease when W_m gets further larger. This makes SALADS select 5 as W_m by default. All the evaluation done by the developers are using $W_m = 5$.

In measuring the performance overhead incurred by SALADS, the developers run the DSSR versions of SPECInt2000 with the official settings. They also run httpd under apache benchmark, opensshl under openssl speed [2], and openssh with a benchmark which uploads 1.5GB test-files using scp [2] for 1000 times. The final results are presented in Fig. 12. On average, the performance overhead introduced by SALADS ranges from 0.2% to 23.5%. In particular, SALADS incurs significant overhead when it protects gzip, gap and twolf. The major reason behind is that these program involves a great many of data structures to manage and access data objects.

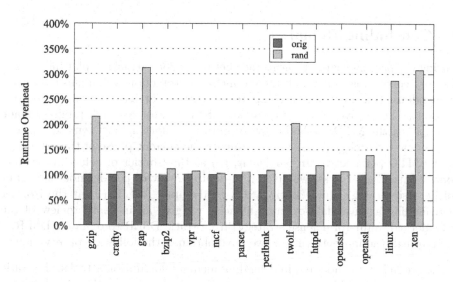

Fig. 12. Performance overhead (from [15] page 15).

Memory Overhead. In addition to performance overhead, the developers also consider the consumption of memory by SALADS. To achieve this measurement, the developers count the use of memory at a large number of random moments during execution of the benchmark programs. The results are demonstrated in Fig. 13. On average, the memory overhead introduced by SALADS to DSSR programs ranged from 0.7% (openssh-2.1.1p4) to 6.1% (twolf).

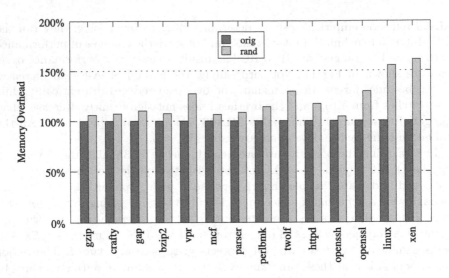

Fig. 13. Memory overhead (from [15] page 15).

4   Concluding Remarks

During the past 25 years, the arms race between cyber attacks exploiting various memory corruption vulnerabilities and memory protection techniques has drawn tremendous attention from the computer security research community. This book chapter seeks to give an in-depth review of the newest research progress made on applying the MTD methodology to protect memory against remote exploits. These new research progress also represents the current phase of the arms race in the MTD perspective. In particular, (a) at the frontier of defending against control-hijacking attacks, we give an in-depth review of the defender side shift (of defense strategy) from static ASLR to dynamic ASLR; (b) at the frontier of defending against data-oriented attacks, we give an in-depth review of the defender side shift (of defense strategy) from static DSLR to dynamic DSLR.

Through the in-depth review, we have obtained the following observations:

- Before MTD was adopted in defending memory corruption attacks, the static nature of old fashioned memory protection had provided the attacker with an incredible advantage, leading to ineffective defenses in many cases.
- MTD techniques can significantly increase the resiliency of a computer system when defending against control-hijacking attacks exploiting memory corruption vulnerabilities.
- MTD techniques may also significantly increase the resiliency of a computer system when defending against data-oriented attacks exploiting memory corruption vulnerabilities.
- When defending against control-hijacking attacks, there are a spectrum of MTD techniques which could be employed. Dynamic ASLR techniques are in general more effective than static ASLR techniques.

– When defending against data-oriented attacks, there are a spectrum of MTD techniques which could be employed. Dynamic DSLR techniques are in general more effective than static DSLR techniques.
– Regarding future directions, we believe that ASLR and DSLR techniques will become more and more adaptive. For example, a new study we conducted very recently has shown that feedback control can make dynamic DSLR more cost-effective under zero-day attacks.

References

1. Arrays of length of zero. http://gcc.gnu.org/onlinedocs/gcc/Zero-length.html
2. Openssh benchmark. http://blog.famzah.net/2010/06/11/openssh-ciphers-performance-benchmark/
3. SPEC CPU benchmark suite (2000). http://www.spec.org/cpu2000/
4. Gimple (2015). https://gcc.gnu.org/onlinedocs/gccint/GIMPLE.html
5. Backes, M., Nürnberger, S.: Oxymoron: making fine-grained memory randomization practical by allowing code sharing. In: USENIX Security Symposium (Security 2014) (2014)
6. Bhatkar, S., Duvarney, D.C., Sekar, R.: Address obfuscation: an efficient approach to combat a broad range of memory error exploits. In: USENIX Security Symposium (Security 2003) (2003)
7. Bhatkar, S., Sekar, R., DuVarney, D.C.: Efficient techniques for comprehensive protection from memory error exploits. In: USENIX Security Symposium (Security 2005) (2005)
8. Bigelow, D., Hobson, T., Rudd, R., Streilein, W., Okhravi, H.: Timely rerandomization for mitigating memory disclosures. In: Proceedings of the 22nd Conference on Computer and Communications Security (CCS 2015) (2015)
9. Bittau, A., Belay, A., Mashtizadeh, A., Mazieres, D., Boneh, D.: Hacking blind. In: 2014 IEEE Symposium on Security and Privacy, Oakland (2014)
10. Bletsch, T., Jiang, X., Freeh, V.W., Liang, Z.: Jump-oriented programming: a new class of code-reuse attack. In: ACM Symposium on Information, Computer and Communications Security (ASIACCS 2011) (2011)
11. Castro, M., Costa, M., Harris, T.: Securing software by enforcing data-flow integrity. In: Proceedings of the 7th Symposium on Operating Systems Design and Implementation (OSDI 2006) (2006)
12. Checkoway, S., Davi, L., Dmitrienko, A., Sadeghi, A.-R., Shacham, H., Winandy, M.: Return-oriented programming without returns. In: ACM Conference on Computer and Communications Security (CCS 2010) (2010)
13. Chen, H., Mao, Y., Wang, X., Zhou, D., Zeldovich, N., Kaashoek, M.F.: Linux kernel vulnerabilities: state-of-the-art defenses and open problems. In: Proceedings of the Second Asia-Pacific Workshop on Systems (2011)
14. Chen, P., et al.: What you see is not what you get! thwarting just-in-time ROP with chameleon. In: 2017 47th Annual IEEE/IFIP International Conference on Dependable Systems and Networks (DSN), pp. 451–462. IEEE (2017)
15. Chen, P., Xu, J., Lin, Z., Xu, D., Mao, B., Liu, P.: A practical approach for adaptive data structure layout randomization. In: Pernul, G., Ryan, P.Y.A., Weippl, E. (eds.) ESORICS 2015. LNCS, vol. 9326, pp. 69–89. Springer, Cham (2015). https://doi.org/10.1007/978-3-319-24174-6_4

16. Chen, S., Xu, J., Sezer, E.C., Gauriar, P., Iyer, R.K.: Non-control-data attacks are realistic threats. In: Proceedings of the 14th Conference on USENIX Security Symposium (Security 2005), vol. 5 (2005)

17. Chen, Y., Wang, Z., Whalley, D., Lu, L.: Remix: on-demand live randomization. In: Proceedings of the 6th ACM Conference on Data and Application Security and Privacy (CODASPY 2016) (2016)

18. Crane, S.: Readactor: practical code randomization resilient to memory disclosure. In: 2015 Symposium on Security and Privacy, Oakland (2015)

19. CVE-2001-0144. SSH CRC-32 compensation attack detector (2001). http://www.securityfocus.com/bid/2347/discuss

20. CVE-2002-0656. Apache openSSL heap overflow exploit (2002). http://www.phreedom.org/research/exploits/apache-openssl/

21. Davi, L., Liebchen, C., Sadeghi, A.-R., Snow, K.Z., Monrose, F.: Isomeron: code randomization resilient to (just-in-time) return-oriented programming. In: Network and Distributed System Security Symposium (NDSS 2015) (2015)

22. Dyninst. Dyninst programmer's guide (2013). www.dyninst.org/sites/default/files/manuals/dyninst/DyninstAPI.pdf

23. Gionta, J., Enck, W., Ning, P.: HideM: protecting the contents of userspace memory in the face of disclosure vulnerabilities. In: Proceedings of the 5th ACM Conference on Data and Application Security and Privacy (CODASPY 2015) (2015)

24. Giuffrida, C., Kuijsten, A., Tanenbaum, A.S.: Enhanced operating system security through efficient and fine-grained address space randomization. In: USENIX Conference on Security Symposium (Security 2012) (2012)

25. Göktas, E., Athanasopoulos, E., Bos, H., Portokalidis, G.: Out of control: overcoming control-flow integrity. In: 2014 IEEE Symposium on Security and Privacy, Oakland (2014)

26. Hiser, J., Nguyen-Tuong, A., Co, M., Hall, M., Davidson, J.W.: ILR: where'd my gadgets go? In: 2012 IEEE Symposium on Security and Privacy, Oakland (2012)

27. Hu, H., Chua, Z.L., Adrian, S., Saxena, P., Liang, Z.: Automatic generation of data-oriented exploits. In: Proceedings of the 24th USENIX Security Symposium (Security 2015) (2015)

28. Hu, H., Shinde, S., Adrian, S., Chua, Z.L., Saxena, P., Liang, Z.: Data-oriented programming: on the expressiveness of non-control data attacks. In: 2016 IEEE Symposium on Security and Privacy (SP), pp. 969–986. IEEE (2016)

29. Kil, C., Jum, J., Bookholt, C., Xu, J., Ning, P.: Address space layout permutation (ASLP): towards fine-grained randomization of commodity software. In: Annual Computer Security Applications Conference (ACSAC 2006) (2006)

30. Lin, Z., Riley, R.D., Xu, D.: Polymorphing software by randomizing data structure layout. In: Flegel, U., Bruschi, D. (eds.) DIMVA 2009. LNCS, vol. 5587, pp. 107–126. Springer, Heidelberg (2009). https://doi.org/10.1007/978-3-642-02918-9_7

31. Lu, K., Nurnberger, S., Backes, M., Lee, W.: How to make ASLR win the clone wars: runtime re-randomization. In: Proceedings of the 22nd Annual Network and Distributed System Security Symposium (NDSS 2016) (2016)

32. Microsoft. A detailed description of the Data Execution Prevention (DEP) feature in Windows XP Service Pack 2 (2008). http://support.microsoft.com/kb/875352

33. Pappas, V., Polychronakis, M., Keromytis, A.D.: Smashing the gadgets: hindering return-oriented programming using in-place code randomization. In: 2012 IEEE Symposium on Security and Privacy, Oakland (2012)

34. Schwartz, E.J., Avgerinos, T., Brumley, D.: Q: exploit hardening made easy. In: USENIX Conference on Security (Security 2011) (2011)

35. Seibert, J., Okhravi, H., Söderström, E.: Information leaks without memory disclosures: remote side channel attacks on diversified code. In: ACM SIGSAC Conference on Computer and Communications Security (CCS 2014) (2014)
36. Shacham, H.: The geometry of innocent flesh on the bone: return-into-libc without function calls (on the x86). In: ACM Conference on Computer and Communications Security (CCS 2007) (2007)
37. Snow, K.Z., Monrose, F., Davi, L., Dmitrienko, A., Liebchen, C., Sadeghi, A.-R.: Just-in-time code reuse: on the effectiveness of fine-grained address space layout randomization. In: 2013 IEEE Symposium on Security and Privacy, Oakland (2013)
38. Song, C., Lee, B., Lu, K., Harris, W.R., Kim, T., Lee, W.: Enforcing kernel security invariants with data flow integrity. In: Proceedings of the 2016 Network and Distributed System Security Symposium (NDSS 2016) (2016)
39. Stanley, D.M., Xu, D., Spafford, E.H.: Improved kernel security through memory layout randomization. In: International Performance Computing and Communications Conference (IPCCC 2013) (2013)
40. Strackx, R., et al.: Breaking the memory secrecy assumption. In: Second European Workshop on System Security (2009)
41. Tang, A., Sethumadhavan, S., Stolfo, S.: Heisenbyte: thwarting memory disclosure attacks using destructive code reads. In: Proceedings of the 22nd ACM SIGSAC Conference on Computer and Communications Security (CCS 2015) (2015)
42. PaX Team. PaX address space layout randomization (ASLR) (2003). http://pax.grsecurity.net/docs/aslr.txt
43. PaX Team. PaX non-executable pages design & implementation (2003). http://pax.grsecurity.net/docs/noexec.txt
44. Wartell, R., Mohan, V., Hamlen, K., Lin, Z.: Binary stirring: self-randomizing instruction addresses of legacy x86 binary code. In: ACM Conference on Computer and Communications Security (CCS 2012) (2012)
45. Xin, Z., Chen, H., Han, H., Mao, B., Xie, L.: Misleading malware similarities analysis by automatic data structure obfuscation. In: Burmester, M., Tsudik, G., Magliveras, S., Ilić, I. (eds.) ISC 2010. LNCS, vol. 6531, pp. 181–195. Springer, Heidelberg (2011). https://doi.org/10.1007/978-3-642-18178-8_16
46. Xu, J., Kalbarczyk, Z., Iyer, R.K.: Transparent runtime randomization for security. In: International Symposium on Reliable Distributed Systems (SRDS 2003) (2003)
47. Zhang, M., Sekar, R.: Control flow integrity for COTS binaries. In: USENIX Conference on Security (Security 2013) (2013)
48. Zhang, Y., Juels, A., Reiter, M.K., Ristenpart, T.: Cross-VM side channels and their use to extract private keys. In: ACM Conference on Computer and Communications Security (CCS 2012) (2012)

Adaptive Cyber Defenses for Botnet Detection and Mitigation

Massimiliano Albanese[1]([✉]), Sushil Jajodia[1], Sridhar Venkatesan[2],
George Cybenko[3], and Thanh Nguyen[4]

[1] George Mason University, Fairfax, VA, USA
{malbanes,jajodia}@gmu.edu
[2] Perspecta Labs, Basking Ridge, NJ, USA
svenkatesan@perspectalabs.com
[3] Dartmouth College, Hanover, NH, USA
george.cybenko@dartmouth.edu
[4] University of Oregon, Eugene, OR, USA
tnguye11@uoregon.edu

Abstract. Organizations increasingly rely on complex networked systems to maintain operational efficiency. While the widespread adoption of network-based IT solutions brings significant benefits to both commercial and government organizations, it also exposes them to an array of novel threats. Specifically, malicious actors can use networks of compromised and remotely controlled hosts, known as botnets, to execute a number of different cyber-attacks and engage in criminal or otherwise unauthorized activities. Most notably, botnets can be used to exfiltrate highly sensitive data from target networks, including military intelligence from government agencies and proprietary data from enterprise networks. What makes the problem even more complex is the recent trend towards stealthier and more resilient botnet architectures, which depart from traditional centralized architectures and enable botnets to evade detection and persist in a system for extended periods of time. A promising approach to botnet detection and mitigation relies on Adaptive Cyber Defense (ACD), a novel and game-changing approach to cyber defense. We show that detecting and mitigating stealthy botnets is a multi-faceted problem that requires addressing multiple related research challenges, and show how an ACD approach can help us address these challenges effectively.

1 Introduction

Organizations increasingly rely on complex networked systems to maintain operational efficiency. While the widespread adoption of network-based IT solutions brings significant benefits to both commercial and government organizations, it also exposes them to an array of novel threats. For instance, advanced

The work presented in this chapter was support by the Army Research Office under grant W911NF-13-1-0421.

S. Jajodia et al. (Eds.): Adaptive Cyber Defense, LNCS 11830, pp. 156–205, 2019.
https://doi.org/10.1007/978-3-030-30719-6_8

persistent threats (APTs) and distributed denial-of-service (DDoS) attacks can bypass traditional defenses by leveraging an arsenal of diverse and sophisticated cyber tools. Specifically, malicious actors can use networks of compromised and remotely controlled hosts, known as botnets, to execute a number of different cyber attacks and engage in criminal or otherwise unauthorized activities. Most notably, botnets can be used to exfiltrate highly sensitive data from target networks, including military intelligence from government agencies and proprietary data from enterprise networks. In a society that has significantly shifted from producer of goods to producer of information-centric services, protecting sensitive and mission-critical data from competitors, state actors, and organized crime has become increasingly critical for the well-being of many commercial and government organizations.

What makes the problem even more complex is the recent trend toward stealthier and more resilient botnet architectures, which depart from traditional centralized architectures and enable botnets to evade detection and persist in a system for extended periods of time. Botnets can achieve resilience through either anti-signature or architectural stealth [40]. Anti-signature stealth entails the capability of manipulating the characteristics of bot-generated traffic to mask features that could be observed by signature-based detectors. On the other hand, architectural stealth entails the capability of establishing an overlay network that minimizes exposure of malicious traffic to detectors. For these reasons, botnets have recently gained significant attention in both the industry and the research community.

One promising approach to botnet detection and mitigation relies on moving-target defense (MTD), a novel and game-changing approach to cyber defense, which is part of the broader trend towards Adaptive Cyber Defense (ACD). MTD has the potential to create asymmetric uncertainty, providing the defender with a tactical advantage over the attacker [18]. Cyber attacks are typically preceded by a reconnaissance phase in which adversaries gather critical information about the target system, including network topology, service dependencies, and unpatched vulnerabilities. System and network configurations are typically static, and do not reconfigure, adapt, or regenerate except in deterministic ways to support maintenance and uptime requirements. In such a static scenario, it is only a matter of time for malicious actors to acquire accurate knowledge about the target system, engineer reliable exploits, and plan their attacks. To address this systemic weakness, MTD techniques are designed to continuously change or shift a system's attack surface [18], which has been formally defined as the *"subset of the system's resources (methods, channels, and data) that can be potentially used by an attacker to launch an attack"* [23]. Continuously reshaping a system's attack surface increases complexity and cost for malicious actors, forcing them to continuously reassess their cyber operations.

In this chapter, we present a holistic, ACD-based approach to botnet detection and mitigation. To dominate the complexity of the problem, we decompose it into three related sub-problems, and tackle them individually. In particular, we presents solutions to (i) optimally deploy a set of detectors,

(ii) identify botnet traffic, and (iii) reduce the overall lifetime of a botnet. We validate our approach through simulation and experiments, and show that our solution is effective in mitigating botnet activity.

The remainder of the chapter is organized as follows. Section 2 discusses related work. Section 3 briefly discusses the threat model and our assumptions, whereas Sect. 4 provides an overview of the research challenges we are addressing. Then, Sects. 5, 7, and 8 discuss the three related challenges and corresponding solutions in detail. Finally, Sect. 9 gives some concluding remarks and indicates directions for future work.

2   Related Work

In response to botnet-borne threats, researchers have developed many different detection mechanisms. The performance of these mechanisms primarily depends on the set of features used to identify malicious traffic. Current research mostly focuses on studying a combination of packet-based, time-based, and behavior-based features to isolate bot traffic from the traffic mix [5]. For instance, BotHunter [16] exploits the sequence of messages between bots and a command and control (C&C) server in a centralized botnet architecture, while Zhang *et al.* exploit a combination of packet-based and time-based features to identify hosts that may potentially belong to a P2P botnet [49]. However, as the accuracy of feature-based detection techniques improves over time, botnets respond with more advanced evasion techniques [38]. On the other hand, architectural stealth techniques aim at building topology-aware botnets to reduce exposure of malicious traffic to detectors. They exploit the fact that detection mechanisms are likely to be deployed on nodes where they can monitor all traffic entering or exiting the network (e.g., network gateways) or significant volumes of internal traffic (e.g., routers). Thus, botmasters can design stealthy communication architectures capable of evading detection techniques such as those described in [16,49] by minimizing observable bot traffic. To this end, Sweeney studied the importance of the physical location of bots (referred to as the *cyber high ground*) to perform stealthy missions such as data exfiltration, and designed a P2P botnet that can effectively exfiltrate data from a network's mission-critical nodes, while maintaining a small network footprint [40].

In the past, researchers have addressed the issue of scalability in Intrusion Detection System (IDS) by modeling it as a zero-sum game between the defender and the attacker [3,20,34,46], where the defender's objective is to optimally place a limited number of monitors to protect a set of target servers. The game-theoretic models in [3,34] develop optimal placement strategies to detect intrusion attempts by considering all possible routes through which the attack can reach a target server from a given set of entry points, while the models in [20,46] develop optimal placement strategies to minimize the attacker's control over the target server.

3   Threat Model and Assumptions

In our threat model, the attacker's ultimate goal is to exfiltrate data from mission-critical nodes, while remaining stealthy and persisting in the system for an extend period of time. To this end, we make the following assumptions, based on previous work by Sweeney [40].

- The attacker can discover the topology of the network, and is aware of what nodes are mission-critical. Reports by Kaspersky labs [19] and Mandiant [1] show that threat actors can infiltrate an organization's network and persist in the system for several years, mapping out the organization and exfiltrating sensitive data and valuable intellectual property.
- Exfiltrating large volumes of data generates abnormally large network flows which in turn may trigger alerts. To avoid detection, the attacker partitions the data to be exfiltrated into m segments $d_1, d_2, \ldots, d_m$, and transmits these segments over a temporal span $\mathcal{T} = \langle t_1, t_2, \ldots, t_m \rangle \subseteq \mathbb{N}^m$, i.e., at each time point t_i, the attacker transmits a data segment d_i to a C&C site. The attacker is said to have *successfully exfiltrated from a mission-critical node* if and only if all the m data segments are exfiltrated by time t_m.
- The attacker is aware of the detector placement strategy employed by the defender.

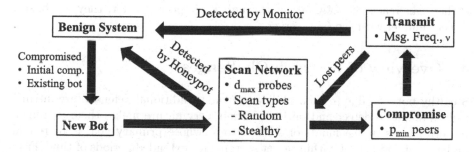

Fig. 1. Lifecycle of a bot

We model the lifecycle of a bot as shown in Fig. 1. It begins when a benign system in the target network is compromised by either an external attacker through a client-side attack or by an existing bot within the network. To construct a resilient botnet, a new bot scans the network to discover benign systems to attack. Here, we assume that all the machines within the network are vulnerable and the corresponding exploits are available to the attacker. A bot can perform two types of scans: *worm-like* or *stealthy scan*. In the worm-like scanning strategy, the bot sends random discovery probes to systems within its subnet, similar to the strategy employed by worms to propagate through a network [45]. These discovery probes include ICMP ping packets and incomplete

TCP handshakes to determine whether a system is hosted at a given IP address and also to learn the configuration of the system, including OS version, services, etc. Due to the randomness of these scans, the bots may send discovery probes to machines that may raise red flags. For example, if a bot on a client machine sends discovery probes to another client machine, this activity may be flagged as anomalous in an enterprise network. In the stealthy scanning strategy, on the other hand, bots first enumerate active connections of the underlying hosts and then send discovery probes only to these machines. Several classes of malware employ this mechanism to move laterally through the network [2]. As one of the attacker's goals is to be stealthy, independent of the scanning strategy, we can reasonably assume the existence of an upper bound d_{max} on the number of discovery probes that a bot would send over a given period of time.

After enumerating target machines, a bot compromises these machines and adds them to its list of peers. As mentioned above, we assume that all machines are vulnerable and can be successfully exploited. We also assume that, in order to build a resilient botnet, each bot needs a minimum number p_{min} of peers. Upon recruiting new machines, the bot begins exchanging update messages with its peers. These messages inform the attacker about the status of each bot within the network and also include data stolen from the corresponding host machine. When an infected host is detected by the defender, it is restored to its original state. If the number of active peers of a bot drops below the predefined threshold p_{min}, then the bot returns to the scanning state to recruit additional machines. Finally, to facilitate remote control by an attacker, the bots periodically check if they can reach the C&C server through their peers. If not, they establish a direct channel with the C&C server.

4   Overview of Research Challenges

Stealthy botnets, due to their ability to evade traditional defenses, are intrinsically difficult to detect and mitigate. Their very nature makes them extremely powerful tools in the hands of APT actors, whose primary goal is to remain undetected and persist within target systems for extended periods of time. From a defensive perspective, the problem of detecting and mitigating stealthy botnets can be broken down into three closely related challenges – captured in the infographic of Fig. 2 – which can be addressed separately, yet in a coordinated fashion, to dominate the complexity of the problem.

In real-world scenarios, it is unfeasible to monitor all network activity in depth. Thus, the first challenge is to deploy a limited number of detectors so as to maximize the likelihood of intercepting botnet-related activity. The second challenge is to analyze traffic collected by deployed detectors in order to isolate malicious data flows and identify bots responsible for those flows. This capability would enable the defender to take down bots and restore compromised hosts to a secure state. Finally, the third challenge is to reduce the overall lifetime of a botnet. Taking down some of the bots in a botnet is only a temporary solution, as residual bots can compromise additional machines to restore the full functionality

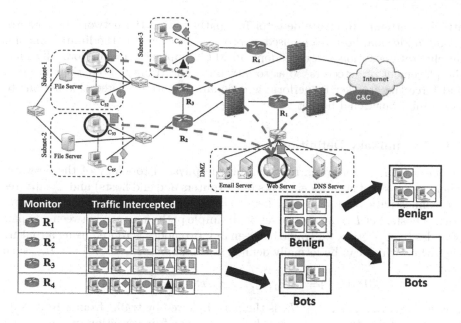

Fig. 2. Botnet detection and mitigation in a sample network scenario

of the botnet. In practice, the defender can claim victory only when every bot has been removed from the system. To achieve this goal, we need to develop a process that iterates through multiple cycles of data collection, analysis, and response, until no further botnet activity can be detected.

5 Detector Placement

As mentioned earlier, collecting and analyzing all traffic traversing every router in a complex network would prove to be a daunting task. Whether the analysis and detection capabilities are distributed or centralized, this solution would not only incur a significant computational cost, but could also increase false positive rates. To address this challenge, we have developed several heuristic detector placement strategies that select subsets of a network's nodes based on their centrality [41]. To this aim, we model a network as a graph, where nodes correspond to hosts and network devices, and edges represent the connectivity between them. A centrality measure captures important properties of a graph to determine how important or central each node is with respect to a given function or mission, which in our case is the botnet's mission to exfiltrate data from the target network to an external server. Centrality measures have found application in a wide range of domains, from social networks to citation ranking, and a prime example is PageRank, the algorithm used by Google to measure the importance of webpages.

Architecturally stealthy botnets are aware of the target network's topology and can potentially discover the location of detectors. Based on this information, attackers attempt to create detector-free paths within the network by compromising additional hosts to be used as proxies. To overcome the limitations of a purely static solution, we can adopt an MTD approach and periodically alter the placement of detectors, so as to introduce uncertainty about their location and force the attacker to perform additional, potentially detectable actions to maintain a functional botnet.

5.1   Preliminary Definitions

Let $G = (V, E)$ be a graph representing the physical topology of the network, where V is a set of network elements (e.g., routers and end hosts) and E captures the connectivity between them. Let $N = \{h_1, h_2, ..., h_n\}$ be a set of mission-critical hosts. Let Π_G denote the set of all simple paths $\pi(v_i, v_j)$ between any pair of nodes $(v_i, v_j) \in V \times V$. Traffic between any two nodes is routed using a routing algorithm, which can be formally defined as a mapping $R_A : V \times V \to \Pi_G$, such that

$$R_A(u, v) = \langle u, z_1, z_2, \ldots, z_r, v \rangle, \forall (u, v) \in V \times V$$

where $\langle u, z_1, z_2, \ldots, z_r, v \rangle \in \Pi_G$ is the path followed by traffic from u to v. Note that we slightly abuse notation and, for the sake of presentation, we may treat a path $\pi \in \Pi_G$ as a set of nodes. Although most routing algorithms attempt to route traffic along the shortest path from source to destination, our approach does not rely on the assumption that traffic is routed along the shortest path, but rather on the more general assumption that we can predict what paths the algorithm will select for routing traffic. However, for the sake of presentation, and without limiting the generality of our approach, we do assume that the networks being studied implement a shortest path routing algorithm.

In order to exfiltrate data from the set N of mission-critical nodes, the attacker compromises a set $B \subseteq V$ of network nodes – referred to as *bots* and such that $B \cap N \neq \emptyset$ – and creates an overlay network to forward captured data to a remote C&C server.

Definition 1 (Exfiltration Path). Given the set $N \subseteq V$ of mission-critical nodes for a network $G = (V, E)$ and a set $B \subseteq V$ of nodes controlled by the attacker, an *overlay path* is a sequence $\pi_o(b_0, \text{C\&C}) = \langle b_0, b_1, b_2, \ldots, b_r, \text{C\&C} \rangle$ of bots – with $b_0 \in N \cap B$ and $b_i \in B$ for each $i \in [1, r]$ – chosen by the attacker to forward traffic from mission-critical node b_0 to a remote C&C site. The *exfiltration path* corresponding to an overlay path $\pi_o(b_0, \text{C\&C})$ is the sequence of nodes in V traversed by traffic exfiltrated through $\pi_o(b_0, \text{C\&C})$, and it is defined as:

$$\pi_e(b_0, \text{C\&C}) = \langle b_0, v_1^0, v_2^0, \ldots, v_{l_0}^0, b_1, v_1^1, v_2^1, \ldots, v_{l_1}^1, b_2, \ldots, b_r, v_1^r, v_2^r, \ldots, v_{l_r}^r, \text{C\&C} \rangle$$

where $R_A(b_i, b_{i+1}) = \langle b_i, v_1^i, v_2^i, \ldots, v_{l_i}^i, b_{i+1} \rangle, \forall i \in [0, r-1]$ is the routing path from b_i to b_{i+1} and $R_A(b_r, \text{C\&C}) = \langle b_r, v_1^r, v_2^r, \ldots, v_{l_r}^r, \text{C\&C} \rangle$ is the path from b_r to C&C.

Example 1. In the example of Fig. 3, if $v_5 \in B$ is a bot and the attacker chooses to exfiltrate traffic through the overlay path $\pi_o(v_1, \mathsf{C\&C}) = \langle v_1, v_5, \mathsf{C\&C} \rangle$, then the corresponding exfiltration path is $\pi_e(v_2, \mathsf{C\&C}) = \langle v_1, v_3, v_5, v_8, \mathsf{C\&C} \rangle$.

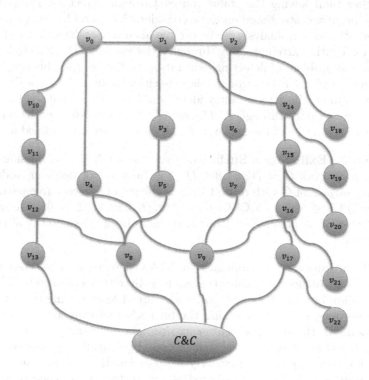

Fig. 3. Example of network graph

For a given set of mission-critical nodes N, the defender's objective is to intercept and detect exfiltration traffic. In order to monitor the network for botnet activity, the defender can deploy detectors on a subset of nodes $D \subset V$. One of several botnet detection mechanisms can be used to detect botnet activity [15,16,48,49]. These detection mechanisms leverage the fact that bots need to communicate with their peers or the $\mathsf{C\&C}$ server to relay captured data.

Definition 2 (Detection). Given the set $N \subseteq V$ of mission-critical nodes for a network $G = (V, E)$ and the set $B \subseteq V$ of bots controlled by the attacker, an exfiltration attempt over an exfiltration path $\pi_e = \langle v_0, v_1, v_2, \ldots, v_r, \mathsf{C\&C} \rangle$, with $v_0 \in N \cap B$, is said to be *detected* iff the exfiltrated traffic traverses a detector node, that is, $\exists d \in D$ s.t. $d \in \pi_e$. A botnet is said to be *stealthy* with respect to N, iff no exfiltration path between nodes in $N \cap B$ and a $\mathsf{C\&C}$ site can be detected.

Unfortunately, existing detection mechanisms suffer from false positives and false negatives, therefore exfiltration attempts may go undetected even when a detector is placed on a node along the exfiltration path. However, a prudent attacker will opt for creating more bots in order to establish a detector-free path, rather than having the traffic routed through detectors, irrespective of their false negative rate. Based on these considerations, and in order to simplify the presentation of our analysis, we ignore the accuracy of detectors and assume that any exfiltration attempt going through a detector node is detected. We will reconsider the problem of detecting exfiltration traffic later in this chapter.

In order to exfiltrate data from a mission-critical node $h \in N$ to a C&C site in a stealthy manner, the attacker must identify a detector-free path $\pi_e^*(h, \text{C\&C}) \in \Pi_G$, and forward data through it. The set of all detector-free paths represents the *exfiltration surface* of the network, which can be formally defined as follows.

Definition 3 (Exfiltration Surface). Given the set $N \subseteq V$ of mission-critical nodes for a network $G = (V, E)$, let $D \subset V$ be a set of detector nodes. The *exfiltration surface* of G with respect to D is the set of detector-free paths $\psi_D = \{\pi_e(h, \text{C\&C}) \mid h \in N \wedge \pi_e(h, \text{C\&C}) \in \Pi_G \wedge \pi_e(h, \text{C\&C}) \cap D = \emptyset\}$. We use Ψ to denote the set of all possible exfiltration surfaces from mission-critical nodes N to C&C sites.

In [42], we proposed an approach to deploy detectors on selected network nodes, so as to reduce the exfiltration surface by either completely disrupting communication between bots and C&C nodes, or at least forcing the attacker to create more bots, thereby increasing the botnet's footprint and the likelihood of detection. As the detector placement problem is intractable, we proposed heuristics based on several centrality measures. Specifically, we showed that the *iterative mission-betweenness centrality strategy* yields the best results. In this strategy, after a node has been selected as a detector, the mission-betweenness centrality of all non-detector nodes is recomputed, and the node with the highest centrality is chosen for placing an additional detector. In practice, this approach prevents two or more detectors from being placed on the same high-centrality path. Although this strategy significantly increases an attacker's effort, the resulting exfiltration surface is static. Therefore, a persistent attacker can gather enough information to precompute the exfiltration surface of the target system and identify a detector-free path to exfiltrate data. We overcome this limitation by designing detector placement strategies that *dynamically* change the exfiltration surface by continually altering the placement of detectors, as discussed in the following subsections.

5.2   Defender's Model

In our defender's model, we consider a resource-constrained setting where the defender can only deploy k detectors. In practice, an upper bound on the number of detectors can be determined by considering the number of systems in the network that can perform detection tasks without impacting the performance of applications running on them. A dual problem is that of minimizing the number of detectors needed to satisfy predefined security requirements. In the following, we formally define the notion of *detector placement*.

We assume that the defender is aware of the location of potential C&C sites. For an enterprise network, C&C locations could include any destination outside the network perimeter. Similarly, for an ISP network, potential C&C sites could be located outside the administered domain. Furthermore, it has been shown that certain IP address ranges are known to participate in malicious campaigns [8,28]. This information can be leveraged to identify potential C&C locations, but, due to the conservative estimate on the location of potential C&C sites, simply blacklisting traffic to these locations would adversely affect legitimate users.

Definition 4 (k-placement). Given a network $G = (V, E)$, a *k-placement* over G is a mapping $pl : V \to \{0, 1\}$ such that $\sum_{v \in V} pl(v) = k$. Vertices v such that $pl(v) = 1$ are called *detector nodes*. We will use $\mathscr{P}_k$ to denote the set of all possible k-placements.

To address the limitations of a static placement and increase the probability of detection, we can continually shift the exfiltration surface by dynamically changing the location of detectors. In our analysis, we discretize time as a finite sequence of integers $\mathscr{T} = \langle t_1, t_2, \ldots, t_m \rangle \subseteq \mathbb{N}^m$, with $m \in \mathbb{N}$, such that for all $1 \le i < m$, $t_i < t_{i+1}$, and model how placements can evolve over time.

Definition 5 (Temporal k-placement). A *temporal k-placement* is a function $tp : \mathscr{T} \to \mathscr{P}_k$. We will use $\mathscr{P}_k^T$ to denote the set of all possible temporal k-placements.

Intuitively, for each time point in $\mathscr{T}$, a temporal k-placement deploys detectors on k network nodes. In order to create uncertainty for the attacker with respect to the location of detectors, we choose temporal k-placement functions by using a probability distribution over all temporal k-placements.

Definition 6 (Temporal probabilistic k-placement). A *temporal probabilistic k-placement* (*tp-k-placement*) is a function $\tau : \mathscr{P}_k^T \to [0, 1]$ such that $\sum_{tp \in \mathscr{P}_k^T} \tau(tp) = 1$.

Example 2. Figure 4 shows an example of temporal probabilistic k-placement τ for the graph of Fig. 3 and for $k = 2$. Each table in the figure represents a different temporal k-placement tp[1]. Note that $\sum_{tp \in \mathscr{P}_k^T} \tau(tp) = 1$. For any given temporal k-placement tp, the i-th column in the corresponding table – with $i \in \{1, 2, \ldots, m\}$ – represents the k-placement pl that tp associates with time point t_i. Note that, for each k-placement pl, $\sum_{v \in V} pl(v) = k$. This example assumes that only certain nodes, namely v_3, v_4, v_5, and v_6, can host detectors.

---

[1] For the sake of presentation, we assume that those shown are the only possible temporal k-placements in $\mathscr{P}_k^T$).

$$\tau \begin{pmatrix} \begin{array}{c|c|c|c|c|c} & t_1 & t_2 & t_3 & \dots & t_m \\ \hline v_3 & 1 & 1 & 0 & \dots & 1 \\ v_4 & 1 & 0 & 1 & \dots & 1 \\ v_5 & 0 & 0 & 1 & \dots & 0 \\ v_6 & 0 & 1 & 0 & \dots & 0 \end{array} \end{pmatrix} = 0.3 \quad \tau \begin{pmatrix} \begin{array}{c|c|c|c|c|c} & t_1 & t_2 & t_3 & \dots & t_m \\ \hline v_3 & 0 & 1 & 1 & \dots & 1 \\ v_4 & 1 & 0 & 0 & \dots & 0 \\ v_5 & 0 & 1 & 0 & \dots & 0 \\ v_6 & 1 & 0 & 1 & \dots & 1 \end{array} \end{pmatrix} = 0.2 \quad \tau \begin{pmatrix} \begin{array}{c|c|c|c|c|c} & t_1 & t_2 & t_3 & \dots & t_m \\ \hline v_3 & 0 & 0 & 0 & \dots & 1 \\ v_4 & 0 & 1 & 1 & \dots & 1 \\ v_5 & 1 & 1 & 0 & \dots & 0 \\ v_6 & 1 & 0 & 1 & \dots & 0 \end{array} \end{pmatrix} = 0.5$$

Fig. 4. Example of temporal probabilistic k-placement

Let the indicator random variable $I_{t_i}^v$ be associated with the event that node v is chosen as a detector at time t_i. Given a temporal probabilistic k-placement τ, the probability with which a node $v \in V$ will be chosen as a detector at time t_i can be derived as

$$pr_{t_i}^v = Pr(I_{t_i}^v = 1 \mid \tau) = \sum_{tp \in \mathscr{P}_k^T \ s.t. \ (\exists pl \in \mathscr{P}_k)(tp(t_i)=pl \wedge pl(v)=1)} \tau(tp) \qquad (1)$$

Thus, at time t_i, the defender selects k nodes for detector placement by sampling from the distribution defined by Eq. 1. We denote such a strategy as $\mathscr{D}_{t_i} \sim \{pr_{t_i}^v\}_{v \in V}$.

5.3 Metrics

To evaluate the performance of a defender strategy, we present two metrics: the *minimum detection probability* and the *attacker's uncertainty*. The minimum detection probability provides a theoretical lower bound on the probability that an exfiltration activity is detected due to the detector placement strategy. On the other hand, the attacker's uncertainty is measured as the entropy in the location of the detectors from the attacker's point of view: the higher the entropy, the higher the attacker's effort required to discover the location of detectors.

5.3.1 Minimum Detection Probability

As mentioned earlier, to be succssfull, the attacker needs to exfiltrate data segments $d_1, d_2, \dots, d_m$ over a temporal span $\mathscr{T} = \langle t_1, t_2, \dots, t_m \rangle \subseteq \mathbb{N}^m$, while remaining undetected. At each time point t_i, the defender chooses a strategy, $\mathscr{D}_{t_i} \sim \{pr_{t_i}^v\}_{v \in V}$ and samples k nodes without replacement. Let D_{t_i} denote the set of detectors at time t_i. Following defender's placement of detectors, the attacker begins exfiltrating data segment d_i. For a chosen overlay path $\pi_o(h, \mathsf{C\&C})$, the traffic will traverse the corresponding exfiltration path $\pi_e(h, \mathsf{C\&C}) = \langle h, v_{i_1}, v_{i_2}, \dots, v_{i_l}, \mathsf{C\&C} \rangle$, with $h \in N$. Therefore, the probability that the attacker's exfiltration of data segment d_i is detected is given by:

$$detectPr(\mathscr{D}_{t_i}, d_i, \pi_e(h, \mathsf{C\&C})) = 1 - \prod_{v \in \pi_e(h, \mathsf{C\&C}) \setminus \{h, \mathsf{C\&C}\}} \left(1 - pr_{t_i}^v\right) \qquad (2)$$

Algorithm 1. $minimumDetectionProb(G, \mathscr{D}_{t_i}, N, \text{C\&C})$

Input: a connectivity graph $G(V, E)$, a defender strategy, $\mathscr{D}_{t_i} \sim \{pr_{t_i}^v\}$, a set $N \subseteq V$ of mission-critical nodes, a potential C&C location

Output: the minimum detection probability of strategy $\mathscr{D}_{t_i}$ at time t_i for graph $G(V, E)$ with respect to mission-critical nodes N and the potential C&C location

1: $H(V', E') \leftarrow$ dual graph of $G(V, E)$, where $V' = E$ and $(e, f) \in E'$ iff e and f share a common vertex $v \in V$
2: $b \leftarrow \epsilon$ // an arbitrarily small value
3: **for all** $(e, f) \in E'$ **do**
4:      $v \leftarrow$ the common vertex of e and f in V
5:      **if** $pr_{t_i}^v < 1$ **then**
6:          $W'(e, f) \leftarrow \log_b(1 - pr_{t_i}^v)$
7:      **else**
8:          $W'(e, f) \leftarrow \infty$
9:      **end if**
10: **end for**
11: // $\forall v \in V$, let $\mathscr{E}(v)$ denote the set $\{e \mid e \in E \wedge e \text{ is incident on } v \in V\}$
12: **for all** h in N **do**
13:      **for all** e in $\mathscr{E}(h)$ **do**
14:          **for all** c in $\mathscr{E}(\text{C\&C})$ **do**
15:              $S \leftarrow$ length of the shortest path from e to c in H
16:              $detectPr(h, e, c) \leftarrow 1 - b^S$
17:          **end for**
18:      **end for**
19:      $detectPr(h) \leftarrow \min\limits_{(e,c) \in \mathscr{E}(h) \times \mathscr{E}(\text{C\&C})} (detectPr(h, e, c))$
20: **end for**
21: **return** $\min\limits_{h \in N}(detectPr(h))$

A rational attacker – who is aware of the defender's strategy – will choose a path that minimizes the probability of detection. Therefore, the path chosen by the attacker to exfiltrate d_i is:

$$\pi_e^{i^*}(h, \text{C\&C}) = \underset{\pi_e(h, \text{C\&C})}{\text{argmin}} \; (detectPr(\mathscr{D}_{t_i}, d_i, \pi_e(h, \text{C\&C}))) \qquad (3)$$

In other words, Eq. 3 can be used to compute the minimum detection probability that a defender strategy $\mathscr{D}_{t_i}$ can guarantee at time t_i. Finally, an exfiltration activity is said to be *detected* when *any* of the m data flows is detected. Therefore, the minimum probability with which a strategy $\mathscr{D}_{t_i}$ detects an exfiltration activity is given by

$$eDetectPr\left(\{\mathscr{D}_{t_i}\}_{i \in [1,m]}\right) = 1 - \prod_{d_i}\left(1 - \min_{\pi_e}(detectPr(\mathscr{D}_{t_i}, d_i, \pi_e))\right) \qquad (4)$$

Given a graph $G(V, E)$, the minimum detection probability of a strategy $\mathscr{D}_{t_i}$ at time t_i – i.e., $\min\limits_{\pi_e}(detectPr(\mathscr{D}_{t_i}, d_i, \pi_e))$ – can be computed using Algorithm 1. At a high-level, the algorithm transforms the graph $G(V, E)$ into a weighted dual graph $H(V', E')$ in which the edge weights are a function of the probability that the corresponding vertex in $G(V, E)$ does not host a detector. Specifically, at time t_i, the path detection probability over any path $\pi_e(u, v)$ (given by Eq. 2) can be re-written as $1 - b^S$, where $S = \left(\sum\limits_{x \in \pi_e(u,v)} \log_b\left(1 - pr_{t_i}^x\right)\right)$ and b is an arbitrarily small value. Here, b^S is the upper bound on the probability that the path $\pi_e(u, v)$ will be free of detectors. Therefore, each edge in E' corresponding to a node $v \in V$ is assigned a weight $\log_b(1 - pr_{t_i}^v)$. Following this assignment, the algorithm determines the shortest path length between the vertices in V' that correspond to edges incident on the mission-critical and C&C vertices in V. The shortest path length represents the maximum probability that data exfiltration is not detected, and the vertices in V corresponding to edges on this shortest path form the path $\pi_e^{i^*}(h, C\&C)$.

In particular, after generating the dual graph $H(V', E')$ on Line 1, Algorithm 1 assigns weights to all the edges $(e, f) \in E'$ based on the probability $pr_{t_i}^v$ that the corresponding vertex $v \in V$ is chosen for detector placement (Lines 3–10). If a detector is placed on vertex $v \in V$ with probability 1, then any exfiltration over a path that contains v will be detected. A rational attacker will avoid such paths and hence the algorithm sets the weight of the corresponding edges in E' as ∞ (Line 8). On the other hand, if the probability is less than 1, then the corresponding edge is assigned a weight $\log_b(1 - pr_{t_i}^v)$ (Line 6). Next, Line 15 computes the length of the shortest path between vertices e and c in V', which correspond to the edges in E that are incident on mission-critical nodes and C&C, respectively. Finally, Line 16 computes the minimum detection probability over all the paths from a mission-critical node $h \in N$ to C&C that traverse edges e and c. Line 19 computes the minimum detection probability for each mission-critical node h by considering all the paths to C&C. Finally, the minimum detection probability with respect to all mission-critical nodes in N for graph $G(V, E)$ is computed on Line 21.

In the worst case, Algorithm 1 takes $O(|E|^2)$ time to generate the edge-dual graph $H(V', E')$ as all pairs of edges in E are checked for a common vertex. As a result, in the worst case $|E'| = O(|E|^2)$. Lines 3–10 run in time $O(|E'|)$ and Line 11 can be computed in time $O(|V|^2)$ by traversing the adjacency matrix of G. To compute the shortest paths between vertices in H (Line 15) that correspond to mission-critical node h and C&C in G, we can leverage the Fibonacci heap implementation of Dijkstra's single-source shortest path algorithm [10]. The complexity of computing the shortest path lengths for a node $h \in N$ (Lines 13–19) is given by $O(\mathscr{E}(h) \cdot (|E'| + |V'| \log |V'|))$. Therefore, in the worst case, the time complexity for computing the shortest path lengths for all mission-critical nodes is $O(|E| \cdot (|E|^2 + |E| \cdot \log |E|))$.

As the time complexity of the algorithm is dominated by the shortest paths computation, the time complexity is $O(|E| \cdot (|E|^2 + |E| \cdot \log |E|))$. The worst-case time complexity for computing the minimum detection probability is $O(|V|^6)$. However, for practical network topologies, our simulation results indicate that the processing time does not exceed $O(|V|^3)$.

5.4   Attacker's Uncertainty

Probabilistic deployment of detectors introduces uncertainty for the attacker with respect to the location of the detectors. Depending upon the nature of the deployed detector (either active or passive), the attacker may progressively learn the location of these detectors through probing. For instance, if an enterprise network deploys an active IDS, a simple probing strategy could consist in sending malicious packets to a node suspected of hosting a detector and, depending on whether the node accepts or rejects the packets, the attacker can determine the node's detection state. In an ISP network, an attacker can leverage probing strategies described by Shinoda *et al.* [35], and by Shmatikov and Wang [36] to identify the presence of detectors in a network.

The uncertainty introduced by a dynamic placement strategy can be quantified by measuring the entropy in locating the detectors at any time t_i. Let $X_{t_i}^-$ be the random variable that maps the set V of potential locations to the corresponding probability of being chosen for detector placement. Therefore, the entropy due to a strategy $\mathscr{D}_{t_i} \sim \{pr_{t_i}^v\}_{v \in V}$ is given by:

$$H(X_{t_i}^- \mid \mathscr{D}_{t_i}) = -\sum_{x \in V} P(X_{t_i}^- = x) \log(P(X_{t_i}^- = x)) \qquad (5)$$

where $\log(P(X_{t_i}^- = x)) = 0$, when $P(X_{t_i}^- = x) = 0$. Note that, based on the above definition of entropy, higher entropy translates into a greater advantage for the defender over the attacker.

5.5   Defender's Strategies

To illustrate the effectiveness of different defender strategies, consider again the network in Fig. 3, which includes mission-critical nodes $N = \{v_0, v_1, v_2\}$. The attacker's objective is to exfiltrate data from any node in N to a C&C server. To protect mission-critical nodes from data exfiltration, we consider the following strategies for placing k detectors.

- *Static Iterative Centrality Placement.* In this strategy, the defender chooses nodes based on the iterative mission-betweenness centrality algorithm proposed in [42]. The defender first computes the mission-betweenness centrality of a node v as $C_M(v) = \displaystyle\sum_{(s,t) \in N \times \text{C\&C s.t. } v \neq t} \frac{\sigma_{st}(v)}{\sigma_{st}}$, where σ_{st} is the number of shortest paths between s and t and $\sigma_{st}(v)$ is the number of those paths that go through v. Upon computing the mission-betweenness centrality for all the nodes, the defender chooses the node with the highest centrality for detector placement. For each subsequent detector placement, the centrality $C_M(v)$ of all non-detector nodes is re-computed and the node with the highest centrality among the non-detector nodes is picked for placing the next detector. In the example of Fig. 3, assume that the defender can place $k = 2$ detectors. Then, node v_9 (or v_8) will be chosen to the place the first detector followed by v_8 (or v_9) to place the second detector.
- *Uniform Random Placement.* The static nature of the above strategy enables an attacker to pre-compute the location of detectors and compromise nodes along a detector-free path. Therefore, in order to create uncertainty about the exact location of detectors, in this strategy, the defender chooses k nodes to place detectors uniformly at random.

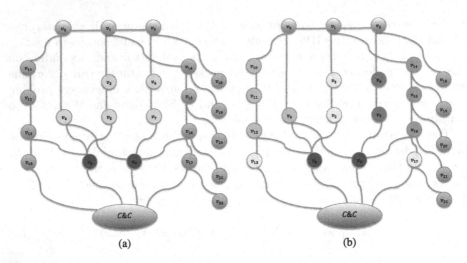

Fig. 5. Candidate detector locations for the network of Fig. 3, based on the (a) centrality-weighted strategy, and (b) expanded centrality-weighted strategy

- *Centrality-Weighted Placement.* Although the uniform random strategy introduces uncertainty for an attacker, it may consider nodes that do not lie on any simple path between mission-critical nodes and C&C. As a result, the uniform strategy may provide a low minimum detection probability. In this strategy, to improve detection guarantees, the defender places k detectors by randomly choosing nodes according to a probability distribution that weights nodes based on their mission-betweenness centrality, i.e., nodes with higher values of $C_M(v)$ have more chances of being chosen for detector placement. For the example shown in Fig. 3, the nodes colored in brown in Fig. 5a are the only nodes considered for detector placement by this strategy (the darkness is proportional to the relative weight of the corresponding node).
- *Expanded Centrality-Weighted Placement.* One of the major drawbacks of the centrality-weighted strategy is that coverage of the exfiltration surface is limited. In fact, that strategy considers only the nodes on the set of all shortest paths between the mission-critical nodes and C&C. In this strategy, the coverage of the exfiltration surface is expanded by considering all the nodes on paths that are up to δ times longer than the shortest paths. Let Π_e be the set of all such paths. The revised centrality of a node v is computed as
$$C_E(v) = \sum_{(s,t)\in N \times C\&C \; s.t. \; v \neq t} \frac{\sigma'_{st}(v)}{\sigma'_{st}}, \text{ where } \sigma'_{st} \text{ is the number of simple paths in}$$
Π_e between s and t and $\sigma'_{st}(v)$ is the number of those paths that go through v. In the example of Fig. 3, when $\delta = 0.25$, the nodes colored in brown in Fig. 5b will be considered for randomizing the placement.

5.6   Simulation Results

We evaluated the proposed strategies using real ISP network topologies obtained from the Rocketfuel dataset [37] and synthetic topologies generated using graph-theoretic properties of typical ISP networks. The Rockfuel dataset provides router-level topologies for 10 ISP networks. For each network, Table 1 provides a summary of the total number of routers within the network and the number of external routers (located outside the ISP) to which the ISP routers are connected. As connections to external routers are outside the monitored domain, we considered a worst-case scenario in our simulations and assumed that all the external routers are potentially routing traffic to C&C servers.

Table 1. Summary of ISP networks from [37]

ASN	Name	No. of routers	No. of ext. routers	ASN	Name	No. of routers	No. of ext. routers
1221	Telstra	2998	329	3356	Level3	3447	1827
1239	Sprintlink	8341	1004	3967	Exodus	895	520
1755	Ebone	605	310	4755	VSNL	121	80
2914	Verio	7102	2432	6461	Abovenet	2720	2066
3257	Tiscali	855	444	7018	AT& T	10152	722

To study the influence of network topology on the performance of a strategy, we evaluated these strategies using simulated medium-scaled ISP networks comprising 3,000 nodes. At the router level, such networks are known to exhibit scale-free network properties wherein the degree distribution follows a power-law distribution. In order to accurately capture the connectivity of an ISP network at the router level, the BRITE network topology generator [26] was used to generate these networks. Ten such networks were considered, with mission-critical nodes varying between 10% and 30% of the network size and 500 C&C locations chosen at random for each network.

For the ISP networks from the Rocketfuel dataset, we varied the size of the detector set as a fraction of the number of mission-critical nodes, whereas, for synthetic topologies, we varied the size of the detector set as a fraction of the network size. These simulations were intended to study the impact on the amount of resources that a network administrator might be willing to invest (proportional to either the number of nodes that need to be protected or the size of the network) to detect exfiltration. In all our simulations, we set $\delta = 0.5$ for the expanded centrality-weighted strategy and tested the statistical significance of the results using paired t-tests at 95% confidence interval. For the sake of presentation, we show the results for a subset of the topologies from the Rocketfuel dataset.

5.6.1   Minimum Detection Probability

As illustrated in Figs. 6 and 7, the probability of detecting exfiltration attempts increases linearly with the number of detectors. We observed that variations in the detection probability for different synthetic networks were less than 1% and hence, for the sake of presentation, we only show mean values.

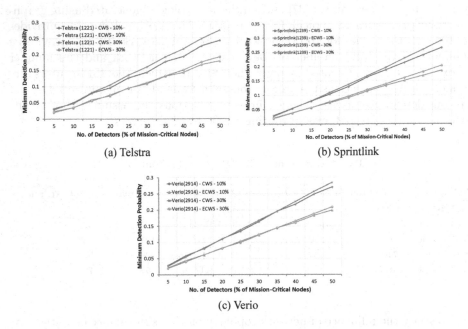

(a) Telstra

(b) Sprintlink

(c) Verio

Fig. 6. Minimum detection probability for different networks using centrality-weighted (CWS) and expanded centrality-weighted (ECWS) strategies

It can be observed that, independently of the number of mission-critical nodes and detectors, the centrality-weighted strategy outperforms the expanded centrality-weighted strategy. This trend can be attributed to the scale-free nature of the topology in which most of the paths traverse a small portion of the nodes. As the expanded strategy considers a larger number of paths and distributes the placement probability across the nodes on these paths, the nodes with high centrality will be chosen with a lower probability than in the case of the centrality-weighted strategy. In these simulations, we observed that the static iterative centrality strategy could not detect exfiltration of data segments in any of the networks as there was at least one detector-free path between one of the mission-critical nodes and a C&C location.

5.6.2 Attacker's Uncertainty

As mentioned earlier, among the detector placement strategies, the static iterative centrality strategy does not introduce any uncertainty for the attacker, whilst the uniform random strategy introduces the highest uncertainty in the location of detectors. To study the attacker's uncertainty in the location of detectors due to the proposed strategies, we computed the relative entropy

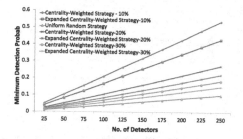

Fig. 7. Minimum detection probability for different strategies using synthetic topologies

introduced by the centrality-weighted and the expanded centrality-weighted strategies w.r.t. the uniform random strategy. As shown in Fig. 8, the centrality-weighted strategy and its expanded version create a level of uncertainty that lies in-between the two ends of the entropy spectrum. In particular, as the expanded strategy potentially considers more nodes, the number of combinations of detector locations, and hence the uncertainty introduced by it is higher than the centrality-weighted strategy. Therefore, for ISP networks, the choice of different centrality-weighted strategies offers a trade-off between entropy and detection probability.

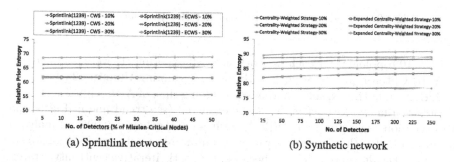

(a) Sprintlink network   (b) Synthetic network

Fig. 8. Relative increase in entropy for the attacker introduced by different strategies

5.6.3 Processing Time

In this section, we evaluate the performance of Algorithm 1 in computing the minimum detection probability and the performance of various detector placement strategies as a function of the network size. For each network size, we generated 10 ISP-type topologies, with 10% of the nodes being mission-critical and 500 C&C locations chosen randomly. We varied the number of detectors from 3% to 7% of the network size and observed similar trends in the processing time. For the sake of presentation, we only show the results when the total number of detectors is set to 3% of the network size. The processing time was averaged

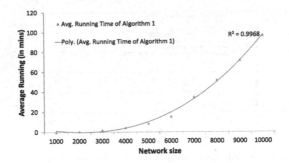

Fig. 9. Processing time for computing the minimum detection probability using Algorithm 1.

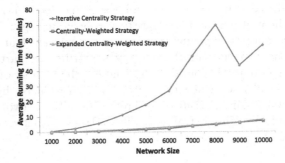

Fig. 10. Processing time for different detector placement strategies

over the 10 topologies for each network size. All experiments were conducted on an AMD Opteron processor with 4 GB memory running Ubuntu 12.04.

Although, in theory, the worst-case processing time of Algorithm 1 is $O(|V|^6)$, for practical network settings, it can be observed (see trend line in Fig. 9) that the execution time grows as $O(|V|^3)$ with an R^2 value of 0.9968. Finally, as shown in Fig. 10, the dynamic strategies compute the detector locations faster than its static alternative. This is because, the static iterative centrality strategy has to re-compute the shortest paths multiple times to determine the location of the detectors.

6   A Game-Theoretic Approach to Detector Placement

In this section, we consider a game-theoretic approach to design effective detection placement strategies. We formulate the botnet defense problem as a Stackelberg security game, thus accounting for the strategic response of attackers to deployed defenses. We consider two formulations of data exfiltration: (i) *uni-exfiltration*, where the source bot routes the stolen data along a single path designated by the attacker; and (ii) *broad-exfiltration*, where each bot propagates the received stolen data to all other bots in the network.

We propose algorithms to compute defense strategies for these data exfiltration formulations: ORANI (Optimal Resource Allocation for uni-exfiltration Interception) and ORABI (Optimal Resource Allocation for Broad-exfiltration Interception). Both ORANI and ORABI employ the double-oracle method [25] to control exploration of the exponential strategy spaces available to attacker and defender. Our main algorithmic contributions lie in defining mixed-integer linear programs (MILPs) and greedy heuristics for implementing the defender and attacker's best-response oracles.

6.1   Game Model: Uni-Exfiltration

Our game model for uni-exfiltration is built on the botnet model introduced in Sect. 5. We model the botnet defense problem as a *Stackelberg security game* (SSG) [21]. In such a game, the defender commits to a mixed (randomized) strategy to allocate limited security resources to protect important targets. The attacker then optimally chooses targets with respect to the distribution of defender allocations.

In the botnet exfiltration game, the attacker attempts to steal sensitive network data. Compromising a mission-critical node enables the attacker to steal data from that node. Compromising other nodes in the network helps the attacker relay the stolen data to a C&C server outside the network, which he controls, through a sequence of compromised nodes (bots) forming an overlay path. Routing between consecutive bots on this paths is beyond the attacker's control, and the sequence of all nodes traversed by exfiltrated traffic is referred to as an *exfiltration path*, as defined Sect. 5.1. In this game model, we consider the case in which the attacker does not divide stolen data into multiple segments before relaying it to C&C.

In our Stackelberg game model, the defender moves first by allocating detection resources, and the attacker responds with a plan for compromise and exfiltration to evade detection. The defender placement of detectors is randomized, so any attack plan succeeds with some probability. As formalized in Definition 2, an exfiltration attempt is detected if there is a detector on the exfiltration path.

Definition 7 (Strategy Space). The strategy spaces of the players are as follows:
Defender: The defender has $K^d < |V|$ detection resources available for deployment on network nodes. We denote by $D = \{D_i \mid D_i \subseteq V, |D_i| \leq K^d\}$ the set of all *pure defense strategies* of the defender. Let $x = \{x_i\}$ be a *mixed strategy* of the defender where $x_i \in [0, 1]$ is the probability that the defender plays D_i, and $\sum_i x_i = 1$.
Attacker: The attacker can compromise up to $K^a < |V|$ nodes. We denote by $A = \{A_j = (B_j, \Pi_j) \mid B_j \subseteq V, |B_j| \leq K^a, \Pi_j = \{\pi_j(c, C\&C) \mid c \in B_j \cap N\}\}$ the set of all pure strategies of the attacker. Each pure strategy A_j consists of: (i) B_j: a set of compromised nodes; and (ii) Π_j: a set of exfiltration paths over B_j.

A simple scenario of the botnet defense game is shown in Fig. 11. The model specification is completed by defining the payoff structure, which is zero-sum.

Definition 8 (Game Payoff). Each mission-critical node $c \in N$ is associated with a value, $r(c) > 0$, representing the importance of data stored at that node. Successfully exfiltrating data from c yields the attacker a payoff $r(c)$, and the defender receives a payoff $-r(c)$. For prevented exfiltrations, both players receive zero.

Note that the maximum achievable payoff for a defender is zero, obtained by preventing all exfiltration attempts. In general terms, let $U^a(D_i, A_j)$ denote the payoff to the attacker if the defender plays D_i and the attacker plays A_j. Since the game is zero-sum, the defender payoff $U^d(D_i, A_j) \equiv -U^a(D_i, A_j)$. The payoff can be decomposed across mission-critical nodes, which is formulated as follows:

$$U^a(D_i, A_j) \equiv \sum_{c \in N} r(c)h(c) \tag{6}$$

where $h(c)$ indicates if the attacker successfully exfiltrates data from node $c \in N$:

$$h(c) = \begin{cases} 1 & \text{if } c \in B_j \text{ and } D_i \cap \pi_j(c, C\&C) = \emptyset \\ 0 & \text{otherwise} \end{cases} \tag{7}$$

The expected utility for the attacker when the defender plays mixed-strategy x is

$$U^a(x, A_j) = \sum_i x_i U^a(D_i, A_j)$$

which is negated to obtain the expected defender payoff $U^d(x, A_j)$.

A defender mixed strategy that maximizes $U^d(x, A_j)$, given that the attacker plays a best response and breaks ties in favor of the defender, constitutes a *Strong Stackelberg Equilibrium* (SSE) of the game.

6.2   ORANI: An Algorithm for Uni-Exfiltration Games

In zero-sum games, the first mover's SSE strategy is also a maximin strategy [22]. Therefore, finding an optimal mixed defense strategy can be formulated as follows:

$$\max_x U_*^d \tag{8}$$

$$\text{s.t. } U_*^d \leq U^d(x, A_j), \ \forall j \tag{9}$$

$$\sum_i x_i = 1, \ x_i \geq 0, \ \forall i, \tag{10}$$

where U_*^d is the defender's utility for playing mixed strategy x when the attacker best-responds. Constraint (9) ensures the attacker chooses an optimal action

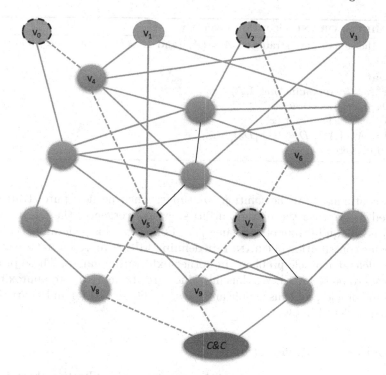

Fig. 11. An example scenario of the botnet exfiltration game with $(K^a = 4, K^d = 1)$. Four mission-critical nodes are $N = \{0, 1, 2, 3\}$. A possible pure strategy of the attacker A_j can be: (i) a set of compromised nodes $B_j = \{0, 2, 5, 7\}$; and (ii) a set of exfiltration paths $\Pi_j = \{\pi_j(0), \pi_j(2)\}$ to exfiltrate data from stealing bots 0 and 2 to the attacker's server $C\&C$. These exfiltration paths $\pi_j(0) = P(0, 5) \cup P(5, C\&C)$ and $\pi_j(2) = P(2, 7) \cup P(7, C\&C)$ relay stolen data via relaying bots 5 and 7 respectively, where $P(0, 5) = (0 \to 4 \to 5)$, $P(5, C\&C) = (5 \to 8 \to C\&C)$, $P(2, 7) = (2 \to 6 \to 7)$ and $P(7, C\&C) = (7 \to 9 \to C\&C)$ are routing paths fixed by the network system. If the defender allocates its one detector to node 9, the attacker fails at exfiltrating data from node 2 since $9 \in \pi_j(2)$ but succeeds from node 0 since $9 \notin \pi_j(0)$.

against x, that is, $U_*^d = \min_j U^d(x, A_j) = \max_j U^a(x, A_j)$. Solving (8–10) is computationally expensive due to the exponential number of pure strategies of the defender and the attacker. To overcome this computational challenge, ORANI applies the *double-oracle* method [17, 25]. Algorithm 2 presents a sketch of ORANI.

ORANI starts by solving a maximin sub-game of (8–10) by considering only small seed subsets D and A of pure strategies for the defender and attacker (Line 2). Solving this sub-game yields a solution (x^*, a^*) with respect to the strategy subsets. ORANI iteratively adds new best pure strategies D_o and A_o to the current strategy sets D and A (Lines 3–5). These strategies D_o and A_o are chosen by the oracles to maximize the defender and attacker utility, respectively, against the current (in iteration) counterpart solution strategies a^* and x^*. This iterative

---

Algorithm 2. ORANI Algorithm Overview

Initialize the sets of pure strategies: $A = \{A_j\}$ and $D = \{D_i\}$ for some j and i;

1: **repeat**
2:     $(x^*, a^*) = \text{MaximinCore}(D, A)$
3:     $D_o = \text{DefenderOracle}(a^*)$
4:     $A_o = \text{AttackerOracle}(x^*)$
5:     $A = A \cup \{A_o\}, D = D \cup \{D_o\}$
6: **until** converge

---

process continues until the solution *converges*: when no new pure strategy can be added to improve the players' utilities. At convergence, the latest solution (x^*, a^*) is an equilibrium of the game [25]. Following this general methodology, the specific contribution of ORANI is in defining MILPs representing the attacker and the defender oracle problems in botnet exfiltration games. These problems are proved to be NP-hard. We thus introduce greedy heuristics to approximately solve these oracle problems significantly faster. These MILPs and heuristics are described in detail in [30].

6.3   Data Broad-Exfiltration

In the botnet defense game model with respect to uni-exfiltration (Sect. 6.1), for each stealing bot, the attacker is assumed to only select a single exfiltration path from that bot to exfiltrate data. In this section, we study the botnet defense game model with respect to an alternative data broad-exfiltration. In particular, for every stealing bot, the attacker is able to broadcast the data stolen by this bot to all other compromised nodes via corresponding routing paths. Once receiving the stolen data, each compromised node continues to broadcast the data to all other compromised nodes. The game model for broad-exfiltration is motivated by the botnet models studied by Rossow *et al.* [32]. Overall, there is a higher chance that the attacker can successfully exfiltrate network data with broad-exfiltration compared to uni-exfiltration. In the following, we briefly describe the botnet defense game model with data broad-exfiltration. The corresponding algorithm, ORABI, to compute an optimal mixed defense strategy is built based upon the double oracle methodology. The algorithm's computation and complexity is described in detail in [30].

In the game model with data broad-exfiltration, the strategy space of the defender remains the same as shown in Sect. 6.1. On the other hand, since the attacker now can broadcast the data, we can abstractly represent each pure strategy of the attacker as a set of compromised nodes $A_j \equiv B_j$ only. Given a pair of pure strategies (D_i, B_j), we need to determine payoffs the players receive. Note that in the case of broad-exfiltration, given (D_i, B_j), the attacker succeeds in exfiltrating the stolen data from a stealing bot if there is an exfiltration path among all the possible exfiltration paths over the compromised set B_j from this bot to $C\&C$ which is not blocked by D_i. Therefore, the players receive a payoff computed as in (6) where the binary indicator $h(c)$ for each mission-critical node $c \in N$ is now determined as:

$$h(c) = \begin{cases} 1 & \text{if } \exists c \in B_j \ \& \ \exists \pi_j(c, C\&C) \\ & \text{s.t. } D_i \cap \pi_j(c, C\&C) = \emptyset \\ 0 & \text{otherwise} \end{cases}$$

In fact, when players plays (D_i, B_j), we can determine if there is an exfiltration path from a stealing bot $c \in B_j \cap N$ which is not blocked by D_i by using depth or breath-first search over the compromised set B_j, which runs in polynomial time.

6.4   Experiments

We evaluate both solution quality and runtime performance of our algorithms compared with previously proposed defense policies. We conduct experiments based on two different datasets: (i) synthetic network topologies generated using JGraphT[2], capturing scale-free properties [9] of many real-world networks; and (ii) real-world network topologies derived from the Rocket-fuel dataset [37]. Each data point in our results is averaged over 50 different samples of network topologies.

6.4.1   Synthetic Network Topology

Data Uni-Exfiltration. We compare six different algorithms: (i) ORANI – both exact oracles; (ii) ORANI-AttG – exact defender oracle and greedy attacker oracle; (iii) ORANI-G – both greedy oracles; (iv&v) Centrality-Weighted Placement (CWP) and Expanded Centrality-Weighted Placement (ECWP) – heuristics proposed in Sect. 5 to generate a mixed defense strategy based on the centrality values of nodes in the network; and (vi) Uniform – generating a uniformly-mixed defense strategy. We consider CWP, ECWP, and Uniform as the three baseline algorithms.

In the first experiment (Fig. 12(a)), we examine solution quality of the algorithms with varying number of nodes. In Fig. 12(a), the x-axis is the number of nodes. The y-axis is the averaged expected utility of the defender obtained by the evaluated algorithms. The data value associated with each mission-critical node is generated uniformly at random within $[0, 1]$. Intuitively, the higher averaged expected utility an algorithm gets, the better the solution quality of the algorithm is. Figure 12(a) shows that all of our algorithms, ORANI, ORANI-AttG, and ORANI-G, defeat the baseline algorithms in obtaining a much higher utility for the defender.

In our second experiment (Fig. 12(b)), we examine the convergence of the double oracle used in ORANI. The x-axis is the number of iterations of adding new strategies for both players until convergence. The y-axis is the average of

---

[2] A free Java graph library available at http://jgrapht.sourceforge.net.

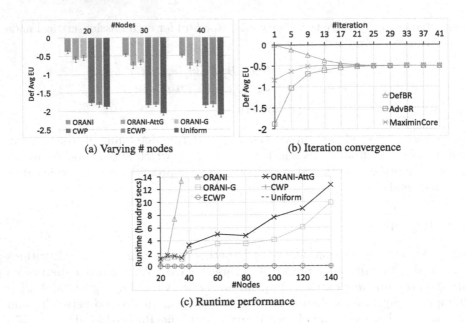

(a) Varying # nodes

(b) Iteration convergence

(c) Runtime performance

Fig. 12. Uni-exfiltration: random scale-free graphs

the defender's expected utility at each iteration with respect to the defender oracle, the attacker oracle, and the Maximin core. The number of nodes in the graph is set to 40. Figure 12(b) shows that ORANI converges quickly, i.e., after approximately 25 iterations. This result implies that there is only a small set of pure strategies involved in the equilibrium despite an exponential number of strategies in total. In addition, ORANI can find this set of pure strategies after a small number of iterations.

In our third experiment (Fig. 12(c)), we investigate runtime performance. In Fig. 12(c), the x-axis is the number of nodes in the graphs and the y-axis is the runtime on average in hundreds of seconds. As expected, the runtime of ORANI grows exponentially when $|V|$ increases. In addition, by using the greedy heuristics, ORANI-AttG and ORANI-G run significantly faster than ORANI. For example, ORANI reaches 1333 s on average when $|V| = 35$ while ORANI-AttG and ORANI-G reach 1266 and 990 s respectively when $|V| = 140$.

Data Broad-Exfiltration. In the case of data broad-exfiltration, we compare eight algorithms: (i) ORABI – both exact oracles; (ii) ORABI-AttG – exact defender oracle and greedy attacker oracle; (iii) ORABI-G – both greedy oracles; (iv) ORABI-AttG-Mul – exact defender oracle and greedy-multi attacker oracle; (v) ORABI-G-Mul – both greedy-multi oracles; and (vi, (vii, (viii) CWP, ECWP, and Uniform.

Our experimental result on solution quality is shown in Fig. 13(a). Figure 13(a) shows that all of our five evaluated algorithms obtain a much higher averaged expected utility for the defender compared to the baseline algorithms.

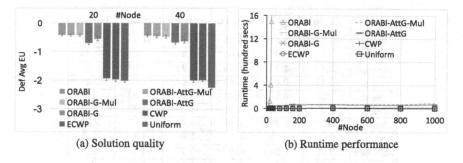

(a) Solution quality   (b) Runtime performance

Fig. 13. Broad-exfiltration: random scale-free graphs

By adding multple new strategies at each iteration, both ORABI-AttG-Mul and ORABI-G-Mul perform approximately as well as ORABI while outperforming ORABI-AttG, and ORABI-G.

In addition, Fig. 13(b) shows that our algorithms with greedy heuristics can scale up to large graphs. For example, when $|V| = 1000$, the runtime of ORABI-AttG-Mul, ORABI-G-Mul, ORABI-AttG, and ORABI-G reaches 89, 20, 71, and 2 s respectively. We conclude that ORABI is the best algorithm for small graphs while ORABI-AttG-Mul and ORABI-G-Mul are proper choices for large-scale graphs.

Finally, we investigate the benefit to the attacker from broad-exfiltration compared to uni-exfiltration. We run ORANI and ORABI on the same set of 50 scale-free graph samples generated by uniformly at random with 20, 30, 40 nodes in each graph respectively. Among all the samples, there are only 58%, 72%, and 52% of the 20-node, 30-node, and 40-node graphs respectively for which the attacker obtains a strictly higher utility by using broad-exfiltration. This result shows that the attacker does not always benefit from broad-exfiltration. Despite broad-exfiltration, the data exchange between any pairs of compromised nodes must follow fixed routing paths specified by the network system, thus constraining the data exfiltration possibilities.

6.4.2 Real-World Network Topology

Our third set of experiments is conducted on real-world network topologies from the Rocketfuel dataset [37]. Overall, the dataset provides router-level topologies of 10 different ISP networks: Telstra, Sprintlink, Ebone, Verio, Tiscali, Level3, Exodus, VSNL, Abovenet, and AT&T. In this set of experiments, we mainly focus on evaluating the solution quality of our algorithms compared with the three baseline algorithms. For each of our experiments, we randomly sample fifty 40-node sub-graphs from every network topology using random walk. In addition, we assume that all external routers located outside the ISP can potentially route data to the attacker's server. Each data point in our experimental results is averaged over 50 different graph samples. The defender's averaged expected utility obtained by the evaluated algorithms is shown in Figs. 14 and 15.

Dataset	ORANI	ORANI-AttG	ORANI-G	CWP	ECWP	Uniform
Telstra	-0.42	-0.44	-0.45	-1.9	-1.94	-2.38
Sprintlink	-0.43	-0.45	-0.45	-1.84	-1.89	-2.36
Ebone	-0.72	-0.73	-0.73	-1.71	-1.75	-2.32
Verio	-0.47	-0.47	-0.47	-1.84	-1.84	-2.25
Tiscali	-0.59	-0.62	-0.61	-1.97	-1.97	-2.2
Level3	-0.63	-0.64	-0.65	-1.85	-1.89	-2.25
Exodus	-0.68	-0.68	-0.68	-1.44	-1.47	-2.34
VSNL	-0.67	-0.68	-0.68	-1.69	-1.78	-2.3
Abovenet	-0.62	-0.64	-0.62	-1.77	-1.77	-2.3
AT&T	-0.31	-0.32	-0.33	-1.91	-1.96	-2.3

Fig. 14. Uni-exfiltration: defender's average utility

Dataset	ORABI	ORABI-AttG-Mul	ORABI-G-Mul	ORABI-AttG	ORABI-G	CWP	ECWP	Uniform
Telstra	-0.41	-0.41	-0.41	-0.41	-0.42	-1.72	-1.78	-2.27
Sprintlink	-0.41	-0.41	-0.41	-0.43	-0.42	-1.72	-1.78	-2.21
Ebone	-0.71	-0.71	-0.71	-0.72	-0.73	-1.58	-1.66	-2.32
Verio	-0.47	-0.47	-0.47	-0.5	-0.5	-1.81	-1.85	-2.26
Tiscali	-0.51	-0.51	-0.51	-0.56	-0.56	-1.88	-1.95	-2.2
Level3	-0.67	-0.67	-0.67	-0.69	-0.68	-1.99	-2.03	-2.37
Exodus	-0.74	-0.74	-0.74	-0.75	-0.75	-1.58	-1.63	-2.37
VSNL	-0.73	-0.73	-0.73	-0.73	-0.73	-1.67	-1.76	-2.38
Abovenet	-0.67	-0.67	-0.68	-0.69	-0.68	-1.81	-1.88	-2.41
AT&T	-0.34	-0.34	-0.34	-0.35	-0.38	-1.88	-1.94	-2.28

Fig. 15. Broad-exfiltration: defender's average utility

Figures 14 and 15 show that all of our algorithms obtain higher defender expected utility than the three baseline algorithms. Further, the greedy algorithms – ORANI-AttG, ORANI-G, and ORABI-AttG, ORABI-G – are shown to consistently perform well on all the ISP network topologies compared to the optimal ones – ORANI and ORABI respectively. In particular, the average expected defender utility obtained by ORANI-G is only ≈ 3% lower than ORANI on average over the 10 network topologies.

7 Bot Identification

To identify and remove bots, we have developed a novel network-based detection scheme, called DeBot, which can identify traffic flows potentially associated with data exfiltration attempts. The proposed solution intercepts traffic from different monitoring points and leverages differences in the network behavior of botnets and benign users to identify suspicious flows. After deploying a number of detectors or monitors as described earlier, we analyze flow characteristics to identify suspicious hosts and use periodogram analysis to identify malicious flows. The fundamental assumption behind the use of periodogram analysis is that exfiltration traffic tends to be relatively more periodic than normal or benign traffic. This approach has been evaluated against different architecturally stealthy botnets in the CyberVAN testbed [7] – developed and maintained by Vencore Labs – and its performance has been compared to two state-of-the-art detection techniques, which we refer to as Stealthy P2P Detector and BSampling. The results indicate that DeBot is effective in detecting botnet activity and mapping out the botnet's architecture, and it outperforms existing solutions with respect to false positive rates. As shown in Fig. 16, the proposed approach to detect exfiltration by stealthy botnets can be divided into four phases: *preprocessing, observation, refinement,* and *analysis.*

In the *preprocessing phase,* we compute the rate at which traffic snapshots should be captured at different monitoring points within the network. We will refer to this rate as the *snapshot rate.* In DeBot, the monitoring period T (e.g., 24 h) is divided into smaller epochs, Δt (e.g., 30 mins). At each epoch in the *observation phase,* the detection mechanism randomly chooses a monitoring point based on the snapshot rates and inspects traffic traversing it during that epoch. During the monitoring period, DeBot maintains a score for each host, which is updated based on the similarity of the host's traffic pattern with other hosts within its neighborhood. At the end of the observation phase, DeBot aggregates the scores for each host based on $\frac{T}{\Delta t}$ traffic snapshots captured from different vantage points. The aggregated score is then used to identify suspicious hosts H_B by comparing the score of each host with the scores of other hosts within its neighborhood.

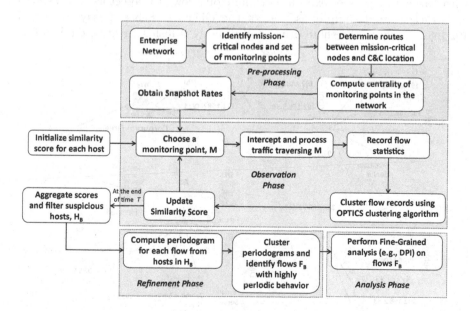

Fig. 16. Overview of DeBot

In the *refinement phase*, DeBot identifies flows corresponding to bot traffic by exploiting the periodic communication pattern between bots. For each host in H_B, it uses periodogram analysis to identify flows that are relatively more periodic than other flows and marks them as suspicious. Then, in the *analysis phase*, suspicious flows are further analyzed using fine-grained analysis techniques such as Deep Packet Inspection.

7.1   Preprocessing Phase

The objective of an exfiltration campaign is to periodically transfer sensitive data to a remote attacker-controlled server. Typically, in an enterprise network, sensitive data is confined to a few servers which we refer to as *mission-critical* servers. Exfiltrated data traverses several intermediate forwarding devices, such as switches, routers and gateways, before reaching the remote server. Any of these internal devices can be used as a monitoring point. In the proposed detection scheme, traffic is mirrored from these devices to a central location for analysis. In large enterprises, a mirroring mechanism is is usually already in place to remotely monitor performance.

Given the sparseness of malicious flows, it is critical to identify monitoring points that are likely to capture such flows. For instance, consider the enterprise network in Fig. 17, with the sensitive data stored in the file servers in Subnet-1 and Subnet-2. The file servers host redundant copies of the data. To exfiltrate this data, an attacker may choose one of several botnet communication architectures with varying degrees of exposure of malicious flows to detectors. For example, an attacker could compromise one of the clients in Subnet-1, say h_1 with IP 192.168.1.2, mount the share drive, and directly transfer sensitive data to C&C.

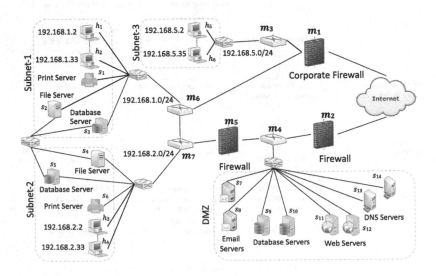

Fig. 17. Enterprise network example

Due to internal routing policies, the traffic traverses two intermediate devices – router m_6 and firewall m_1 – before exiting the network. For the purpose of presentation, we denote this communication architecture as a path, $h_1 \rightarrow m_6 \rightarrow m_1 \rightarrow$ C&C. Other possible exfiltration paths include, but not limited to: $h_3 \rightarrow m_7 \rightarrow m_6 \rightarrow m_1 \rightarrow$ C&C, where h_3 is a compromised host in Subnet-2, $h_1 \rightarrow m_6 \rightarrow m_7 \rightarrow m_5 \rightarrow m_4 \rightarrow s_9 \rightarrow m_4 \rightarrow m_2 \rightarrow$ C&C, in which a database server s_9 was compromised and used as a relay. It can be observed that the percentage of malicious flows intercepted by different monitoring points depends on the internal routing policy and the attacker's choice of communication architecture.

Table 2. Number of flows vs. number of bot flows

Monitoring point	No. of unique flows	No. of unique bot flows	% of bot flows
m_1	5,248	0	0
m_2	149,392	3,451	2.31
m_3	106,448	1,724	1.62
m_4	913,680	7,766	0.85
m_5	690,748	4,352	0.63
m_6	126,156	1,728	1.37
m_7	149,580	3,455	2.31
Total	1,146,784	11,124	0.97

Detecting malicious traffic within large networks calls for a scalable detection mechanism. Although capturing traffic from all monitoring points would ensure that all malicious flows are intercepted, such approach is not scalable. Therefore, it is crucial to identify the most effective set of monitoring points so as to limit the amount of data to be analyzed, whilst ensuring that a sufficient number of malicious flows are intercepted for the detection mechanism to be able to distinguish malicious flows from benign flows.

To understand the impact of different monitoring points on processing time and accuracy of a detection mechanism, we simulated the network scenario of Fig. 17 in the CyberVAN testbed [7], which can generate benign user traffic. In the network of Fig. 17, we consider a stealthy botnet with a communication architecture composed of a server in the DMZ and four compromised hosts, two in Subnet-1 and two in Subnet-2. The bots exfiltrate data from the file server and forward it to the server in the DMZ, which aggregates data and relays it to C&C. Table 2 shows the number of flows intercepted at different monitoring points during a 12-h monitoring period: when m_4 is chosen as a monitoring point, the detection mechanism processes 6 times more records than m_7, while intercepting only twice as many bot flows as m_7. This example shows that the relationship between the volume of traffic monitored and the number of malicious flows intercepted is not linear.

To improve the likelihood of intercepting malicious flows in resource-constrained environments, Venkatesan *et al.* [41] proposed a dynamic monitoring strategy that exploits graph-theoretic properties of the network, which is modeled as a graph $G(V, E)$, with a subset of nodes $M_c \subseteq V$ identified as mission-critical. For each potential monitoring point $m \in M \subset V$, they compute a new centrality measure, known as the *mission-betweenness centrality* $C_M(m)$, which is a function of the fraction of shortest paths between mission-critical nodes and C&C that traverse m. The time horizon is divided into smaller observation epochs and in each epoch a monitoring point m is chosen with probability $\frac{C_M(m)}{\sum_{m' \in M} C_M(m')}$. As the above strategy only considered monitoring points on shortest paths, to improve coverage the authors proposed the *expanded centrality-weighted strategy*, which also considers monitoring points on paths that are δ times longer than the shortest paths.

In this work, we adopt the principle behind the dynamic strategy mentioned above – i.e., choosing monitoring points with high centrality – and modify it to account for internal routing policies. In [41], the authors assumed that traffic between systems is routed through the shortest path. However, an enterprise network is segmented into subnets and the route between any two systems depends on the routing policies at different monitoring points. Such policies are influenced by several factors such as network load and security policies. We use the tracert tool to identify the routes traversed by traffic between two systems s and t and in turn a set of monitoring points that can intercept that traffic. We use $\mathscr{R}$ to denote the set of all routes between systems in a network.

As mentioned earlier, stealthy botnets reduce exposure to detectors by compromising additional systems and using them as proxies to relay traffic to C&C. In an enterprise network, most communication patterns follow a client-server model. Thus, to avoid suspicious patterns, compromised servers could

---

Algorithm 3. $computeSnapshotRates(M, M_c, S, \mathscr{R})$

Input: a set M of potential monitoring points, a set M_c of mission-critical nodes, a set S of potential
proxy servers, and a set R of routes between pairs of nodes in M
Output: the snapshot rate $P(m)$ for each monitoring point $m \in M$
1: **for all** $m_c \in M_c$ **do**
2:     $\mathscr{R}'_{m_c} \leftarrow \emptyset$
3:     **for all** $s \in S$ **do**
4:         $\mathscr{R}'_{m_c} \leftarrow \mathscr{R}'_{m_c} \cup \{R_1 || R_2 \mid (R_1, R_2) \in \mathscr{R}_{m,s} \times \mathscr{R}_{s,\text{C\&C}}\}$
5:     **end for**
6: **end for**
7: $C_B(m) \leftarrow 0, \forall m \in M$ // Initialize mission-betweenness centrality
8: **for all** m_c in M_c **do**
9:     $\sigma(m) \leftarrow 0, \forall m \in M$
10:     **for all** $R \in \mathscr{R}'_{m_c}$ **do**
11:         **for all** $m \in M \cap R$ **do**
12:             $\sigma(m) \leftarrow \sigma(m) + 1$
13:         **end for**
14:     **end for**
15:     $C_B(m) \leftarrow C_B(m) + \frac{\sigma(m)}{|\mathscr{R}'_{m_c}|}, \forall m \in M$
16: **end for**
17: $P(m) \leftarrow \frac{C_B(m)}{\sum_{m' \in M} C_B(m')}, \forall m \in M$
18: **return** P

---

take the role of a proxies for bots within the network. However, we make a conservative assumption that all systems (both clients and servers) can act as proxies for the botnet. This assumption creates a worst-case scenario for the defender, thereby making it challenging to design a detection mechanism. To collect large traffic samples of such botnets, we first compute the centrality of all the monitoring points, similarly to the expanded centrality-weighted strategy in [41].

We use algorithm *computeSnapshotRates* (Algorithm 3) to compute the snapshot rates. For each mission-critical node m_c and potential proxy s, the algorithm first determines all the paths through s by concatenating routes from m to s, $\mathscr{R}_{m,s}$, with routes from s to C&C. Here, we conservatively assume that any destination outside the network is a potential C&C server. The resulting set of routes $\mathscr{R}'_m$ is used to compute the mission-betweenness centrality of each monitoring point (lines 7–16). Finally, the snapshot rate for each monitoring point is computed on line 17. Assuming that the topology of the network remains static during the entire monitoring period, this is a one-time computation. The snapshot rate of a monitoring point $m \in \mathscr{M}$ is the probability that m is chosen by DeBot for analyzing traffic traversing it during an epoch. Randomness introduces uncertainty for the attacker and increases the cost and complexity of establishing a stealthy botnet architecture.

7.2 Observation Phase

DeBot identifies suspicious hosts by comparing their network characteristics with other hosts within their neighborhood. The neighborhood of a host is the set of hosts that are expected to exhibit similar network characteristics in the absence of malicious activity. Hosts whose characteristics deviate from their neighboring hosts are classified as suspicious. In this chapter, without loss of generality, we assume that hosts in the same subnet exhibit similar behavior, thus a host's neighborhood is represented by its subnet. Prior to starting this phase, DeBot initializes the similarity scores of host pairs, which quantify the similarity in the network behavior of any two hosts. At the beginning of each observation epoch $\Delta t_i, i \in \left[1, \frac{T}{\Delta t}\right]$, DeBot selects a monitoring point m_i based on the snapshot rates computed in the preprocessing phase. Traffic traversing m_i during Δt_i is intercepted and statistics of each flow are recorded. In this work, a flow is uniquely identified by the tuple (src, dst, sport, dport, protocol). Flow statistics are used as features to cluster flows, and subsequently update the similarity scores of host pairs based on the number of common clusters between them. Finally, at the end of the time horizon, DeBot identifies suspicious hosts by comparing the aggregate behavior of each host with other hosts within its subnet.

For each flow f, DeBot records the median number of packets sent $(pkts_{sent}(f))$ and received $(pkts_{recv}(f))$, and the median number of bytes sent $(bytes_{sent}(f))$ and received $(bytes_{recv}(f))$ during an epoch Δt_i. We refer to the tuple $\langle pkts_{sent}(f), pkts_{recv}(f), bytes_{sent}(f), bytes_{recv}(f) \rangle$ as the statistics of flow f. A TCP session is considered a flow when a SYN packet is acknowledged by a SYN-ACK packet. However, as the traffic snapshot may include incomplete sessions, a TCP session is included in the flow record table $\mathscr{F}_i$ if at least one packet and its acknowledgment are intercepted during the same epoch. For UDP packets, only flows in which a request is followed by a response are considered.

To identify flows that exhibit similar network behavior, the flow records in $\mathscr{F}_i$ are clustered using the OPTICS clustering algorithm [4], a density-based clustering algorithm that, unlike K-means, can identify arbitrarily shaped clusters by grouping closely-spaced records. OPTICS uses a priority queue to linearly order the input records so that records that are closely-spaced are placed together. In OPTICS, a group of records is identified as a cluster if two conditions hold: (i) it includes at least *minPts* records; and (ii) for any two records in the cluster, there is a sequence of records within the cluster such that every pair of consecutive records is within a distance ϵ. Records that do not belong to any cluster are labeled as noise.

As DeBot operates on traffic snapshots, selecting an optimal value for *minPts* is crucial to ensure that intercepted bot flows form a cluster and are not treated as noise. If *minPts* is too high, the traffic snapshot might not have intercepted a sufficient number of bot flows to form a cluster, whereas, if it is too low, it will lead to creation of multiple clusters. Therefore, the choice of *minPts* is influenced by both the frequency ν of bot communication and the length Δt of the observation window. The relationship between the three variables can be approximated as $minPts = \kappa \cdot \frac{\Delta t}{\nu}$ where $0 < \kappa < 1$ is a constant. In order to limit the number of meaningful clusters, *minPts* is fixed and the length of an observation epoch is expressed as a function of ν, i.e., $\Delta t = \frac{minPts}{\kappa} \cdot \nu$. In order words, the choice of Δt bounds the frequency with which bots can send/receive update messages without losing stealth.

As mentioned earlier, DeBot tracks the similarity in network behavior between hosts. Let $\mathscr{N}(h)$ denote the set of hosts in the neighborhood of host h, and let a scoring function $sim(h_i, h_j)$ quantify the similarity between two hosts h_i and h_j. Before the observation phase, the scoring function is initialized as $sim(h_i, h_j) = 1, \forall h_j \in \mathscr{N}(h_i)$. As noted earlier, in our work all hosts that are in the same subnet as host h are considered its neighbors. Let $\mathscr{C}_k$ denote the set of flow clusters at the end of an observation epoch Δt_k, and let $C_{h_i} \subseteq \mathscr{C}_k$ be the subset of clusters containing flows from/to host h_i. The scoring function is updated as follows:

$$sim(h_i, h_j) = \lambda(m_k, h_i) \cdot \left(\frac{|C_{h_i} \cap C_{h_j}|}{|C_{h_i} \cup C_{h_j}|} \right) + (1 - \lambda(m_k, h_i)) \cdot sim(h_i, h_j) \quad (11)$$

where $\lambda(m_k, h_i)$ is a scalar-valued function that models the rate at which the similarity score is updated. To define this function, we first define the *visibility* of a monitoring point m as the set of hosts whose incoming and outgoing traffic traverses m. For instance, in the network of Fig. 17, the visibility of M_3 is the set of hosts in the subnet 192.168.5.0/24. In DeBot, if a host is not visible to the current monitoring point, then the score is updated at a slower rate. In particular, the $\lambda()$ function is defined as:

$$\lambda(m_k, h_i) = \begin{cases} 0.5 \text{ if } h_i \in visibility(m_k) \\ 0.25 \text{ otherwise} \end{cases}$$

At the end of the observation phase, the aggregate network score of a host is computed as the sum of the similarity scores of the host with hosts in its neighborhood, i.e., $agg\_score(h_i) = \sum_{h_j \in \mathcal{N}(h_i)} sim(h_i, h_j)$. A high aggregate score implies that the host exhibited network characteristics similar to the hosts in its neighborhood while a low score implies that the host's network characteristics deviated from the other hosts. Based on this rationale, a host h_i is identified as suspicious if its aggregate score is less than $\mu_{agg}(\mathcal{N}(h_i)) - \sigma_{agg}(\mathcal{N}(h_i))$, where $\mu_{agg}(\mathcal{N}(h_i))$ and $\sigma_{agg}(\mathcal{N}(h_i))$ are, respectively, the mean and standard deviation of the aggregate scores of hosts in the neighborhood of h_i.

7.3   Refinement Phase

A bot participating in an exfiltration campaign regularly communicates with its peer bots or C&C to send or receive updates. Table 3 shows the observed communication frequency of different instances of POS malware. Such periodic behavior has also been observed in botnets that are known for stealing credentials, such as Storm, Waldec, and Zeus [32].

In DeBot, we leverage the periodic communication feature of bots to identify malicious host pairs. To determine whether a host h_i is periodically communicating with another host h_j, the communication pattern between h_i and h_j is treated as a signal in the time domain and transformed to the frequency domain using Discrete Fourier Transform (DFT). After the transformation, the Power Spectrum Density (PSD) of different frequencies is analyzed and compared with the PSD of other connections generated by host h_i to identify periodic communications. Details are provided in the following subsections.

Table 3. Communication frequency of malware

POS malware	Victim	Period
BlackPOS [24]	Target	10 mins
FrameworkPOS [24]	Home Depot	60 mins + random mins
Backoff [24]	UPS	45 s
Punkey [27]	CiCi's Pizza (suspected)	20 mins or 45 mins

7.3.1   Detecting Periods Using Periodogram Analysis

Let $TS_{i,j} = \{ts_1, ts_2, ...\}$ be the set of timestamps at which a connection was initiated from host h_i to h_j. The monitoring period $[0, T]$ is divided into equally-spaced time points $T_{i,j} = \{t_1, t_2, ..., t_N\}$, where $t_{k+1} - t_k = \Delta s$ and $N = \frac{T}{\Delta s}$. When traffic between two hosts h_i and h_j is continuously monitored, the corresponding connection pattern is treated as a signal that has been sampled at evenly-spaced time intervals, $X_{h_i,h_j}(t_k), \forall t_k \in T_{i,j}$, defined as:

$$X_{h_i,h_j}(t_k) = \begin{cases} 1, \exists ts_l \in TS_{i,j}, \ ts_l \in (t_{k-1}, t_{k+1}) \\ 0, \text{otherwise} \end{cases}$$

A Discrete Fourier Transform (DFT) converts a signal in the time domain to the frequency domain by expressing the signal as a sum of sinusoidal components using the equation:

$$F_{h_i,h_j}(\omega) = \sum_{k=1}^{N} X_{h_i,h_j}(t_k)e^{-i\omega t_k} \tag{12}$$

where $\omega = 1, ..., N$ and $e^{i\theta} = cos(\theta) + i \cdot sin(\theta)$. Essentially, the DFT coefficient, $F_{h_i,h_j}(\omega)$, at frequency ω *correlates* the signal X_{h_i,h_j} with a sequence of sine and cosine waves at frequency ω – the higher the coefficient value, the greater the similarity. The strength of each frequency in the signal is computed by the power spectrum density. Several methods exists to estimate the power spectral density [39]. In this work, we use the periodogram method as it is computationally less expensive than other methods. The periodogram of the time series X_{h_i,h_j} is given by:

$$P_{h_i,h_j}(\omega) = \frac{1}{N}|F_{h_i,h_j}(\omega)|^2 = \frac{1}{N}\left[\left(\sum_{k=1}^{N} X_{h_i,h_j}(t_k)cos\ \omega t_k\right)^2 + \left(\sum_{k=1}^{N} X_{h_i,h_j}(t_k)sin\ \omega t_k\right)^2 \right] \tag{13}$$

Figure 18 shows the periodogram of a sample of Zeus traffic obtained from a public repository [13]. In this network trace, the bot connected with its peer bot every 60 s, which, in the periodogram, is represented by the frequency corresponding to the highest power. Employing the above approach directly in DeBot, however, presents several limitations:

- *Unevenly-spaced observations*: Equation 13 assumes that the traffic being analyzed was sampled at equally-spaced time intervals. However, DeBot employs a dynamic monitoring strategy which only captures snapshots of traffic from different monitoring points. Thus, there may be long periods of unobserved connection patterns between pairs of hosts, and the above periodogram analysis may not accurately estimate the power of different frequencies in the signal.
- *Detecting periodicity*: DFT treats every discrete time series as periodic. Thus, labeling a connection pattern as periodic based on high peaks in the

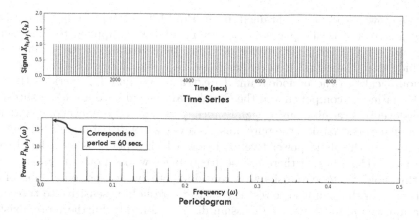

Fig. 18. Communication pattern and periodogram of a Zeus bot

periodogram will lead to a large number of false positives. Furthermore, *random fluctuations* due to noisy data and *spectral leakage* due to finite-length sampling may also produce peaks at frequencies that do not correspond to the true frequency of the signal.

In addition to addressing the above limitations, the detection mechanism should be robust in the following two scenarios, which we also address in the following subsections.

- *Random perturbations*: As malicious flows are detected based on their periodicity, bots can evade detection by introducing random perturbations to the connection pattern.
- *False positives*: Legitimate applications, such as software updates and email clients, also generate periodic flows, which may be misclassified as malicious.

7.3.2 Lomb-Scargle Periodogoram

The dynamic monitoring strategy results in sampling traffic between two hosts h_i, h_j at time points $t_k, k = 1, 2, ..N$ that are not evenly spaced. To study the periodicity of an unevenly-spaced discrete time series, we use the Lomb-Scargle periodogram to estimate the power spectrum [33]. The Lomb-Scargle periodogram modifies the classical periodogram given in Eq. 13 by introducing a time translation parameter τ:

$$P_{h_i,h_j}(\omega) = \frac{1}{2} \cdot \left[\frac{\left(\sum_{k=1}^{N} X_{h_i,h_j}(t_k) \cos \omega(t_k - \tau) \right)^2}{\sum_{k=1}^{N} \cos^2 \omega(t_k - \tau)} + \frac{\left(\sum_{k=1}^{N} X_{h_i,h_j}(t_k) \sin \omega(t_k - \tau) \right)^2}{\sum_{k=1}^{N} \sin^2 \omega(t_k - \tau)} \right]$$

$$(14)$$

where

$$\tau = (1/2\omega) \tan^{-1} \left[\left(\sum_{k=1}^{N} \sin(2\omega t_k) \right) \bigg/ \left(\sum_{k=1}^{N} \cos(2\omega t_k) \right) \right]$$

The power spectrum obtained from Eq. 14 is shown to be statistically equivalent to the least squares fit of a sinusoidal wave applied to the discrete time series [33]. High peaks in the resulting periodogram are not sufficient to conclude that the signal is periodic. Noise in a signal can also produce large spurious peaks in the periodogram. To extract the candidate periods that are due to harmonic components in the signal – and not due to noise – a threshold power-level is determined using significance tests. For a given level of confidence, a significance test models the pure noise as a Gaussian distribution $\mathcal{N}(\mu, \sigma)$ and determines a threshold power-level z_0 below which a power is considered to be pure noise [11]. Thus, if there are no frequencies whose power is greater than the threshold z_0, then the signal is considered to be non-periodic. One of the limitations of the significance test is that the threshold is sensitive to the choice of parameters μ and σ for the Gaussian distribution [11]. Furthermore, existing non-parametric methods [44] are applicable to evenly-spaced time series and, thus, cannot be directly adopted in our setting.

7.3.3 Relative-Periodicity

In this work, instead of checking whether the connection pattern from host h_i to host h_j is periodic using significance test, we determine if it is *relatively* periodic by comparing its periodogram with that of other connection patterns generated by h_i during the monitoring period. Thus, we use the system's typical network behavior (instead of white noise) as the baseline to check for periodicity.

Let $\mathcal{P}_{h_i} = \{P_{h_i,h_j}(\omega)\}$ be the set of periodograms of connection patterns originating from host h_i (obtained using Eq. 14). To determine which periodogram exhibits an anomalously higher periodicity, the periodograms in $\mathcal{P}_{h_i}$ are clustered using agglomerative clustering. While clustering, the difference in periodic structures between two periodograms is assessed using the power distance metric [44]. The power distance between P_{h_i,h_j} and P_{h_i,h_k} is computed by first identifying the set of frequencies $\omega_{i,j}$ with the K-highest powers in P_{h_i,h_j}. The power distance is then defined as:

$$pDist = ||P_{h_i,h_j}(\omega_{i,j}) - P_{h_i,h_k}(\omega_{i,j})||$$

In our evaluation, we set $K = 1000$. To ensure that the total energy is constant, before computing $pDist$, the powers are normalized as follows:

$$X(t) = \frac{X(t) - \frac{1}{N}\sum_{i=1}^{N}X(i)}{\sqrt{\sum_{i=1}^{N}\left(X(t) - \frac{1}{N}\sum_{i=1}^{N}X(i)\right)}}, t = 1, 2, ...N$$

In the proposed hierarchical cluster analysis of periodograms, the linkage criteria between sets of periodograms was computed using the Ward's method. After building a hierarchical structure of the periodograms, clusters are formed by the set of periodograms whose pairwise distance is less than a threshold γ.

The value of γ was set to $0.95 \cdot max_d$ where max_d is the maximum distance between any two sets of periodograms as determined by the Ward's method. Finally, if a cluster contains only one periodogram, the connection pattern of the corresponding host pair is considered to be relatively periodic. The rationale behind this approach is that connection patterns corresponding to bot flows are anomalously more periodic than other connection patterns from the same hosts, thus the corresponding periodogram will form an individual cluster. The host pairs (h_i, h_j) that are identified as relatively periodic are marked as suspicious for further analysis.

Finally, in the analysis phase, flows generated by host pairs marked as suspiciously periodic can be analyzed using fine-grained tools such as Deep Packet Inspection or submitted for manual inspection to the security operations center.

8 Botnet Lifetime

Ultimately, the defender's objective is to eradicate a botnet from the network. Enterprise-scale solutions require protection mechanisms that are both proactive in preventing the propagation of botnets and reactive in detecting and responding to bots already present within the network. To address this need and solve the third challenge mentioned earlier, we propose to deploy—in a defense-in-depth approach—a mix of two classes of countermeasures, namely honeypots and network-based detectors. While honeypots are used to detect intrusion attempts—including a bot's attempt to compromise another machine—network-based detection mechanisms can identify (through behavioral analysis) bots that coexist with benign machines. Both honeypots and network-based detectors can be treated as resources available to the defender in limited supply due to cost constraints.

8.1 Reinforcement Learning Model

To optimally and dynamically deploy these mechanisms in an iterative fashion, and consequently reduce the lifetime of botnets, we developed a solution based on a reinforcement learning model [43]. Reinforcement learning (RL) is an algorithmic method for solving sequential decision-making problems wherein an agent (or decision maker) interacts with the environment to learn how to respond under different conditions [14]. Formally, the agent seeks to discover a policy that maps the system state to an optimal action. In our work, the agent learns a policy that maximizes the total number of bots detected and removed over time. As the location of bots is unknown prior to the deployment of defense mechanisms, the agent estimates the short-term and long-term rewards of an action by monitoring network activity within different network segments. In particular, the agent monitors the behavior of hosts with respect to potential attack indicators (e.g., scanning activity and number of outgoing sessions) to inform the next iteration of detector placement. This approach was compared to three other strategies, namely:

- a static strategy, which does not modify detector placement over time;
- an MTD centrality-based strategy, which periodically alters detector placement based on topology-driven centrality measures;
- a myopic strategy, which makes placement decisions to optimize short-term benefits based on feedback from the operating environment.

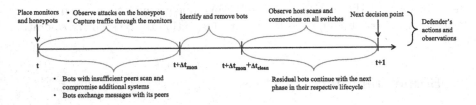

Fig. 19. Timeline of defender's and attacker's actions and observations

The defender's objective is to maximize the number of bots detected and removed using a limited number of resources (honeypots and monitors). In an enterprise, any machine that connects to the target network is susceptible to compromise and subsequent recruitment as a bot. Hence, determining the locations for placing defense mechanisms is critical to detect bots and curb their spread within the network. Furthermore, as bots can propagate through the network, the placement of these defenses must also dynamically change to detect bots in different subnetworks. Due to the evolving nature of the threat, we propose a reinforcement learning approach to guide the defender's sequential decision-making process of placing monitors and honeypots over time.

In our model, we consider an infinite horizon wherein the agent makes decisions on a periodic basis. The time between two consecutive decisions is referred to as an epoch. A timeline with the sequence of events that occur between consecutive decisions is shown in Fig. 19. At each decision point, the agent determines the network segments that will be monitored during the next epoch. At the beginning of an epoch, as described in Sect. 3, bots perform one of two detectable activities, depending on the stage in their respective lifecycle: (i) scanning and subsequently compromising machines within the network (these bots are referred to as *scanning bots*), or (ii) exchanging update messages with their peers and the C&C server (referred to as *transmission bots*). The agent observes the network activity for a time period $\Delta t_{mon} \in [0, 1]$—i.e., for a fraction of the epoch – during which (i) honeypots may be scanned and compromised by scanning bots, and (ii) traffic through the monitors is captured for analysis by a centralized bot detection mechanism. At time $t + \Delta t_{mon}$, the detector processes captured traffic and identifies a set of potential bots. We assume that the network-based detection mechanism is imperfect, with a known true positive rate, while inference based on network activity on honeypots is assumed to be perfect[3], with a true positive rate of 1.

After identifying potential bots, the defender removes them by restoring the corresponding machines to their pristine state. Let $\Delta t_{clean} \in [0,1]$ be the time[4] taken by the detector to process the captured traffic and subsequently remove the identified bots. In a resource-constrained setting with an imperfect detection mechanism, the defender may not have detected all the bots in the network. As a result, undetected bots continue with the next stage in their respective lifecycle. Bots with an insufficient number of peers will scan the network while bots with enough peers will exchange messages. The basic elements of the model are defined below.

Decision Variable. Given N potential monitoring points, for each point the agent may choose one of the following actions, denoted with symbols m, h, b, and e respectively: (m) passively monitor traffic traversing that monitoring point; (h) place a honeypot; (b) place both defense mechanisms; or (e) do nothing. Then, the set of decisions at time t is represented as a vector $x_t = (x_1^t, x_2^t, ..., x_N^t)$, where $x_i^t \in \{m, h, b, e\}$. Note that placing multiple monitors on the same monitoring point does not provide any additional benefit.

System State. The state of the system should capture the location of bots within the network. However, as the location of bots is unknown prior to the placement of defense mechanisms, we derive the state of the system by observing attack indicators in different segments of the network. Anomalous behaviors – such as a large number of unsuccessful login attempts, increase in the number of host scans, and a large number of outgoing sessions – are some of the most common symptoms of an ongoing attack [47]. Thus, in our model, we determine the potential locations of bots by observing anomalous behaviors in different segments of the network. In particular, to estimate the number of bots in different segments of the network, we track the total number of host scans and the total number of sessions that were recorded since the latest removal of bots from the network, i.e., in the time period $[t + \Delta t_{mon} + \Delta t_{clean}, t + 1)$ in Fig. 19. These features can be observed at all monitoring points – not just those where defense mechanisms have been deployed – with very low overhead.

In a network with N monitoring points, the state S_t of the system at any time t can be defined as a $2N$-dimensional vector $(\psi_1^h, \psi_1^s, \psi_2^h, \psi_2^s, ..., \psi_N^h, \psi_N^s)$, where ψ_i^h and ψ_i^s are, respectively, the host scans state and the sessions state of monitoring point i, with $i \in [1, N]$. In this work, we model the host scans state and the sessions state of each monitoring point as either LOW, MEDIUM or HIGH. In the presence of benign network activity, determining the accurate state of each feature (host scans or sessions) at different monitoring points is challenging. To address this issue, the defender must first establish a baseline behavior for each feature, for example by counting how many times a feature is observed during the time period $[t + \Delta t_{mon} + \Delta t_{clean}, t + 1)$. If μ_i^f and σ_i^f are

---

[3] Attempted access to a honeypot can be assumed an indicator of malicious activity.
[4] Both Δt_{mon} and Δt_{clean} are defined as a fraction of an epoch.

the mean and standard deviation of each feature $f \in \{h, s\}$ at monitoring point i, then the state at any given time t could be defined as:

$$\psi_i^f(t) = \begin{cases} HIGH, & \text{if } Total_i^f(t) \geq \mu_i^f + \sigma_i^f \\ MED, & \text{if } Total_i^f(t) \in (\mu_i^f - \sigma_i^f, \mu_i^f + \sigma_i^f) \\ LOW, & \text{if } Total_i^f(t) \leq \mu_i^f - \sigma_i^f \end{cases} \quad (15)$$

where, $Total_i^f(t)$ is total number of observations of feature f that were recorded during the time period $[t + \Delta t_{mon} + \Delta t_{clean}, t + 1)$. In the following, when t is clear from the context, we will use ψ_i^f instead of $\psi_i^f(t)$. The intuition behind Eq. 15 is that any large deviation from the expected behavior is considered to be anomalous. It must be noted that the objective of this work is *not* to design a specific bot detector, but rather to develop a strategy for placing defense mechanisms to enable enterprise-scale botnet detection and mitigation. While fine-tuning the definition of ψ_i^f will yield more accurate results, it is beyond the scope of this work.

Reward Function. In an RL model, the choice of an optimal action is influenced by the immediate reward $R(S_t, x_t)$ of an action. Here, the reward of an action is defined as the number of bots that are correctly identified. However, taking an action x_t at time t, when the system is in state S_t, yields a reward that is measured at a later time, $t + \Delta t_{mon} + \Delta t_{clean}$. This is a class of time-lagged information acquisition problems, where we do not know the value of the current state until it is updated after the uncertainty in the bot activity is revealed. Therefore, the immediate reward of an action is *estimated* by using information from recent observations. Such problems occur in real world, such as when travel and hotel reservation decisions are done today for a future date and the value of making such decisions is unknown until the date has occurred [12,29,31].

We estimate the number of bots in a network segment by determining the number of hosts that have deviated from the expected behavior. Similar to the motivation behind deriving the state of the system, the defender first establishes a baseline for the network activity of each machine across all monitoring points. Let $\mu_{mc,i}^f$ and $\sigma_{mc,i}^f$ be the mean and standard deviation of feature f for machine mc when observed from monitoring point i. We consider a simple threshold scheme to decide whether a machine is a potential bot. Given any machine mc and a monitoring point i, if $Total_{mc,i}^f(t)$ is the total number of observations of feature f that were recorded during the time period $[t + \Delta t_{mon} + \Delta t_{clean}, t + 1)$, then machine mc is considered a potential bot if and only if $(\exists i \in [1, N])(Total_{mc,i}^f(t) \geq \mu_{mc,i}^f + 3 \cdot \sigma_{mc,i}^f)$. It should be noted that this rule to identify suspicious machines can be modified based on the specific settings of the target network and does not limit the generality of the proposed RL model.

Post-decision System State. The post-decision system state, S_t^x, is the state to which the system transitions after the decision x_t is taken. Similar to the reward function, the change in the state of the system can only be observed at a later time, in this case $t + 1$. Therefore, we *estimate* the post-decision state of the system by determining the *expected* effect of a decision.

Our estimation is based on the rationale that the objective of placing a defense mechanism at a monitoring point is to remove bots from that portion of the network by identifying machines that exhibit anomalous behaviors. In particular, suppose that machines $mc_1, mc_2, ..., mc_k$ are identified as potential bots from the monitoring point i due to deviations w.r.t. a feature f. Then, placing a defense mechanism at time t (a honeypot if f is the host scans count, or a network-based detector if f is the sessions count) is expected to confirm, after a monitoring period Δt_{mon}, whether the suspected bots are actually bots and, if so, restore the corresponding machines $mc_j, j \in [1, k]$ to their pristine state. As a result of the cleaning process, the agent expects to record $\hat{\mu}^f_{mc_j,i}, \forall j \in [1, k]$ observations of feature f at the monitoring point i during the time $[t + \Delta t_{mon} + \Delta t_{clean}, t + 1)$. Assuming that the machines that are not expected to be affected by this decision continue with their latest recorded behavior (i.e., the behavior exhibited during $[t - 1 + \Delta t_{mon} + \Delta t_{clean}, t))$, then the new post-decision state of monitoring point i for a feature f can be obtained by using Eq. 15, where the estimated total number of observations of feature f is given by $\widehat{Total}^f_i(t + 1) = \sum_{j \in [1,k]} \hat{\mu}^f_{mc_j,i} + \sum_{j \notin [1,k]} \mu^f_{mc_j,i}$, with $\hat{\mu}^f_{mc_j,i}$ being the estimated behavior of machine mc_j after the placement of a defense mechanism. It must be noted that, since the baseline values $(\mu^f_{mc_j,i}, \sigma^f_{mc_j,i})$ of all machines at different monitoring points are established as a preprocessing step, the post-decision state reached by a system due to an action can be obtained immediately.

Exogenous Information. The exogenous information, or uncertainty, B_{t+1} is the information from the environment that is acquired after decision x_t. The uncertainty is attributed to the co-existence of benign and malicious behavior within the network, making it challenging to model the evolution of bots. In the RL model, the uncertainty is captured by observing network activity and extracting features from different monitoring points.

State Transition Function. The state transition function, defined as $S_{t+1} = \tau(S_t, x_t, B_{t+1})$, captures how the system state evolves. However, due to the absence of a model to predict B_{t+1}, the state transition probabilities are unknown. Hence, a reinforcement learning based approach is used to study the evolution of the system state.

Objective Function. The objective function is defined as the long-run total discounted value of the states $V^j(S)$ as the iteration index $j \to \infty$, which is derived using the recursive Bellman's optimality equation [6] below (Eq. 16). Here, $V^j(S)$ is the cumulative sum of discounted $R(S_t, x_t)$ rewards for the learning phase, whose iterations are indexed from 1 to j. In this work, we consider a 365-day cycle in which decisions are made at the start of each day. The learning phase goes through several iterations (indexed with j) of 365-day cycles. As the value of a state is measured in terms of number of correctly identified bots, the objective function will be to maximize the long-run total discounted value of the states $V^j(S)$: the higher the value of $V^j(S)$, the better the system state. The model strives to transition from one good state to another by making a decision that is guided by the highest value of the estimated future states that are reachable at any given time t.

8.2 Phases of Reinforcement Learning

RL achieves the objective through three phases, namely, exploration, learning, and learned. The recursive Bellman's optimality equation that updates the value of the states is given as follows:

$$V^j(\hat{S}^x_{t-1}) = (1 - \alpha^j)V^j(\hat{S}^x_{t-1}) + \alpha^j \nu^j \tag{16}$$

$$\nu^j = \left[\max_{x_t \in \mathscr{X}} \left\{R(S_t, x_t) + \beta V^j(\hat{S}^x_t)\right\}\right] \tag{17}$$

where $V^j(S)$ denotes the value of state S at the j-th iteration, $\hat{S}^x_t$ is the estimated post-decision state reached by the system at state S_t under the action x_t, α^j is the learning parameter that is decayed gradually, $\mathscr{X}$ is the set of all feasible decisions from which the model will choose a decision at every iteration, and β is the fixed discount factor that allows the state values to converge in a long run. It should be noted that the value of the estimated post-decision state $\hat{S}^x_{t-1}$ is updated at time t (in Eq. 17) using the estimated reward function and the value of the estimated post-decision states that can be reached under different actions. In a classical RL formulation, the immediate real rewards and the immediate value of the post-decision states at time t are known; hence, the value of the post-decision state at time $t-1$ can be updated using Eqs. 16 and 17. However, as both the rewards and post-decision states are estimated, we update the value of the post-decision state only after the real reward for an action is observed.

A snapshot of the state-transition diagram is shown in Fig. 20, in which U_t denotes the uncertainty (before and after removing bots) after taking an action, and O_{t+1} denotes the features observed at different monitoring points after removing the bots. The bot removal stage is denoted by E_{t+1}. In this model, during the learning phase, the choice of an action x_t when the system is in state S_t is determined by the estimated reward function $R(S_t, x_t)$. After taking the action x_t, the uncertainty U_{t+1} unfolds, transitioning the system to the state S_{t+1}. As the uncertainty U_{t+1} unfolds (shown in trapezoidal box in

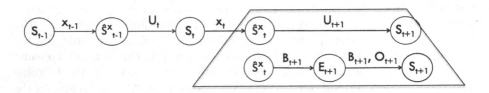

Fig. 20. State transition diagram

Algorithm 4. Exploration and Learning Phase

Input: Baseline values ψ_i^f of each feature $f \in \{h, s\}$ for each monitoring point i, baseline values $\psi_{mc,i}^f$ for each machine mc, decision space $\mathscr{X}$, initial learning parameter $\alpha^0 = 0.8$ at time $t = 0$, discount parameter $\beta = 0.95$, number of iterations for learning $J = 1000$.

Output: State value function, $V(S), \forall S$

```
 1:  V(S) ← 0, ∀S
 2:  for all j = {1, ..., J} do
 3:      if j ≤ 0.3 · J then
 4:          Phase ← Exploration
 5:      else
 6:          Phase ← Learning
 7:      end if
 8:      for all t = {1, ..., 365} do
 9:          Observe features from the monitoring points and determine state St
10:          if Phase = Exploration then
11:              Choose a random defense placement decision, xt
12:          else
13:              Estimate immediate reward R(St, xt) and post-decision state Ŝtx, as described in
                 Section 8.1, ∀xt ∈ X
14:              Choose the action xt' that gives the maximum value in Eq. 17
15:          end if
16:          if t > 2 then
17:              Observe the real reward at t + Δtmon + Δtclean
18:              Decay the learning parameter, αj = αj/(1+e), where e = j²/(1.25·10¹⁴+j)  // see [14]
19:              Update value of post-decision state Vj(Ŝt-1x) using Eq. 16 and Eq. 17 with the real
                 reward and the value of αj
20:          end if
21:      end for
22:  end for
```

Fig. 20), bots are removed at stage E_{t+1} and the agent observes the real reward which is then used to update the value of the estimated post-decision state $\hat{S}_{t-1}^x$. The three phases of learning are described below.

Exploration Phase. In this phase, the RL algorithm explores several non-optimal decisions and acquires the value of the system states that are visited. As described in Algorithm 4, Eq. 16 is used without the max operator in Eq. 17 by taking random decisions for placing defense mechanisms, and the values of $V^j(S_t^x)$ and $V^j(\hat{S}_{t-1}^x)$ are taken from the previously stored values, if the state was visited, or set to 0 otherwise. Since the algorithm begins with $V^0(S) = 0, \forall S$ at $j = 0$, exploration helps to populate the values of some of the states that are visited. Exploration is stopped after a certain number of iterations, which depends on the size of the state-space and the number of iterations planned for the learning phase. In our simulation, we stopped exploration after 30% of the total number of iterations. The idea here is to explore as many states as possible during the learning phase. A low value of this parameter would imply not enough states being explored, whereas a high value would lead to non-convergence of state values during the learning phase. Thus, our choice is reasonable and quite common for this class of problems.

Learning Phase. In this phase, the algorithm takes near-optimal decisions at time t, which are obtained from Eq. 17 with the max operator (lines $13 - 14$ of Algorithm 4). The value of the post-decision state at time $t - 1$ is updated at time $t + 1$ as per Eq. 16 with the real rewards. After several iterations, learning is stopped when convergence of the value of the states is achieved, as measured in terms of the mean-square error of the stochastic gradient [31].

Learned Phase. This is the implementation phase of the RL. The inputs to this phase include the value of the states at the time when learning was terminated and the estimated reward function. In this phase, the RL algorithm takes optimal decisions at each time t, which is obtained from Eq. 17 with the max operator. The algorithm then evaluates all its feasible actions and chooses an action that takes the system to the post-decision state with the highest value.

8.3   Simulation Results

Differently from the myopic strategy, the reinforcement learning strategy also considers long-term benefits of possible detector placements. Figure 21 shows how the number of bots in the network changes over time for each of the four strategies considered. If the number of bots reaches 0, then the botnet has been completely removed from the system. As expected, the static placement strategy exhibits the worst performance: once a botnet is established, bots that are not observable by the static detectors can persist in the network indefinitely. MTD strategies, on the other hand, provide significantly better protection as they introduce uncertainty about the location of detectors. The myopic strategy shows

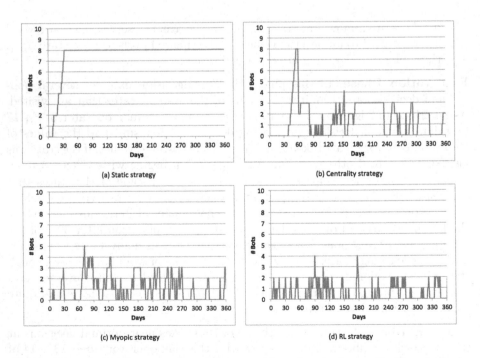

Fig. 21. Comparison between the reinforcement learning strategy and other strategies

a significant improvement over the centrality strategy in reducing the lifetime of the botnet, because it also considers information obtained from the network. Finally, among the MTD strategies, the RL approach shows the largest reduction in the botnet's lifetime.

9   Conclusions and Future Work

Stealthy botnets pose significant threats due to their ability to evade traditional defenses and persist in the target system for extended periods of time. Detecting and mitigating these threats is a multifaceted problem that calls for novel solutions to address the several interrelated challenges that stealthy bots introduce. We have shown how the problem can be decomposed into relatively simpler problems that can be tackled separately, while keeping the big picture in mind. Essentially, we need to understand where and when to monitor for potentially suspicious activity, how to look at observed traffic to identify potentially compromised machines, and how to ensure that each and every bot has been removed from the network. Moving Target Defense (MTD) has proved to be a viable and promising approach in tackling these challenges, enabling us to achieve some interesting results. Of course, the security benefits of deploying MTD techniques come at a cost for the defender, which can be measured in terms of increased overhead to maintain availability for legitimate users. The tradeoff between security and cost can generally be controlled by configuring the parameters of an MTD technique.

In this chapter, we have presented the key findings of our work on disrupting stealthy botnets through the use of a novel moving target defense approach. Specifically, we have targeted botnets that are being used for exfiltrating sensitive data from mission-critical systems. Defending against such botnets is challenging, as research has shown how they have become increasingly sophisticated and have the capability of operating in stealth mode by minimizing their footprint.

In order to defeat exfiltration attempts by modern botnets, we have proposed a moving target defense approach for placing detectors across the network – in a resource-constrained environment – and dynamically and continuously changing the placement of detectors over time. Specifically, we have proposed several strategies based on centrality measures that capture important properties of the network. Our objective is to increase the attacker's effort and likelihood of detection by creating uncertainty about the location of detectors and forcing the botmaster to perform additional actions in an attempt to create detector-free paths through the network.

We have presented two metrics to evaluate the proposed strategies – namely the *minimum detection probability* and the *attacker's uncertainty* – and an algorithm to compute the minimum detection probability. We validated our approach through simulations, and the results confirmed that the proposed solution can effectively reduce the likelihood of successful exfiltration campaigns.

As part of our work on optimal detector placement, we also proposed a Stackelberg game model that accounts for the strategic response of attackers to deployed defenses. We proposed two double-oracle based algorithms, ORANI and ORABI, to compute optimal defense strategies with respect to data uni-exfiltration and broad-exfiltration formulations, respectively. We also provided greedy heuristics to approximate the defender and the attacker best-response oracles. We conducted experiments based on both random scale-free graphs and real-world ISP network topologies, demonstrating the advantages of our game-theoretic solution over previous strategies.

To identify and remove bots, we have developed a novel network-based detection scheme, called DeBot, which can identify traffic flows potentially associated with data exfiltration attempts. The proposed solution intercepts traffic through deployed detectors and leverages differences in the network behavior of botnets and benign users to identify suspicious flows. We analyze the characteristics of traffic flows to identify suspicious hosts and use periodogram analysis to identify malicious flows. The fundamental assumption behind the use of periodogram analysis is that exfiltration traffic tends to be relatively more periodic than normal or benign traffic. This approach has been evaluated against different architecturally stealthy botnets in the CyberVAN testbed and its performance has been compared to two state-of-the-art detection techniques. The results indicate that DeBot is effective in detecting botnet activity and outperforms existing solutions with respect to false positive rates.

Finally, to achieve the defender's ultimate objective of eradicating a botnet from the network, we proposed to deploy – in a defense-in-depth approach – a mix of two classes of countermeasures, namely honeypots and network-based detectors. While honeypots are used to detect intrusion attempts – including a bot's attempt to compromise another machine – network-based detection mechanisms can identify (through behavioral analysis) bots that coexist with benign machines. Both honeypots and network-based detectors can be treated as resources available to the defender in limited supply due to cost constraints. To optimally and dynamically deploy these mechanisms in an iterative fashion, and consequently reduce the lifetime of botnets, we developed a solution based on a reinforcement learning model.

Our future plans include but are not limited to: (i) introducing a probabilistic model to account for false negatives in the deployed detectors; (ii) defining and evaluating the performance of the proposed detector placement strategies against more sophisticated attacker's strategies; and (iii) casting the model in a game-theoretic framework to study the Nash equilibria and dominant strategies.

References

1. APT1: Exposing one of China's cyber espionage units. Technical report, Mandiant, February 2013
2. Lateral movement: how do threat actors move deeper into your network? Technical report, Trend Micro (2013)
3. Alpcan, T., Başar, T.: An intrusion detection game with limited observations. In: Proceedings of the 12th International Symposium on Dynamic Games and Applications (ISDG 2006), Sophia-Antipolis, France, July 2006

4. Ankerst, M., Breunig, M.M., Kriegel, H.P., Sander, J.: OPTICS: ordering points to identify the clustering structure. In: Proceedings of the 1999 ACM SIGMOD International Conference on Management of Data (SIGMOD 1999), pp. 49–60. ACM, Philadelphia, May 1999
5. Beigi, E.B., Jazi, H.H., Stakhanova, N., Ghorbani, A.A.: Towards effective feature selection in machine learning-based botnet detection approaches. In: Proceedings of the IEEE Conference on Communications and Network Security (IEEE CNS 2014), pp. 247–255. IEEE, San Francisco, October 2014
6. Bellman, R.E.: Dynamic Programming. Princeton University Press, Princeton (1957)
7. Chadha, R., et al.: CyberVAN: a cyber security virtual assured network testbed. In: Proceedings of the 2016 IEEE Military Communications Conference (MILCOM 2016), pp. 1125–1130. IEEE, Baltimore, November 2016
8. Collins, M.P., Shimeall, T.J., Faber, S., Janies, J., Weaver, R., Shon, M.D., Kadane, J.B.: Using uncleanliness to predict future botnet addresses. In: Proceedings of the 7th ACM SIGCOMM Internet Measurement Conference (IMC 2007), pp. 93–104. ACM, San Diego, October 2007
9. Faloutsos, M., Faloutsos, P., Faloutsos, C.: On power-law relationships of the Internet topology. ACM SIGCOMM Comput. Commun. Rev. **29**(4), 251–262 (1999)
10. Fredman, M.L., Tarjan, R.E.: Fibonacci heaps and their uses in improved network optimization algorithms. J. ACM (JACM) **34**(3), 596–615 (1987)
11. Frescura, F.A.M., Engelbrecht, C.A., Frank, B.S.: Significance tests for periodogram peaks, June 2007. https://arxiv.org/abs/0706.2225
12. Ganesan, R., Jajodia, S., Shah, A., Cam, H.: Dynamic scheduling of cybersecurity analysts for minimizing risk using reinforcement learning. ACM Trans. Intell. Syst. Technol. **8**(1) (2016)
13. García, S., Grill, M., Stiborek, J., Zunino, A.: An empirical comparison of botnet detection methods. Comput. Secur. **45**, 100–123 (2014)
14. Gosavi, A.: Simulation-Based Optimization: Parametric Optimization Techniques and Reinforcement Learning, Operations Research/Computer Science Interfaces, vol. 55, 2nd edn. Springer, New York (2003)
15. Gu, G., Perdisci, R., Zhang, J., Lee, W.: BotMiner: clustering analysis of network traffic for protocol- and structure-independent botnet detection. In: Proceedings of the 17th USENIX Security Symposium (USENIX Security 2008), pp. 139–154. USENIX Association, San Jose, July 2008
16. Gu, G., Porras, P., Yegneswaran, V., Fong, M., Lee, W.: BotHunter: detecting malware infection through IDS-driven dialog correlation. In: Proceedings of the 16th USENIX Security Symposium (USENIX Security 2007), pp. 167–182. USENIX Association, August 2007
17. Jain, M., Korzhyk, D., Vaněk, O., Conitzer, V., Pěchouček, M., Tambe, M.: A double oracle algorithm for zero-sum security games on graphs. In: Proceedings of the 10th International Conference on Autonomous Agents and MultiAgent Systems (AAMAS 2011), pp. 327–334. IFAAMAS, Taipei, May 2011
18. Jajodia, S., Ghosh, A.K., Swarup, V., Wang, C., Wang, X.S. (eds.): Moving Target Defense: Creating Asymmetric Uncertainty for Cyber Threats, Advances in Information Security, vol. 54. Springer, New York (2011). https://doi.org/10.1007/978-1-4614-0977-9
19. Kaspersky Labs: Kaspersky lab and ITU research reveals new advanced cyber threat, May 2012. http://usa.kaspersky.com/about-us/press-center/press-releases/kaspersky-lab-and-itu-research-reveals-new-advanced-cyber-threat

20. Khalil, K., Qian, Z., Yu, P., Krishnamurthy, S., Swam, A.: Optimal monitor placement for detection of persistent threats. In: Proceedings of the IEEE Global Communications Conference (IEEE GLOBECOM 2016). IEEE, Washington, DC, December 2016
21. Kiekintveld, C., Jain, M., Tsai, J., Pita, J., Ordóñez, F., Tambe, M.: Computing optimal randomized resource allocations for massive security games. In: Proceedings of the 8th International Conference on Autonomous Agents and Multi-Agent Systems, pp. 689–696. IFAAMAS, Budapest, May 2009
22. Korzhyk, D., Yin, Z., Kiekintveld, C., Conitzer, V., Tambe, M.: Stackelberg vs. Nash in security games: an extended investigation of interchangeability, equivalence, and uniqueness. J. Artif. Intell. Res. **41**(2), 297–327 (2011)
23. Manadhata, P.K., Wing, J.M.: An attack surface metric. IEEE Trans. Softw. Eng. **37**(3), 371–386 (2011)
24. Marschalek, M., Kimayong, P., Gong, F.: POS malware revisited - look what we found inside your cashdesk. Cyphort labs special report, Cyphort, Inc. (2014)
25. McMahan, H.B., Gordon, G.J., Blum, A.: Planning in the presence of cost functions controlled by an adversary. In: Proceedings of the 20th International Conference on Machine Learning (ICML 2003), pp. 536–543. AAAI Press, Washington DC, August 2003
26. Medina, A., Lakhina, A., Matta, I., Byers, J.: BRITE: an approach to universal topology generation. In: Proceedings of the 9th International Symposium on Modeling, Analysis and Simulation of Computer and Telecommunication Systems, pp. 346–353. IEEE, Cincinnati, August 2001
27. Merritt, E.: New POS malware emerges - Punkey, April 2015. https://www.trustwave.com/Resources/SpiderLabs-Blog/New-POS-Malware-Emerges--Punkey/
28. Moreira Moura, G.C.: Internet Bad Neighborhoods. Ph.D. thesis, University of Twente, The Netherlands, March 2013
29. Nascimento, J.M., Powell, W.B.: An optimal approximate dynamic programming algorithm for the lagged asset acquisition problem. Math. Oper. Res. **34**(1), 210–237 (2009)
30. Nguyen, T.H., Wellman, M.P., Singh, S.: A stackelberg game model for botnet data exfiltration. In: Rass, S., An, B., Kiekintveld, C., Fang, F., Schauer, S. (eds.) GameSec 2017. LNCS, vol. 10575, pp. 151–170. Springer, Vienna (2017). https://doi.org/10.1007/978-3-319-68711-7_9
31. Powell, W.B.: Approximate Dynamic Programming: Solving the Curses of Dimensionality, 2nd edn. Wiley, Hoboken (2011)
32. Rossow, C., Andriesse, D., Werner, T., Stone-Gross, B., Plohmann, D., Dietrich, C.J., Bos, H.: SoK: P2PWNED - modeling and evaluating the resilience of peer-to-peer botnets. In: Proceedings of the 2013 IEEE Symposium on Security and Privacy (S&P 2013), pp. 97–111. IEEE, Berkeley (2013)
33. Scargle, J.D.: Studies in astronomical time series analysis. ii-statistical aspects of spectral analysis of unevenly spaced data. Astrophys. J. **263**, 835–853 (1982)
34. Schmidt, S., Alpcan, T., Albayrak, Ş., Başar, T., Mueller, A.: A malware detector placement game for intrusion detection. In: Lopez, J., Hämmerli, B.M. (eds.) CRITIS 2007. LNCS, vol. 5141, pp. 311–326. Springer, Heidelberg (2008). https://doi.org/10.1007/978-3-540-89173-4_26
35. Shinoda, Y., Ikai, K., Itoh, M.: Vulnerabilities of passive internet threat monitors. In: Proceedings of the 14th USENIX Security Symposium (USENIX Security 2005), pp. 209–224. USENIX Association, Baltimore, August 2005

36. Shmatikov, V., Wang, M.H.: Security against probe-response attacks in collabo-
    rative intrusion detection. In: Proceedings of the 2007 Workshop on Large Scale
    Attack Defense, pp. 129–136. ACM, Kyoto, August 2007
37. Spring, N., Mahajan, R., Wetherall, D., Anderson, T.: Measuring ISP topologies
    with Rocketfuel. IEEE/ACM Trans. Netw. **12**(1), 2–16 (2004)
38. Stinson, E., Mitchell, J.C.: Towards systematic evaluation of the evadability of
    bot/botnet detection methods. In: Proceedings of the 2nd USENIX Workshop on
    Offensive Technologies. USENIX Association, San Jose, July 2008
39. Stoica, P., Moses, R.L.: Introduction to Spectral Analysis, 1st edn. Prentice Hall,
    Upper Saddle River (1997)
40. Sweeney, P.J.: Designing effective and stealthy botnets for cyber espionage and
    interdiction: finding the cyber high ground. Ph.D. thesis, Thayer School of
    Engineering, Darthmouth College, August 2014
41. Venkatesan, S., Albanese, M., Cybenko, G., Jajodia, S.: A moving target defense
    approach to disrupting stealthy botnets. In: Proceedings of the 3rd ACM Workshop
    on Moving Target Defense (MTD 2016), pp. 37–46. ACM, Vienna, October 2016
42. Venkatesan, S., Albanese, M., Jajodia, S.: Disrupting stealthy botnets through
    strategic placement of detectors. In: Proceedings of the 3rd IEEE Conference
    on Communications and Network Security (IEEE CNS 2015), pp. 55–63. IEEE,
    Florence, September 2015. Best Paper Runner-up Award
43. Venkatesan, S., Albanese, M., Shah, A., Ganesan, R., Jajodia, S.: Detecting
    stealthy botnets in a resource-constrained environment using reinforcement learn-
    ing. In: Proceedings of the 4th ACM Workshop on Moving Target Defense (MTD
    2017), pp. 75–85. ACM, Dallas, October 2017
44. Vlachos, M., Yu, P., Castelli, V.: On periodicity detection and structural periodic
    similarity. In: Proceedings of the 5th SIAM International Conference on Data
    Mining (SDM 2005), pp. 449–460. SIAM, Newport Beach, April 2005
45. Wang, Y., Wen, S., Xiang, Y., Zhou, W.: Modeling the propagation of worms in
    networks: a survey. IEEE Commun. Surv. Tutorials **16**(2), 942–960 (2014)
46. Wellman, M.P., Prakash, A.: Empirical game-theoretic analysis of an adaptive
    cyber-defense scenario (preliminary report). In: Poovendran, R., Saad, W. (eds.)
    GameSec 2014. LNCS, vol. 8840, pp. 43–58. Springer, Cham (2014). https://doi.
    org/10.1007/978-3-319-12601-2_3
47. West, M.: Preventing system intrusions. In: Network and System Security, pp.
    29–56, , 2nd edn. Syngress (2014)
48. Zeng, Y., Hu, X., Shin, K.G.: Detection of botnets using combined host- and
    network-level information. In: Proceedings of the IEEE/IFIP International Con-
    ference on Dependable Systems and Networks (DSN 2010), pp. 291–300. IEEE,
    Chicago, June 2010
49. Zhang, J., Perdisci, R., Lee, W., Luo, X., Sarfraz, U.: Building a scalable system
    for stealthy P2P-botnet detection. IEEE Trans. Inf. Forensics Secur. **9**(1), 27–38
    (2014)

Optimizing Alert Data Management Processes at a Cyber Security Operations Center

Rajesh Ganesan[1]([✉]), Ankit Shah[1], Sushil Jajodia[1], and Hasan Cam[2]

[1] Center for Secure Information Systems, George Mason University,
Mail Stop 5B5, Fairfax, VA 22030-4422, USA
{rganesan,ashah20,jajodia}@gmu.edu
[2] Army Research Laboratory, 2800 Powder Mill Road,
Adelphi, MD 20783-1138, USA
hasan.cam.civ@mail.mil

Abstract. Alert data management is one of the top functions performed by a Cyber Security Operation Centers (CSOC). This chapter is focused on the development of an integrated framework of several tasks for alert data management. The tasks and their execution are sequenced as follows: (1) determining the regular analyst staffing of different expertise level for a given alert arrival/service rate, and scheduling of analysts to minimize risk, (2) sensor clustering and dynamic reallocation of analysts-to-sensors, and (3) measuring, monitoring, and controlling the level of operational effectiveness (LOE) with the capability to bring additional analysts as needed. The chapter presents several metrics for measuring the performance of the CSOC, which in turn drives the development of various optimization strategies that optimize the execution of the above tasks for alert analysis. It is shown that the tasks are highly inter-dependent, and must be integrated and sequenced in a framework for alert data management. For each task, results from simulation studies validate the optimization model and show the effectiveness of the modeling and algorithmic strategy for efficient alert data management, which in turn contributes to optimal overall management of the CSOCs.

1 Introduction

The desiderata of a CSOC enterprise can broadly be structured into the following major elements: (1) all alerts must be investigated in a timely manner, (2) resources (analysts) must be optimally managed, and (3) desired performance must be achieved. Alert data management of a CSOC consists of several tasks that influence the above elements, and it is imperative that the tasks are optimized to achieve the best CSOC performance. Among the different tasks at a CSOC, this chapter presents three most important tasks, and shows how they are inter-dependent, integrated, sequenced, and optimized. These tasks include (1) determining the regular analyst staffing of different expertise level for a given

© Springer Nature Switzerland AG 2019
S. Jajodia et al. (Eds.): Adaptive Cyber Defense, LNCS 11830, pp. 206–231, 2019.
https://doi.org/10.1007/978-3-030-30719-6_9

alert arrival/service rate, and scheduling of analysts to minimize risk, (2) sensor clustering and dynamic reallocation of analysts-to-sensors, and (3) measuring, monitoring, and controlling the level of operational effectiveness (LOE) with the capability to bring additional analysts as needed.

The framework for alert data management system is shown in Fig. 1. The framework consists of the CSOC alert analysis block, which is central to the system. In this block the alerts are analyzed as per Fig. 2, which is described in Sect. 2. The inputs to the central block are analyst-to-sensor allocation and the alerts per sensor generated by the IDS and any backlog from the previous shift of operation. The outputs include the determination of innocuous and significant alerts, which are further processed as shown in Fig. 2. Additionally, the backlog of unanalyzed alerts is also measured and monitored, which indicates the LOE status of the CSOC. There are three optimization models that are integrated in Fig. 1 and each of them perform an important task. The tasks briefly presented next while the details are presented later in this chapter. The tasks are sequenced in the following order which serves as the road map of the integrated model framework shown in Fig. 1 that is described in this chapter.

Task 1 determines (1) the staffing levels for regular analysts who are categorized into junior, intermediate and senior analysts based on their level of expertise, and (2) the analyst schedule over a two-week (14-day) work cycle that provides adequate analyst expertise mix to handle the alerts in every shift of operation. Under normal operating condition, the above staffing and scheduling of analysts would maintain the level of operational effectiveness of a CSOC and the risk (% of unanalyzed alerts per shift of operation) is minimized and kept at the desired level of performance. Regular analyst levels can be determined based on the alert arrival rate for the given number of sensors, and the alert service rate of each analyst of a particular expertise level using queueing theory [2], which maintains a baseline queue of alerts at any given time (the number of unanalyzed alerts which constitutes an acceptable or desired baseline risk). Regular analyst levels also determine one portion of the analyst budget for alert analysis at a CSOC. Ideally, a zero baseline risk would be desired but as per queueing theory, a zero queue length would need many analysts that could be impractical from a budgetary standpoint. Hence, the queue length for a given arrival rate of alerts and service rate by hired analysts (number of hires within the budget) is deemed to be acceptable, against which the CSOC's LOE performance is measured. An optimization model as shown at the bottom in Fig. 1 achieves the scheduling of regular analysts, which is described in [3]. Under continuous CSOC operation, Task 1 also receives input from Task 3 for any additional analysts that are needed when the queue length exceeds the desirable level (risk increases and the LOE degrades). The additional analysts are also scheduled by the optimization algorithm for scheduling as shown in Fig. 1. The additional analysts constitute another portion of the analyst budget for alert analysis at a CSOC. In summary, the schedule optimization block considers the regular and additional analyst staff and produces a shift schedule for the analysts who must report to work for the immediate following day (24 h or 2 shifts).

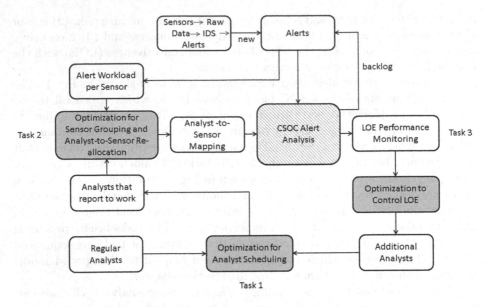

Fig. 1. Framework for alert data management

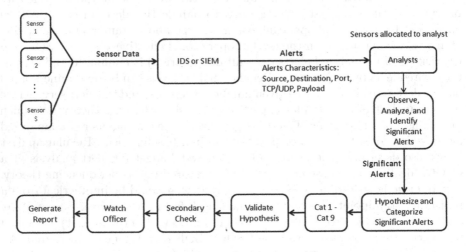

Fig. 2. Alert analysis process [1].

Task 2 performs sensor clustering and dynamic reallocation of analysts-to-sensors as shown in Fig. 1. The sensor grouping and dynamic allocation block is another optimization model that considers the alert workload expected per sensor for the next shift of operation (new alerts and backlog alerts per sensor) and the available number of analysts of various expertise levels that report to work (output of Task 1) in order to generate groups of sensors and the analyst-to-sensor allocation. An optimization model determines the analyst to sensor mapping such that the tooling and credential expertise to analyze alerts from a sensor are met along with capability to investigate the rate of alert generation by the sensor. The output of Task 2 and the alerts per sensor generated by the IDS and the backlog of alerts are passed into the central analysis block where alert investigation happens based on the process laid out in Fig. 2. The LOE performance is monitored as the output of the CSOC alert analysis process, which leads to Task 3.

Task 3 uses the LOE status as the input or trigger to the LOE optimization, which outputs the number of additional analysts (on-call) required to handle the backlog that is above the normal or baseline backlog queue. The optimization algorithm is a reinforcement learning model which operates in a dynamic mode to determine the additional number of analysts per shift of operation. The output of Task 3 is one of the inputs to Task 1 for scheduling the analysts as shown in Fig. 1. The three tasks are integrated and loop over in order to achieve effective alert data management of a CSOC.

The chapter is organized as follows. Section 2 describes the major elements that provide context for the alert data management of a CSOC, which includes the alert analysis process along with three major characteristics: alert, performance (LOE), and resource characteristics. In Sect. 3, the description of an integrated framework of three optimization models one for each of the above tasks, and their respective roles in effective alert data management are presented. Section 4 presents the related literature. Section 5 concludes the chapter with the major contributions.

2   Alert Analysis Process of a CSOC

Alerts are generated and analyzed by cyber security analysts as shown in Fig. 2. In the current system, the number of analysts that report to work remains fixed, and sensors are pre-assigned to analysts. A 12 h shift cycle is used, and analysts work six days on 12 h shift and one day on 8 h shift, thus working a total of 80 h during a two-week period. There is a very small overlap between shifts to handover any notes and the work terminal or workstation to the analyst from the following shift. The type and the number of sensors allocated to an analyst depend upon the experience level of the analysts. The experience level of an analyst further determines the amount of workload that they can handle in an operating shift. The workload for an analyst is captured in terms of the number of alerts/hr that can be analyzed based on the average time taken to analyze an alert. In this chapter, three types of analysts are considered (senior L3, intermediate L2, and junior L1 level analysts), and their workload value is proportional to their level of expertise.

A cybersecurity analyst must do the following: (1) observe all alerts from the IDS such as SNORT or a Security Information and Event Management (SIEM) tool such as ArcSight [4], (2) thoroughly analyze the alerts that are identified as significant alerts that are pertinent to their pre-assigned sensors, and (3) hypothesize the severity of threat posed by a significant alert and categorize the significant alert under Category 1–9. The description of the categories are given in Table 1 [5]. If an alert is hypothesized as a very severe threat and categorized under Cat 1, 2, 4, or 7 (incidents) then the watch officer for the shift is alerted and a report is generated (see Fig. 2). The Level of Operation Effectiveness (LOE) of a CSOC is measured at the end of every day of operation.

Table 1. Alert categories [5]

Category	Description
1	Root Level Intrusion (Incident): Unauthorized privileged access (administrative or root access) to a DoD system
2	User Level Intrusion (Incident): Unauthorized non-privileged access (user-level permissions) to a DoD system. Automated tools, targeted exploits, or self-propagating malicious logic may also attain these privileges
3	Unsuccessful Activity Attempted (Event): Attempt to gain unauthorized access to the system, which is defeated by normal defensive mechanisms. Attempt fails to gain access to the system (i.e., attacker attempts valid or potentially valid username and password combinations) and the activity cannot be characterized as exploratory scanning. Can include reporting of quarantined malicious code
4	Denial of Service (DOS) (Incident): Activity that impairs, impedes, or halts normal functionality of a system or network
5	Non-Compliance Activity (Event): This category is used for activity that, due to DoD actions (either configuration or usage) makes DoD systems potentially vulnerable (e.g., missing security patches, connections across security domains, installation of vulnerable applications, etc.). In all cases, this category is not used if an actual compromise has occurred. Information that fits this category is the result of non-compliant or improper configuration changes or handling by authorized users
6	Reconnaissance (Event): An activity (scan/probe) that seeks to identify a computer, an open port, an open service, or any combination for later exploit. This activity does not directly result in a compromise
7	Malicious Logic (Incident): Installation of malicious software (e.g., trojan, backdoor, virus, or worm)
8	Investigating (Event): Events that are potentially malicious or anomalous activity deemed suspicious and warrants, or is undergoing, further review. No event will be closed out as a Category 8. Category 8 will be re-categorized to appropriate Category 1–7 or 9 prior to closure
9	Explained Anomaly (Event): Events that are initially suspected as being malicious but after investigation are determined not to fit the criteria for any of the other categories (e.g., system malfunction or false positive)

2.1   Alert Characteristics

Alert Generation. The network data collected by the sensors is analyzed by an IDS or a SIEM, which automatically analyses the data and generates alerts. Most of the alerts are deemed insignificant by the IDS or SIEM, and about 1% of the alerts generated are classified as significant alerts.[1] The significant alerts are those with a different pattern in comparison to previously known alerts. The significant alerts must be further investigated by cybersecurity analysts and categorized.

Based on the past alert generation rate per day, a historical daily average alert generation rate can be derived, which is used as a baseline for determining a static workforce size, their expertise levels, and their daily work schedule. In reality, the number of alerts generated per sensor per hour varies throughout the day. On days when the number of alerts generated exceeds the above historical daily average alert generation rate, the static workforce size cannot cope with the additional workload, which will result in many alerts that will not be thoroughly investigated. Consequently, the backlog also increases (LOE is reduced). Hence, dynamic scheduling of cybersecurity analysts is a critical part of cybersecurity defense, which includes both the static workforce and a dynamic (on-call) workforce to meet the everyday varying demands on the workforce for alert investigation. In this chapter, the alert generation is modeled as a Poisson distribution, whereas the variation in alert generation per sensor is modeled as a Poisson distribution. The sum of the above distributions taken together will generate the historical daily-average alert generation per day (referred as the baseline alert generation rate). The parameters of the above distributions can be altered as needed based on historical patterns in alert generation, and the dynamic programming model presented in this chapter will adapt and converge to find the optimal dynamic schedules for the analysts that minimizes the backlog, which is the metric to measure the LOE.

Alert Prediction. The uncertainty in the alert generation rate is the primary driver for modeling a dynamic (on-call) workforce in addition to the static workforce that report to work daily. In order to determine the size and expertise composition of the static workforce, the historical daily-average for alert generation is used. However, to determine the size of the dynamic (on-call) workforce on a daily basis, one of the key inputs to the stochastic dynamic programming model is the number of additional alerts (over and above historical daily-average) estimated per sensor for the next day. It should be noted that the dynamic scheduling of analysts is required not only due to the dynamic increase in alert traffic generation rate of the sensors but also the detection of very important attacks/exploits/vulnerabilities such as the first-time detection of zero-day attacks and vulnerabilities (e.g., heartbleed vulnerability and exploit), which

---

[1] We arrived at the 1% figure based on our literature search and numerous conversations with cybersecurity analysts and Cybersecurity Operations Center (SOC) managers. Our model treats this value as a parameter that can be changed as needed.

could trigger an increase in alert generation rates for the shifts and days following the attack or requires additional monitoring as explained below. When a new zero-day attack is detected or reported in the news, additional dynamic (on-call) analysts are required to determine (i) whether such (zero-day) attacks have already exploited any vulnerability in the network, (ii) what defensive mechanisms such as new signatures (or attack detection rules) must be developed and used to detect (zero-day) attacks, and (iii) what and how attack detection should be reported to upper level management and other agencies. Hence, workload of cybersecurity analysts is increased significantly when zero-day attacks are detected or reported in the industry, even if the traffic rate of sensors during this period may not have necessarily increased. This type of significant event is expected to increase the workload between shifts and the team work of analysts includes not only thorough inspection of events but also preparing and sharing reports, and developing new attack detection rules if needed. In this research, a one-day (one-shift) look-ahead on-call analyst selection model will be run every day (shift) at an appropriate time such that there is sufficient time for the dynamic force to report to work prior to the starting of their shift.

The chapter assumes a Poisson distribution for the baseline average hourly rate of alert generation and a Poisson distribution to introduce variability and spikes in the hourly rate of alert generation. A prediction model for alert estimation using real-world data collected by the CSOC can replace the Poisson distribution in practice. To use the dynamic programming model in practice, the cyber-defense organization could develop statistical models to analyze their data patterns, and replace the distributions that are used in this chapter for making hourly alert predictions for each day of operation. The chapter assumes that the organization has developed a statistical model for alert prediction using historical actual alert generation data, and has determined that the alert generation rate comprises of two distributions. Since, real alert data was not available, the chapter assumes another stream of data to mimic the actual alert generation rate that draws a single random number using only a Poisson distribution whose average is the sum of average of the Poisson distributions that was used to generate the predicted stream of data. In summary, in the real-world, the actual alert rate will come from the intrusion detection system itself and the predicted alert rate will come from the statistical alert prediction model developed by the organization. The avgTTA/hr (LOE status) is estimated using the above rate of alert generation a explained next.

2.2   Performance Characteristics

Performance Metric: LOE. The readiness level of a CSOC is paramount to achieving the above mission successfully. The readiness level must be quantified and measured so that it provides a manager with full understanding of the impact of the interdependencies between various factors that affect the dynamics of the CSOC operations, and take corrective actions as needed. Some of these factors include (1) backlog of alerts that depends on the alert generation and processing rates, (2) the false positive and negative rates of analysts, (3) the optimal allocation of analysts to sensors, (4) optimal scheduling of the analysts with the right expertise mix in a shift, (5) grouping of sensors, (6) triaging of alerts, (7) the availability of tooling and credentials of analysts in a shift, and (8) effective team formation with highest collaborative scores among the analysts. In this chapter the readiness of the CSOC is defined as the level of operational effectiveness (LOE) of a CSOC, which is a color-coded scheme that indicates the timely manner in which an alert was investigated at the CSOC [6]. The LOE is continuously monitored for every hour of the work shift. Among the factors given above that affect the LOE of a CSOC, this chapter investigates two factors, namely, (1) the dynamic optimal scheduling of CSOC analysts to respond to the uncertainty in the day-to-day demand for alert analysis, and (2) the dynamic optimal allocation of CSOC analyst resources to the sensors that are being monitored. Thus, the objective of this research is to maintain the LOE of a CSOC at the desired level through the dynamic optimal scheduling and allocation of CSOC analyst resources.

In this chapter, the LOE of a CSOC is monitored as follows. The chapter identifies a common metric that is influenced by the disruptive factors that affect the normal operating condition of a CSOC, and this metric is the total time for alert investigation (TTA) for an alert after its arrival in the CSOC database. Any delay in data transmission between the IDS and the CSOC is ignored, and is not part of the TTA metric. In this chapter, it is assumed that an alert will be immediately queued after it arrives in the CSOC database. The TTA of an alert consists of the sum of two parts as shown in Fig. 3: (1) waiting time in queue, and (2) time to investigate an alert, after it has been drawn for investigation by the analyst. Clearly, when the rate of alert generation increases or a new alert pattern decreases the throughput of the system or when the CSOC capacity is reduced by analyst absenteeism the immediate impact is felt in terms of the delays experienced by the alerts waiting in the queue for investigation. Since all the alerts must be investigated, the queue length could become very long. The above means that the alerts stay much longer in the system and the average TTA calculated for each hour (avgTTA/hr) of operation of the CSOC increases.

The avgTTA/hr is calculated at the end of each hour of CSOC operation by using the individual values of TTA for all the alerts that completed investigation during that hour. A baseline value for avgTTA/hr is established for normal operating condition of the CSOC as shown in Fig. 4. It is a requirement of the CSOC that the avgTTA/hr remain within a certain upper-bound (four hours, for example), which is referred as the threshold value for avgTTA/hr. If the avgTTA/hr is maintained below the threshold during any given hour of CSOC operation then the LOE is said to be *optimal*, however, if the avgTTA is maintained at the baseline value then the LOE is said to be *ideal*. Different tolerance bands are created both below and above the threshold value of avgTTA to indicate a color-coded representation of LOE status (see Fig. 4).

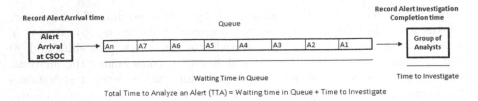

Fig. 3. Total time for alert investigation (TTA) [6]

Performance Metric: Notion of Risk per Sensor. A formal definition for the risk per sensor used in this chapter is presented below. Recently, an optimal scheduling for the cybersecurity analysts was published to minimize risk, where risk was measured as a single overall metric for all the sensors by computing the % of significant alerts that remained unanalyzed by the analysts at the end of a shift [1,3]. However, the context of the problem in this chapter is different. In this chapter, the sensors generate a certain number of alerts per shift, however, in some instances the % of significant alerts among all the alerts generated could vary between the sensors. In another instance, the % of significant alerts between two sensors could be the same but the significant alerts of one sensor takes more time than the other due to a new alert. The consequence of the above is that the number of alerts per sensor that remain unanalyzed (constitutes risk per sensor) is uneven among all the sensors at the end of the shift. This condition results in an imbalance among the risk values obtained from each sensor. In order to balance the number of unanalyzed alerts among all sensors, this research uses a modified notion of risk r_s per sensor s at any time of observation as follows:

$$r_s = V_s - c_s \quad \forall s \tag{1}$$

where r_s is the number of unanalyzed alerts per sensor, which is also the unanalyzed alert queue length for a sensor, V_s is the number of alerts generated from the start of the shift till the time of observation, and c_s (alert coverage) is defined

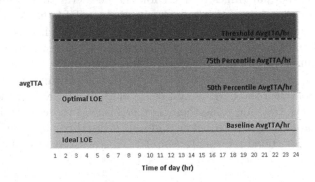

Fig. 4. Color-coded representation of (LOE) [6]

as the number of alerts thoroughly analyzed by the analysts from the start of the shift till the time of observation.

It should be clearly noted that at any observation time during the progress of a shift, the risk per sensor is measured from the number of unanalyzed alerts that have queued up for that sensor (observable to a shift manager), and the queue might contain some significant alerts that are detected upon alert investigation. In other words, one of the metrics to initiate sensor grouping and reallocation decision at any time during a shift is based upon the magnitude of the unevenness in the number of alerts that remain unanalyzed per sensor, which is captured by the alert queue statistics of each sensor. The other metric is analyst utilization. Furthermore, it should also be noted that the adaptive reallocation model is both reactive and proactive in the analyst to sensor decision making process. The model is reactive to the alert queue statistics that is observed for each sensor. The model assumes that the causes for an imbalance in alert queue length (risk per sensor) between sensors would persist after reallocation. Hence, the reallocation model is proactive because it uses the above assumption to compute both the expected alert and significant alert rates for the remainder time in the shift. Using the above reactive and proactive computations, the model determines a new analyst to sensor reallocation decision that will balance the risk per sensor among all sensors as the shift progresses.

In general, a shift manager would observe the length of the unanalyzed alert queue that builds up for each sensor, which is defined as r_s. An imbalance among r_s for all sensors is used as a metric to perform reallocation.

It is true that unless an alert is thoroughly analyzed, its category or severity is unknown. Also, the time taken to analyze an alert depends on its category or severity, whether or not it is a known or a new pattern of alert, and the expertise level of the analyst. Therefore, at the time of drawing an alert from the queue for investigation, since its category or severity is unknown, the time to analyze an alert in this chapter is based upon an average time from a probability distribution, which can be obtained from historical real world data. The total time needed to thoroughly analyze all the alerts and significant alerts can be compared to the total time available, which is based on the current capacity of the organization (number and expertise mix of analysts), their sensor-to-analyst allocation rules, and shift-schedules, in order to determine the % of significant alerts that would remain unanalyzed (risk). Such a risk metric could be used to initiate actions to build analyst capacity for the organization with optimal number of analysts, expertise mix in a work-shift, sensor-to-analyst allocation, and optimal shift schedules. Hence, the scope of the chapter is focused on capacity building for a cyber-defense organization through the optimal allocation and scheduling of its analysts, regardless of the type of alert (category or severity), using the notion that some alerts will need more time than the others. Several parameters are considered in this chapter to calculate the alert investigating capacity of the organization, which includes number of sensors, an average alert generation rate for the sensors, number of analysts, their expertise level, sensor-to-analyst allocation, analyst time to investigate an alert, and their work-shift schedule. The chapter assumes that all the alerts that were thoroughly investigated were also accurately categorized. It should be noted that as a second metric (quality), once a significant alert has been detected by thorough alert analysis, a different definition of risk can be used to measure the quality of work performed by capturing the true positive and false negative rates. Furthermore, the severity of the threat that an alert poses to the organization, and actions to mitigate the threat can be taken. However, such a definition of risk and the actions to mitigate are beyond the scope of this chapter.

2.3 Resource Characteristics

The analysts have certain characteristics as well. They differ from each other in terms of their expertise levels such as junior, intermediate, and senior analysts. They also have different tooling knowledge and individual credentials (security clearance levels such as confidential, secret, top secret, and so on) to investigate certain types of alerts. From our conversations with CSOC managers, it was learnt that tooling knowledge was correlated to the level of expertise. For example, junior analysts would have access to basic tools while senior analysts would have access to the entire tool-set. Also, their alert service rate, and false positives and negatives rate are not the same among them. Typically, higher expertise is associated with lower false positives and negatives rate. Similarly, optimal matching of analyst's tooling knowledge and credentials with sensor requirements could reduce the number of unanalyzed alerts (backlog). The following are the characteristics of analysts (resources) who investigate alerts.

1. L3 - senior analyst. L3 analysts are assigned 4–5 sensors and they and can handle on average 12 alerts per hour (5 min/alert).
2. L2 - intermediate analyst. L2 analysts are assigned 2–3 sensors and they can handle on average 7–8 alerts per hour (8 min/alert).
3. L1 - junior analyst. L1 analysts are assigned 1–2 sensors and they and can handle on average 5 alerts per hour (12 min/alert).
4. Analysts work in two 12-h shifts, 7 PM–7 AM and 7 AM–7 PM. However, the optimization model can be adapted to 8 h shifts as well.
5. Each analyst on regular (static) schedule works for 80 h in 2 weeks (6 days in 12-h shift and 1 day in 8-h shift)
6. When a group of analysts are allocated to a group of sensors by the optimization algorithm, the alerts generated by that group of sensors are arranged in a single queue based on their arrival time-stamp, and the next available analyst within that group will draw the alerts from the queue based on a first-in-first-out rule.
7. Based on experience, an analyst spends, on average, about the same amount of time to investigate alerts from the different sensors that are allocated, which can be kept fixed or drawn from a probability distribution such as Poisson or Uniform.
8. Analysts of different experience levels can be paired to work on a sensor.

9. Writing reports of incidents and events during shifts is considered as part of alert examining work, and the average time to examine the alert excludes the time to write the report. Analysts spend 80% of their time on alert analysis and the remaining time on training and writing reports.
10. L1 analysts are not scheduled on-call because the purpose of on-call workforce is to schedule the most efficient workforce to handle the additional alerts above the historical daily-average that are generated.
11. Analysts of different experience levels can be paired to work on a sensor.

3 Alert Data Management

The framework for alert data management system is shown in Fig. 1. This section describes the requirements and modeling assumptions, which is followed by the detailed description of the tasks.

3.1 Effective Alert Analysis at a CSOC- Requirements

The requirements of the cybersecurity system can be broadly described as follows. The cybersecurity analyst scheduling system,

1. shall ensure that LOE is maintained at the baseline that is established for normal operating conditions,
2. shall ensure that an optimal number of staff is available and are optimally allocated to sensors to meet the demand to analyze alerts,
3. shall ensure that a right mix of analysts are staffed at any given point in time, and
4. shall ensure that weekday, weekend, and holiday schedules are drawn such that it conforms to the working hours policy of the organization.

3.2 Effective Alert Analysis at a CSOC - Model Assumptions

The assumptions of the optimization model are as follows.

1. At the end of the shift any unanalyzed alert is carried forward into the next shift. The backlog indicates the avgTTA/hr, which in turn indicates the LOE status of the CSOC.
2. All alerts that were thoroughly investigated were also accurately categorized. Hence, false positives and false negatives are not modeled in this chapter.
3. The optimization model is run for 24-h to determine the sensor-to-analyst allocation for that day. Simulation statistics on risk and analyst utilization are calculated at the end of the 24-h day.

In the following, the three tasks are described in detail along with their optimization models and results. Tasks 1, 2, and 3 loop over as in Fig. 1, which achieves the effective alert data management in a CSOC.

3.3   Task 1: Scheduling of Analysts to Minimize Risk

Task Description. The objective of Task 1 is to formulate and test an adaptive and dynamic analyst scheduling strategy for effective cyber-defense that is capable of using an estimate of the varying future alert generation rates and scheduling an optimal number of cybersecurity analysts at different expertise levels that minimizes the risk and maintains risk under a pre-determined upper bound for a set of system defined parameters and constraints.

Simulation and Optimization Model. The scheduling optimization model takes inputs from (1) a static mixed-integer programming model for obtaining the minimum number of analysts (static or regular workforce) for a historical daily-average alert generation rate calculated over the past two-week period, and (2) a dynamic LOE model (Task 3) based on stochastic dynamic programming to obtain the minimum number of additional workforce and their expertise level that is needed (dynamic or on-call workforce) based on the estimated additional alerts per sensor for the next day. The mathematical details of the models, algorithms, and implementation guidelines are available in [1,3].

Scheduler Module: The input to the 14-day static scheduling module is the number of personnel needed per level per day, which is derived from the integer programming optimization module. An optimal schedule for the static workforce can be derived based on the following constraints.

1. Each analyst gets at least 2 days off in a week and every other weekend off.
2. An analyst works no more than 5 consecutive days.
3. An analyst works 80 h per two weeks counted over 14 consecutive days between a Sunday and a Saturday. Both 12 h and 8 h shift patterns are allowed.

The objective of the static workforce scheduling algorithm is to find the best days-off schedule and days-on schedule for both 12 h and 8 h shifts for all analysts in the organization subject to the above scheduling constraints. A mixed integer programming scheduling model is used to obtain the 14-day static schedule. During the 14-day schedule, the dynamic programming algorithm would assign on-call status to those analysts who have the day-off. The number of on-call analysts that actually report to work in a day is drawn from those who have been designated with the on-call status.

Results of a Heuristic for Static and Dynamic Workforce Scheduling. The days-off scheduling heuristic is given in [7]. The minimum number of employees needed W as per the scheduling constraints is given as follows.

$$W_1 \geq \lceil \frac{k_2 max(n_1, n_7)}{k_2 - k_1} \rceil \qquad (2)$$

$$W_2 \geq \lceil \frac{1}{5} \sum_{j=1}^{7} n_j \rceil \qquad (3)$$

$$W_3 \geq max(n_1, \ldots, n_7) \qquad (4)$$

$$W = max(W_1, W_2, W_3) \qquad (5)$$

where k_1 weekends are off in k_2 weekends, and $n_1, \ldots, n_7$ is the number of employees needed on $Sunday, \ldots, Saturday$ respectively. For a sample scenario of 10 sensors and 6 L1, 6 L2, and 8 L3 analysts required per day (split equally in two 12 h shifts), $k_1 = 1$, and $k_2 = 2$, and $n_1, \ldots, n_7 = 20$. The value of W is 40 (12 L1, 12 L2, and 16 L3), which is the number of employees that the organization must hire (be on payroll) to meet the days-off constraints given above. It should be noted that in the above situation, there are no part-time analysts and all full-time analysts work 12 h shifts ($12 * 7 = 84$ h in every 14-day cycle).

Table 2 shows the combined output of the scheduling heuristic for scheduling static and a fixed dynamic workforce in which X represents days-off for analysts, and c indicates the days on which on-call analysts are scheduled at each level of expertise. The issue with fixing the number of people that are on-call per day at the beginning of the 14-day period is that the cyber defense system is no longer adaptable to higher alert generation rates that exceed the alert rates covered by the fixed on-call workforce. In contrast to the above, the dynamic programming algorithm will select the actual number of on-call workforce required for the next day from the available on-call workforce for that day, which provides greater scheduling flexibility and adaptability to varying alert generation rates. L1 (junior) analysts are not scheduled for on-call workforce.

3.4   Task 2: Sensor Clustering and Dynamic Allocation of Analysts-to-Sensors

Task Description. Understanding the importance and relationship between sensors, analysts, and shift characteristics leads to the two essential properties that must be met in performing the grouping of sensors into clusters, and the allocation of analysts to clusters. The following properties serve as objectives for Task 2. Property 1: meeting the cluster's requirement for specific analyst expertise mix of junior, intermediate, and senior analysts, complete tools coverage that allows the analysts to handle the type of alerts generated by the sensors in the cluster, and analyst credentials such as security clearances needed for the cluster. Property 1 ensures that a high quality of work performance is maintained. In other words, a cluster with a sub-optimal mix of expertise, or with analysts lacking credentials or tooling knowledge is said to perform inefficiently. We define quality of work performed in terms of metrics such as minimizing false positive and negative rates. However, it must be noted that the chapter does not measure the quality metrics directly, instead, it expects the quality of alert analysis to be high if the above property is met. Property 2: minimizing and balancing the number of unanalyzed alerts in each cluster at the end of the daily work-shift because a large number or an imbalance of unanalyzed alerts among clusters, due to factors such as lack of analyst credentials or tooling expertise in a cluster, would pose a security risk to the organization whose network is monitored by these sensors. Property 2 deals with quantity of unanalyzed alerts and ensures that the overall number of unanalyzed alerts are minimized. It also ensures that no cluster has been unduly disadvantaged in the grouping and allocation process that has resulted in some clusters having a higher number of unanalyzed alerts over other clusters. The quantity of unanalyzed alerts per cluster is measured in this chapter and the motivation behind this property is explained below.

Simulation and Optimization Model. Figure 5 shows the framework of the adaptive model presented in [8], which consists of an optimization and a simulation model. The optimization model used for grouping of sensors to form clusters, and analyst allocation to the clusters is modeled and solved using mixed integer programming. Analyst and sensor characteristics are provided as inputs to the optimization model. A minimum mix of analyst expertise levels, a complete tool-set coverage, and analyst credential requirements on the clusters are provided as constraints to the model. The outputs of the optimization model are clusters (groups of sensors) and the analyst to cluster allocation, which are then provided as inputs to the simulation model. The simulation model is used for verification and validation in which a CSOC work-shift is simulated with the analyst and sensor characteristics used in the optimization model. The number of unanalyzed alerts that remain per cluster at the end of the shift is measured by replicating the shift several times in order to achieve a 95% confidence interval.

Results. A case with an increase in alert generation rate is considered. For other results refer to [8]. Due to an increase in the number of alerts generated from some sensors in the last few (one or more) shifts, the estimated average alert generation rate on the respective sensors is increased, though the number of scheduled analysts from various expertise levels, at the start of the shift, remain the same as in the nominal case. In this *case study*, the average alert generation rate per day on twenty sensors were increased to a range between 100 to 150 significant alerts (*i.e.* 1% of 10000 to 15000 total alerts).

The outputs for the case with an increase in alert generation rate is shown in Table 3. It is to be noted that the workload (number of alerts generated) has increased while the resource capacity (number of alerts that could be analyzed by analysts) has remained the same as compared to the nominal case. As a result, there are more number of alerts that remained unanalyzed at the end of the shift compared to the nominal case. In order to minimize the maximum number of unanalyzed alerts at the end of the shift among all the clusters, more number of clusters (11) were created compared to the nominal case (8).

The analyst credential requirement $C5$, for sensors S17, S19, S38, and S40 is met by allocation of analyst A12 to the cluster Q1, while analyst credential

Table 2. Scheduling of L1, L2, and L3 level analysts for both static and a fixed dynamic workforce using days-off scheduling heuristics, X- days-off, and c- on-call [3]

Day →		1	2	3	4	5	6	7	8	9	10	11	12	13	14	
Level ↓ Analyst ID ↓	Sat	Sun	Mon	Tue	Wed	Thu	Fri	Sat	Sun	Mon	Tue	Wed	Thu	Fri	Sat	Sun
L3 1	x	c	x	x						c	x			x	x	x
2	x	x	c	x	x					c	x				x	x
3	x	x		c	x	x					x	c			x	x
4	x	x			c	x			x				x	c	x	x
5	x	c	x	x						c	x			x	x	x
6	x	x	x	c	x						x	c			x	x
7	x	x			x	x	c				x	x			c	x
8	x	x				x	c		x				x	x	c	x
9		c	x				x	x	x	c			x			
10		x	c	x			x	x			x	c				
11				c	x	x	x	x			x	c				
12				c	x	x	x	x			x	c				
13			x	x			c	c	x	x			x			
14			x	x	x		c	c			x	x				
15				x	x	x	x	c	c		x	x				
16					x	x	c	c	x		x	x				

Day →		1	2	3	4	5	6	7	8	9	10	11	12	13	14	
Level ↓ Analyst ID ↓	Sat	Sun	Mon	Tue	Wed	Thu	Fri	Sat	Sun	Mon	Tue	Wed	Thu	Fri	Sat	Sun
L2 1	x	c	x	x						c	x			x	x	x
2	x	x	c	x	x					c	x				x	x
3	x	x		c	x	x					x	c			x	x
4	x	x			c	x			x				x	c	x	x
5	x	c	x	x						c	x			x	x	x
6	x	x	x	c	x						x	c			x	x
7	x	x			x	x	c				x	x			c	x
8	x	x				x	c		x				x	x	c	x
9		c	x				x	x	x	c			x			
10		x	c	x			x	x			x	c				
11				c	x	x	x	x			x	c				
12				c	x	x	x	x			x	c				

Day →		1	2	3	4	5	6	7	8	9	10	11	12	13	14	
Level ↓ Analyst ID ↓	Sat	Sun	Mon	Tue	Wed	Thu	Fri	Sat	Sun	Mon	Tue	Wed	Thu	Fri	Sat	Sun
L1 1	x	x	x	x						x	x			x	x	x
2	x	x	x	x	x					x	x				x	x
3	x	x		x	x	x					x	x			x	x
4	x	x			x	x			x				x	x	x	x
5	x	x	x	x						x	x			x	x	x
6	x	x	x	x	x						x	x			x	x
7	x	x			x	x	x				x	x			x	x
8	x	x				x	x		x				x	x	x	x
9		x	x				x	x	x	x			x			
10		x	x	x			x	x			x	x				
11				x	x	x	x	x			x	x				
12				x	x	x	x	x			x	x				

requirement $C4$, for sensors S16, S18, S20, S36, S37, and S39 is met by allocation of analysts A10, A11, and A12 to the clusters Q5, Q6, Q10, and Q11. There is a complete tool-set coverage, and the minimum mix of analyst expertise levels

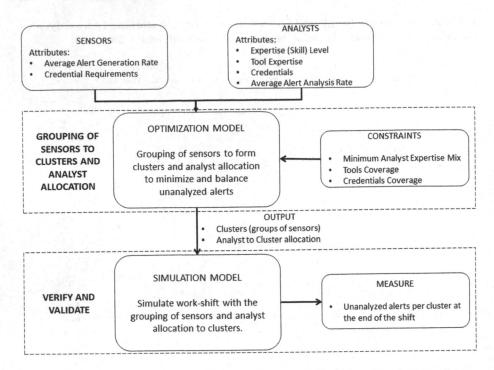

Fig. 5. Adaptive grouping of sensors into clusters and allocation of analysts to clusters model [8].

is maintained on each of the clusters. The average number of unanalyzed alerts per cluster at the end of the shift is balanced as shown in Table 3 with 50 as the maximum average number of unanalyzed alerts that remained on either of the clusters.

The following meta-principles are derived for sensor grouping and allocation to analyst optimization model.

1. It was observed from the clusters that the number of alerts per cluster could vary among them, however, the more important metric is to minimize and balance the number of unanalyzed alerts among the clusters at the end of the shift.
2. It was observed that an analyst allocated to only one cluster may not be a good strategy because if the alert generation rate among the group of sensors in the respective cluster decreases during the shift, the analyst will be idling.
3. The integrated grouping and allocation methodology was able to generate *optimal* sensor to analyst allocation for a given alert generation rate by sensors and known alert service rate by analysts that met the expertise, tooling, and credentials requirements of the cluster.
4. It was observed that the maximum number of clusters that could have been generated (upper-bound, R) is the minimum of the product between the num-

Table 3. Outputs for increase in alert generation rate [8]

Clusters	Groups of sensors	Analysts allocated	Avg. alerts gen.	Avg. alerts unan.
Q1	S17, S19, S38, S40	A2, A4, A6, A12	213	50
Q2	S5, S7	A4, A7, A11	78	50
Q3	S14, S22, S34	A4, A6, A10	122	50
Q4	S1, S13, S21	A1, A6, A11	78	48
Q5	S6, S8, S16, S32	A3, A9, A11, A12	176	50
Q6	S3, S27, S30, S36, S37, S39	A3, A8, A9, A10	297	50
Q7	S2, S11, S23, S31, S35	A1, A9, A12	204	50
Q8	S9, S28	A5, A7, A12	70	50
Q9	S24, S33	A3, A7, A11	93	50
Q10	S4, S10, S20, S25, S29	A5, A6, A12	209	50
Q11	S12, S18, S15, S26	A1, A2, A7, A11	100	50

ber of analysts at each level, and the maximum clusters that could be allocated to an analyst from the respective level.

5. The number of unanalyzed alerts per cluster at the end of the shift can be brought to zero if and only if there are sufficient numbers of analysts hired with various levels of expertise, tooling knowledge, and credentials.

6. With limited analyst resource, the goal is to ensure that the number of unanalyzed alerts is minimized and balanced among the clusters by meeting the expertise, tooling, and credentials requirements of the cluster.

3.5   Task 3: Measuring, Monitoring, and Controlling LOE

Task Description. The objective of Task 3, is to develop an intelligent and adaptive decision support tool for the CSOC manager to take optimal actions (when and how much) to allocate the additional resources in order to maintain an optimal LOE status throughout the 14-day cycle of the CSOC. Due to the dynamic and sequential decision making framework of the 14-day CSOC operation, the chapter presents a reinforcement learning (RL)-based model for representing the manager's decision making process under uncertainty. The RL model takes the continuously monitored LOE status of the CSOC operation as one of its inputs, and takes corrective actions depending on the extent of deviation of the current avgTTA/hr value from the baseline avgTTA/hr value for the CSOC system. The decisions made by the RL model is compared with greedy and rule-based uniformly distributed resource actions to demonstrate the superior decision making ability of the RL model in the face of uncertainties due to disruptive factors. While the rule-based actions are limited in its use of future resources in advance (not adaptive), the greedy actions are myopic in nature that responds to surges in alert backlog by allocating additional resources without any consideration of the future resource needs. For further details refer to [9].

Simulation and Optimization Model. A framework for the dynamic LOE optimization model is provided in Fig. 6. The dynamic optimization model consists of two main blocks - the alert analysis process simulation block, and the RL-based optimization block, which is executed one after another. As explained later, the time over the 14-day work cycle is indexed in 1 h time steps. As shown in Fig. 6, at each time step, the state of the CSOC system is observed, and a decision is made by the RL agent. The decision is then ratified by the CSOC manager, and implemented for the next hour of CSOC operation. The details of the above blocks are presented next.

Simulation Model of Alert Analysis Process: The simulation model of the alert analysis process consists of four main blocks as shown in Fig. 6. They include the CSOC system inputs (system parameters and alert generation by IDS), the uncertain events (both internal and external) that affect the CSOC system inputs, the alert analysis process block in which the work shift is simulated, and the performance metrics block that captures the LOE status of the CSOC at each point in time using the avgTTA/hr metric.

Work-day Simulation: A work-day at the CSOC is simulated and each simulation run corresponds to one operation day of 24 h. Alerts are generated using a Markovian distribution. Analysts are considered as resources, and they investigate alerts from a single queue of alerts populated by the IDSs in a first-come-first-served (FCFS) manner. The time taken to investigate an alert by an analyst, T, is the average time taken based on historical statistics observed in the organization. In this chapter, T is maintained constant except when a new alert pattern causes an increase in T, although it could also be drawn from a probabilistic distribution for each alert. It is assumed that all analysts spend 80% of their effort in a shift toward alert analysis, and the rest of the time is spent on report writing, training, and on generating signatures. Hence, an analyst could increase their effort on alert analysis up to 20% when the need arises, which will increase the service rate of alerts investigated in a day. The alert analysis process of a CSOC is considered to be in steady state under normal operating conditions, which means that the average alert arrival rate per hour, average queue length, average waiting time in queue, and average alert investigation time are all normal. Hence, a baseline avgTTA/hr value for the CSOC system can be established using queueing theory, and the LOE status is ideal as shown in Fig. 4. A threshold value for avgTTA/hr is also established. The scheduled analyst staffing levels are adequate to maintain a pre-determined acceptable avgTTA/hr (and LOE status) of the CSOC.

RL-based Optimization Model for Decision Making: The past hour performance of the CSOC from the real-world alert analysis process (simulation block in this chapter), presents the avgTTA/hr and LOE status to the CSOC manager. Disruptive events, if any, from the past hour are known. It is imperative that the CSOC manager considers the current avgTTA/hr metric and LOE state of the system in order to make a decision to add additional resources or do nothing for the next hour of CSOC operation. The decision is non-trivial because of the uncertainties in the future disruptive events and the limited additional resource that is available to the CSOC manager. Accurate prediction of future disruptive events such as a zero day attack is very hard. However, one can observe from the history of past events, such as resources that were needed to mitigate a past disruptive event and the frequency of occurrences of each type of disruptive events, and build a probability distribution that can simulate the arrival process of real-world uncertain events. By interacting with the unknown environment via simulation and by learning from the past decisions, the goal of the RL-based decision support system is to optimally plan the allocation of additional resources such that in the long run (over several 14-day cycles), the CSOC system with an adaptive RL-based decision performs far better (in terms of its LOE) than making ad hoc or greedy or a rule-based decision.

When a disruptive event occurs, a CSOC manager, in the order of preference as determined through our discussions with CSOC managers at the Army Research Lab, would utilize the remainder 20% of analyst time on alert analysis, spend some of their own time to assist the analysts in clearing the alert backlog, and bring on-call analysts to supplement the regular analyst workforce. The RL model manages the additional resource allocation that follows the CSOC manager's order of preference, and decides the quantity of additional resource and the timing of when to allocate the additional resources. For further details refer to [9].

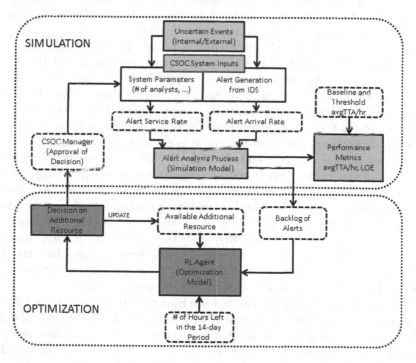

Fig. 6. Dynamic LOE optimization model framework [9].

Results. The following section presents the results from an experiment in which five uncertain disruptive events occurred over the 14-day work cycle. It is reemphasized that in the case of an occurrence of an uncertain event, the additional 20% effort for alert analysis, which is available during every hour of the shift from resources such as analysts and watch officers is utilized first. The on-call analysts are called upon only if needed as a second source of additional resource. Figure 7(a–c) shows only the depletion of on-call additional resource. Where ever the plot is horizontal, it means that the on-call resource was not utilized because (1) either the backlog was 0 (avg. queue length is 1,175 alerts as in the baseline case), or (2) that the backlog was low and was cleared by adding the additional 20% analyst effort for alert analysis.

In the greedy approach as shown in Fig. 7(a), the additional resources were utilized in a myopic manner. An additional resource is called upon only when the workload that is worth the resource's time is accumulated. Such a strategy is commonly employed in various organizations where on-call resources are limited and expensive. For example, an analyst is called upon as soon as four hours of workload is accumulated. As shown in Fig. 7(g), since there is a waiting time for the four hours of workload to accumulate, the LOE (and the alert backlog as shown in Fig. 7(d)) is observed to climb into the yellow zone. As soon as the additional resources are assigned, the LOE is restored into the green zone. However, since this strategy is also myopic (greedy), it can be observed from Fig. 7(g) that there were no additional resources left after ten days, and the LOE climbed into the red zone on the eleventh day.

In the rule-based uniformly distributed approach, the following rules are followed. The CSOC's additional resources are evenly distributed at the start of the 14-day period. As the days progress, any unused resource is rolled over into the following day. Resources allocated to future days cannot be used in advance, which sharply differs from the greedy approach that can exhaust as much resource as needed. The decision to allocate additional resources depends on the available resource on that day, and it is a reaction to the magnitude of the uncertain event that occurs. Figure 7(b) shows that there are unused resources at the end of the 14-day period because resources were evenly distributed and no major event occurred toward the end of the 14 days that consumed all of the remaining resources. Due to the rule that future resources could not be used in advance because they are reserved for future uncertainties, the LOE was found to have higher variance (see Fig. 7(h)) than the greedy and RL approaches as shown in Figs. 7(g) and (i), respectively. Despite being better than greedy in reacting to the uncertainties, the LOE eventually crosses the red band (4-h avgTTA threshold) with the onset of the 4th uncertain event.

There are two critical decisions that are learned with reinforcement learning, (i) when to call the additional resources with respect to the time available in the 14-day work cycle, and (ii) how many additional resources to call upon such that an optimal LOE is maintained over the 14-day work cycle. It can be seen from Fig. 7(c) that very few resources were utilized until the fourth event on the tenth day as compared to Figs. 7(a) and (b) in which the additional on-call resources were exhausted. As a result, there were fluctuations in the avgTTA values but the LOE was maintained in the green zone. With the event on the tenth day ($t = 240$), the majority of the additional on-call resources were called upon to keep the LOE in the green zone. With resources still remaining at the end of the 14-day work cycle as shown in Fig. 7(c), it can be observed from Fig. 7(i) that the LOE was maintained in the green zone throughout the 14-day work cycle.

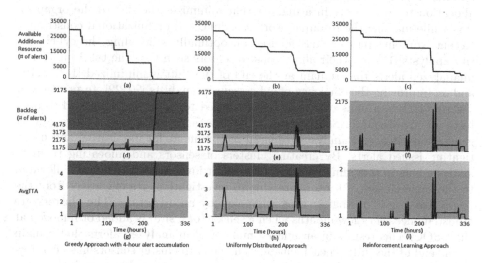

Fig. 7. 5 uncertain events: (a–c) available additional resource, (d–f) backlog, and (g–i) AvgTTA (LOE) [9]

4 Related Literature

D'Amico and Whitley [10] identified six analysis roles of cybersecurity analysts: triage analysis, escalation analysis, correlation analysis, threat analysis, incident response, and forensic analysis. The amount of time an analyst spends in triage analysis is a function of the alert generation rate and is bounded by no more than 80% of the analyst's time (effort) [11]. The rest of the time is spent on writing reports, updating signatures in the IDS from new alert patterns, and training. Triage analysis is the fundamental function of a CSOC. In triage analysis, the large amount of data (alerts) that are generated from IDS using pattern matching techniques [12,13], and automated techniques for malicious behavior [14,15], are investigated to identify suspicious activities (significant alerts). The thorough analysis of a significant alert requires adequate analyst time, which varies between analysts depending on their level of expertise and the category of the alert from the sensor. The alert data received at the CSOC often contains false positive alerts which lead to wastage of time of the analysts. Similarly, there could be false negatives (missing alerts) in the alert data due to unknown vulnerabilities. The experience of an analyst (expertise level) can help in reducing the number of false positives and false negatives in a shift [16]. It is the desiderata of a CSOC to adequately staff analysts such that all the generated alerts are investigated in a timely manner, and to maintain a proper mix of expertise levels among the analysts allocated to groups of sensors such that the number of false negatives and false positives are minimized.

Managing a CSOC requires critical decision making on scheduling the optimal number of cybersecurity analysts of various expertise levels and an optimal allocation to the sensors in a manner that minimizes the risk to the organization while meeting the resource, work schedule, and organizational constraints. Recent work in literature has focused on optimally scheduling the cybersecurity analysts [17] and their allocations to sensors such that the total number of unanalyzed alerts that remain at the end of the shift is minimized [1,3,18,19], and on improving the efficiency of cybersecurity analysts [20–22]. In practice, at several CSOCs, groups of sensors are clustered together and allocated to analysts for investigation. Sensors have attributes such as historical average alert generation rates, and analyst credential requirements for monitoring and investigating issued alerts. By creating clusters of sensors and allocating them to analysts help in providing context during alert investigation. Analysts allocated to the same cluster of sensors are able to investigate alerts efficiently from the respective sensor during an alert campaign by an adversary. The clusters are adjusted once every few months when a sensor (or a site) is added or removed at the CSOC. This results in an uneven number of unanalyzed alerts that remain at the end of the shift on each cluster. The need to cluster sensors has been recognized, but it has not been implemented often enough, and in a unified manner that takes human factors (analyst attributes) into consideration. To the best of the authors' knowledge, a unified model that creates clusters (groups of sensors), and allocates cybersecurity analysts to the clusters by taking into account the unique attributes of both, sensors and analysts, to minimize and balance the work (unanalyzed alerts) that remain at the end of the shift has not been studied or researched in published literature.

A CSOC is a unique amalgamation of people, processes, and technology. A CSOC performs many roles in terms of variety of services offered, which are broadly categorized into reactive, proactive, and security quality management services [23]. Alert management, incident handling, and vulnerability handling are categorized under the reactive services, while intrusion detection services [24] and development of security tools [22] are categorized under the proactive services. D'Amico and Whitley [10] conducted a cognitive task analysis to study the analytical process that transforms data into security situation awareness, which is categorized as a security quality management service provided by a CSOC. A complete list of services offered by a CSOC is given in Killcrece et al. [23].

Real-time work schedule adjustments or reactive-scheduling has been studied for over three decades. Early research work on reactive-scheduling used rolling horizon technology for the job-shop scheduling [25] while real-time schedule adjustment in a call center environment had been studied to provide a high level of customer service [26]. A sequential mixed-integer programming with loose bounds has been shown to achieve higher profit improvement than experienced managers in real-time work schedule adjustment decisions at a quick service restaurant study [27]. In the hospitality and tourism industry, research efforts have focused on using mixed-integer programming to solve tour scheduling problems to minimize labor cost and meet service standards [28,29]. In manufacturing systems, predictive-reactive scheduling has been studied where schedules are revised in response to real-time events such as machine breakdowns and random job arrivals [30]. The relationship between situational constraints (schedule or allocation disruption) and the effect on worker performance has been a topic of interest in organization studies [31] which relates excessive schedule disruptions to worker morale and to worker turnovers.

5 Conclusions

The chapter presented an innovative integrated framework for efficient alert data management, and highlighted three very important tasks and the strategies to optimize them. The desiderata of a CSOC enterprise can be achieved by the integrated model in which all alerts could be investigated in a timely manner, the CSOC resources (analysts) could be optimally managed, and the LOE desired performance could be achieved. The tasks considered in this chapter are (1) determining the regular analyst staffing of different expertise level for a given alert arrival/service rate, and scheduling of analysts to minimize risk, (2) sensor clustering and dynamic reallocation of analysts-to-sensors, and (3) measuring, monitoring, and controlling the level of operational effectiveness (LOE) with the capability to bring additional analysts as needed. The chapter demonstrated the framework under which the inter-dependent tasks can be integrated, sequenced, and optimized, which is very useful for CSOC managers to make shift-to-shift decisions on scheduling analysts, allocation them to sensors, and maintaining the LOE of the CSOC. The algorithmic details of each optimization model are available in the cited references under each task, and the best strategy is to optimize individual tasks such that the outputs of one task are the inputs to another as described in the introduction to the chapter. As on-going and future research, trade-off analysis of competing factors would be studied, along with other tasks such as alert prioritization and team formation in order to increase the fidelity of the integrated model.

Acknowledgment. The authors would like to thank Dr. Cliff Wang of the Army Research Office for the many discussions which served as the inspiration for this research. This work is partially supported by the Army Research Office under grant W911NF-13-1-0421.

References

1. Ganesan, R., Jajodia, S., Cam, H.: Optimal scheduling of cybersecurity analyst for minimizing risk. ACM Trans. Intell. Syst. Technol. 8(4), 52:1–52:32 (2017)

2. Gross, D., Shortle, J., Thompson, J., Harris, C.: Fundamentals of Queuing Theory. Wiley, New York (2008)
3. Ganesan, R., Jajodia, S., Shah, A., Cam, H.: Dynamic scheduling of cybersecurity analysts for minimizing risk using reinforcement learning. ACM Trans. Intell. Syst. Technol. 8(1), 1–21 (2016)
4. Bhatt, S., Manadhata, P.K., Zomlot, L.: The operational role of security information and event management systems. IEEE Secur. Privacy 12(5), 35–41 (2014)
5. CIO: DON cyber crime handbook. Department of Navy, Washington, DC (2008)
6. Shah, A., Ganesan, R., Jajodia, S., Cam, H.: A methodology to measure and monitor level of operational effectiveness of a CSOC. Int. J. Inf. Secur. 17(2), 121–134 (2018)
7. Pinedo, M.: Planning and Scheduling in Manufacturing and Services. Springer, New York (2009). https://doi.org/10.1007/978-1-4419-0910-7
8. Shah, A., Ganesan, R., Jajodia, S., Cam, H.: Optimal assignment of sensors to analysts in a cybersecurity operations center. IEEE Syst. J. 13, 1060–1071 (2018)
9. Shah, A., Ganesan, R., Jajodia, S., Cam, H.: Dynamic optimization of the level of operational effectiveness of a CSOC under adverse conditions. ACM Trans. Intell. Syst. Technol. 9(5), 51:1–51:20 (2018)
10. D'Amico, A., Whitley, K.: The Real Work of Computer Network Defense Analysts. In: Goodall, J.R., Conti, G., Ma, K.L. (eds.) VizSEC 2007. MATHVISUAL. Springer, Heidelberg (2008). https://doi.org/10.1007/978-3-540-78243-8_2
11. West-Brown, M.J., Stikvoort, D., Kossakowski, K.P., Killcrece, G., Ruefle, R.: Handbook for computer security incident response teams (CSIRTs). DTIC Document CMU/SEI-2003-HB-002 (2003)
12. Bejtlich, R.: The Tao of Network Security Monitoring: Beyond Intrusion Detection. Pearson Education Inc., Boston (2005)
13. Crothers, T.: Implementing Intrusion Detection Systems. Wiley, New York (2002)
14. Di Pietro, R., Mancini, L.V. (eds.): Intrusion Detection Systems. Advances in Information Security, vol. 38. Springer, New York (2008)
15. Northcutt, S., Novak, J.: Network Intrusion Detection, 3rd edn. New Riders Publishing, Thousand Oaks (2002)
16. Kott, A., Wang, C., Erbacher, R.F.: Cyber Defense and Situational Awareness. Springer, Cham (2014). https://doi.org/10.1007/978-3-319-11391-3
17. Altner, D.S., Rojas, A.C., Servi, L.D.: A two-stage stochastic program for multishift, multi-analyst, workforce optimization with multiple on-call options. J. Sched. 21, 517–531 (2017)
18. Ganesan, R., Shah, A.: A strategy for effective alert analysis at a cyber security operations center. In: Samarati, P., Ray, I., Ray, I. (eds.) From Database to Cyber Security. LNCS, vol. 11170, pp. 206–226. Springer, Cham (2018). https://doi.org/10.1007/978-3-030-04834-1_11
19. Ganesan, R., Shah, A., Jajodia, S., Cam, H.: A novel metric for measuring operational effectiveness of a cybersecurity operations center. In: Wang, L., Jajodia, S., Singhal, A. (eds.) Network Security Metrics, pp. 177–207. Springer, Cham (2017). https://doi.org/10.1007/978-3-319-66505-4_8D
20. Erbacher, R.F., Hutchinson, S.E.: Extending case-based reasoning to network alert reporting. In: 2012 ASE International Conference on Cyber Security, pp. 187–194 (2012)
21. Sundaramurthy, S.C., et al.: A human capital model for mitigating security analyst burnout. In: Eleventh Symposium on Usable Privacy and Security (SOUPS 2015), pp. 347–359. USENIX Association (2015)

22. Sundaramurthy, S.C., McHugh, J., Ou, X., Wesch, M., Bardas, A.G., Rajagopalan, S.R.: Turning contradictions into innovations or: how we learned to stop whining and improve security operations. In: Twelfth Symposium on Usable Privacy and Security (SOUPS 2016), pp. 237–250. USENIX Association (2016)
23. Killcrece, G., Kossakowski, K.P., Ruefle, R., Zajicek, M.: State of the practice of computer security incident response teams (CSIRTs). Technical report CMU/SEI-2003-TR-001, Software Engineering Institute, Carnegie Mellon University, Pittsburgh, PA, table 9, p. 66 (2003)
24. Scarfone, K., Mell, P.: Guide to intrusion detection and prevention systems (IDPS). Special Publication 800-94, NIST (2007)
25. Nelson, R.T., Holloway, C.A., Mei-Lun Wong, R.: Centralized scheduling and priority implementation heuristics for a dynamic job shop model. AIIE Trans. $9(1)$, 95–102 (1977)
26. Cleveland, B., Mayben, J.: Call Center Management on Fast Forward: Succeeding in Today's Dynamic Inbound Environment. Call Center Press, Annapolis (1997)
27. Hur, D., Mabert, V.A., Bretthauer, K.M.: Real-time work schedule adjustment decisions: an investigation and evaluation. Prod. Oper. Manag. $13(4)$, 322–339 (2004)
28. Love, R.R., Hoey, J.M.: Management science improves fast-food operations. Interfaces $20(2)$, 21–29 (1990)
29. Loucks, J.S., Jacobs, F.R.: Tour scheduling and task assignment of a heterogeneous work force: a heuristic approach. Decis. Sci. $22(4)$, 719–738 (1991)
30. Vieira, G.E., Herrmann, J.W., Lin, E.: Rescheduling manufacturing systems: a framework of strategies, policies, and methods. J. Sched. $6(1)$, 39–62 (2003)
31. O'Connor, E.J., Peters, L.H., Rudolf, C.J., Pooyan, A.: Situational constraints and employee affective reactions: a partial field replication. Group Organ. Stud. $7(4)$, 418–428 (1982)

Online and Scalable Adaptive Cyber Defense

Benjamin W. Priest[1]([✉]), George Cybenko[1], Satinder Singh[2],
Massimiliano Albanese[3], and Peng Liu[4]

[1] Dartmouth College, Hanover, NH, USA
{benjamin.w.priest.th,gvc}@dartmouth.edu
[2] University of Michigan, Ann Arbor, MI, USA
baveja@umich.edu
[3] George Mason University, Fairfax, VA, USA
malbanes@gmu.edu
[4] Penn State, State College, PA, USA
pliu@ist.psu.edu

Abstract. This chapter introduces cyber security researchers to key concepts in the data streaming and sketching literature that are relevant to Adaptive Cyber Defense (ACD) and Moving Target Defense (MTD). We begin by observing the challenges met in the big data realm. Particular attention is paid to the need for compact representations of large datasets, as well as designing algorithms that are robust to changes in the underlying dataset. We present a summary of the key research and tools developed in the data stream and sketching literature, with a focus on practical applications. Finally, we present several concrete extensions to problems related to ACD applications throughout this book, with a focus on improving scalability.

1 Introduction

Resilience in the face of a changing and uncertain environment is a prime concern motivating the study of Adaptive Cyber Defense (ACD). Practical ACD solutions must perform actions quickly, often as a reaction to some environmental stimulus, and usually without a human in the loop. Accordingly, efficient software and hardware implementations of ACD technologies should be fast, cheap to deploy, and scale to the size of their problem domain. Furthermore, they must deliver near-real-time performance. These requirements can be difficult to achieve when data sizes are large. For example, a technique that relies on network communication traffic statistics might need to process millions or even billions of events per hour depending on the size of the underlying network. At this point even data storage becomes tricky, especially on embedded components.

Several ACD technologies discussed in this book depend upon potentially large data inputs. Indeed, network-level techniques that depend on the traffic or topology of a computer network as data inputs are susceptible to many of

© Springer Nature Switzerland AG 2019
S. Jajodia et al. (Eds.): Adaptive Cyber Defense, LNCS 11830, pp. 232–261, 2019.
https://doi.org/10.1007/978-3-030-30719-6_10

the usual challenges that accompany so-called big data. While there may be a large number of events serving as input to an ACD system, often the actionable or "interesting" information is small compared to the entirety of available data. Indeed, in many cases relatively simple statistics of the data suffice to support adaptive decision making. Additional aspects of big data analysis for cyber security can be found in [14].

In particular, consider the problem of detecting changes in network traffic over the gateway between an enterprise network and the outside world. The conventional wisdom is to utilize either a sliding window of event logs or to chunk them into discrete time windows of some fixed length (e.g. 1 h), and compare current statistics to past statistics when attempting to either detect an interesting network event or diagnose one after the fact.

For example, one might want to keep track of the number of distinct remote hosts that send and/or receive traffic from each enterprise host. A significant increase in the number of remote requesters over time could indicate a distributed denial-of-service attack, especially if other enterprise hosts experience a similar increase, and even more so if there is significant overlap among the new remote hosts contributing to each quantity. One could naïvely solve this problem by maintaining a set for each remote host, which allows not only recovery of the quantity in question but also the computation of the intersections between different hosts. However, such an approach does not scale to the size of the Internet, where the memory and time costs become prohibitive. We will describe data structures and techniques that allow approximate recovery of the distinct elements, as well as heuristic recovery of their intersection sizes using exponentially less memory and constant time, a significant improvement over the naïve approach.

It may also be interesting to keep track of the enterprise hosts sending or receiving the most data over the proxy, or alternatively the source-destination pairs that communicate the most data over some time period. Changes in these statistics indicate changes in the underlying dynamics of the flow of web traffic, and could be used to detect or diagnose a malicious exfiltration event, or perhaps a denial of-service attack. Records of the changes in these orderings over time also give forensic analysts good places to start looking when diagnosing network events after-the-fact. However, the look-up table enumerating all sources, destinations, or source-destination pairs to their traffic volume over each, say, hour, tends to be intractably large in practice. Furthermore, sorting such a data structure is an expensive process. We will describe data structures and techniques that allow an approximate recovery of these highest volume or so-called "heavy hitter" elements, which also uses exponentially less memory than the naïve solution.

Time is a critical element in ACD scenarios. The state of a computer network can change, sometimes rapidly, including not only the network topology but also the nature of the traffic flow exchanged across the network. Near-real-time reaction speed demands that data processing update quickly in the face of burgeoning events. Consequently, batch analyses that require total recomputation in the presence of changes to the underlying data are unsuitable in all but the smallest or simplest cases. Consequently, the robustness and efficacy requirements of ACD solutions demand data processing that updates quickly in the presence of changes to the underlying data. So-called "online processing" is the algorithmic solution to this problem, wherein algorithms are assumed to have only sequential access to data elements. In this model, update time and pass complexity become important measurements of algorithmic efficiency. In general single-pass or truly online algorithms are preferred, although algorithms requiring a provably small number of passes may be acceptable in some cases where the underlying data is small enough to be warehoused. Many problems traditionally solved with batch algorithms admit to online generalizations, although sometimes at the expense of additional algorithmic complexity and implementation difficulty. Unfortunately, many problems are known not to admit online algorithms, or such algorithms are unknown.

ACD technologies typically do not rely upon the data itself, but instead on statistics thereof. In fact, a data summary that can be queried for the desired statistics suffices in many cases. Moreover, coarse approximations are often acceptable. Any practical summary should furthermore scale weakly with data size both in terms of time and memory. A further desirable property of such a data structure is that it admit some form of merge procedure, where two data structures accumulated over different streams are combined to produce a data structure that summarizes the concatenation of the two streams. Such summaries, usually called sketches, are an important object of study in the data stream model.

Algorithms in the data stream model obtain only sequential access to elements of the data, possibly in adversarial order, and are typically allowed additional memory at most logarithmic in the data size. A common generalization is the turnstile or dynamic data streaming model, which allows updates to the underlying data structure, e.g. changing the value of an index in a vector, or removing an edge from a graph. Sketches in the turnstile data streaming model are called linear sketches, as they amount to linear transforms of the underlying data. In addition to the measures introduced in the online processing model, the data stream model further emphasizes memory complexity as an efficiency metric. Some relaxations allow a provably small number of passes, or expand the memory complexity to polyloglinear in one dimension of the data. This latter relaxation of often called the semi-streaming model, and is usually required for complex structured data such as matrices, tensors, and graphs. In exchange for small size and low update cost, sketches typically produce approximate results and are only suitable for answering queries of a particular statistic of the data, which must be known ahead of time. The scaling particulars depend upon the desired queries, and some queries are known to be insoluble in sublinear memory.

This chapter will provide a short introduction to streaming and sketching algorithms and discuss their application to ACD problems, including specific applications to technologies discussed in other chapters. Section 2 provides a brief overview of practical and nearly practical sketching technologies. Section 5 discusses some applications of sketching algorithms to ACD technologies, including those discussed in other chapters. Section 4 discusses the problem of identifying important vertices in large graphs used elsewhere in this book, and provides some sketch-based solutions for approximating or estimating them. Finally, Sect. 6 summarizes the contributions of this chapter.

2   Sketching and the Data Stream Model

For the remainder of this chapter we shall refer to data inputs, be they network traffic logs, sensor readings, database records, et cetera, as data streams. Data streams generalize the phenomenon described above, wherein data inputs are sufficiently large and latency demands are sufficiently high that challenges arise in interfacing between the program and the input, computing functions on large segments of the input, and storing the input in working or disk memory. A streaming algorithm interacts with the input sequentially, updating some internal data structure for each input element before discarding it. Consequently, both the additional memory required by an algorithm and the number of passes that it takes over the data input become resources just as important as the number of operations required.

A stream $\sigma = \langle a_1, a_2, \ldots, a_m \rangle$ is a sequence of elements in the universe U, $|U| = n$. We assume throughout that that the hardware has working memory storage capabilities $o(\min\{m, n\})$. Thus, clearly, storing σ in working memory is not an option. A streaming algorithm $\mathcal{A}$ has three primary subroutines: INIT, PROCESS, and OUTPUT. Algorithm 1 summarizes such an algorithm. INIT initializes a data structure $\mathcal{D}$ that $\mathcal{A}$ will maintain. This typically involves both allocating memory and determining randomness, often in the form of sampling hash functions from a suitable hash family. As $\mathcal{A}$ reads over σ, it calls PROCESS on each element, updating $\mathcal{D}$. These elements are discarded after they are read. OUTPUT performs some computation over $\mathcal{D}$, returning the desired statistic. We will sometimes use the notation $\mathcal{D}(\sigma)$ to indicate the data structure state after $\mathcal{A}$ has read over σ. We refer to this process as *accumulating* $\mathcal{D}(\sigma)$.

Algorithm 1. Stream Algorithm Operations

1: **procedure** INIT
2:     Initialize $\mathcal{D}$
3: **procedure** PROCESS($i \in \sigma$)   \\pass over σ
4:     Apply i to $\mathcal{D}$
5: **procedure** OUTPUT
6:     Query $\mathcal{D}$ for output

In order to allow $\mathcal{A}$ to scale to immense data sizes, it is necessary to maintain the constraint $|\mathcal{D}| = o(\min\{m, n\})$. Ideally we need only store a constant number of tokens and counters so that $|\mathcal{D}| = \tilde{O}(1) = O(\log m + \log n)$. Here we use the tilde to suppress logarithmic factors, and will continue to do so throughout the rest of the chapter. The logarithmic memory constraint guarantees that

$\mathcal{D}$'s memory footprint is equivalent to the cost of storing a constant number of elements of σ.

Unfortunately, logarithmic memory constraints are not always possible. For example, detecting whether a particular element of U comprises more than half of a stream σ requires space $\Omega(\min\{m,n\})$! In particular, it is known that many fundamental properties of complex structured data such matrices and graphs require memory linear in some dimension of the data [49,51]. In such cases, the logarithmic requirements of streaming algorithms are sometimes relaxed to $O(n\,\mathrm{polylog}\,n)$ memory. In the case of matrices, here n refers to one of the matrix's dimensions, whereas for graphs n refers to the number of vertices. This is usually known as the *model*, although some authors also use the term to refer to $O(n^{1+\gamma})$ for small γ [28,53]. We will use the former definition in this document.

In addition to restricting the space requirements of $\mathcal{A}$, it is important that its procedures PROCESS and OUTPUT be $o(\min\{m,n\})$, so that updating and querying $\mathcal{D}$ remains fast. Again, $\widetilde{O}(1)$ or even $O(1)$ is the preferred performance. The particular constraints on the efficiency of these subroutines depends on the problem, but in general in real-time systems it is important that PROCESS require time less than the data interarrival time.

By enforcing these constraints on $\mathcal{A}$, we loose the ability to arbitrarily query the data for different statistics or examine individual elements. Furthermore, for most problems we must grant that OUTPUT returns an approximate statistic. Here we give up exact computation in favor of Monte Carlo guarantees, where output is guaranteed to have low error with high probability. However, this loss of freedom, accuracy and precision is a necessary condition for fast, scalable analysis of many practical problems. Fortunately, this is not so great a sacrifice if we know ahead of time what aspects of the input are of interest.

A stream σ is often thought of as updates to a hypothetical frequency vector $f(\sigma)$, which holds a counter for each element in U. We will drop the parameterization of f where it is clear. For example, σ could be a stream of IP addresses querying a server, in which case f is a vector of the number of queries issued from each source IP. Given our assumptions so far, f is clearly too large to store directly, although for some problems a sparse representation might be feasible. Where $\mathcal{A}$ and $\mathcal{D}$ are as above, we instead accumulate $\mathcal{D}$ using $\mathcal{A}$ by reading over σ. $\mathcal{D}$ can be thought of as a lossy compression of f that only preserves some statistic thereof, such as a norm.

Consider the situation where a frequency vector f represents a mutable, persistent object, such as a matrix storing trust scores between machines in a managed network. It would be nice to not have to throw away the accumulated data structure $\mathcal{D}$ and restart $\mathcal{A}$ whenever one of the elements of f changes. Accordingly, we commonly generalize the concept of a frequency vector by allowing that σ applies positive or *negative* updates to f. We call a stream σ a *turnstile* or *dynamic* stream if its elements are of the form (i,c), where i is an index of f (an element of U), $c \in [-L,\ldots,L]$ for some integer L, and (i,c) indicates that $f_i \leftarrow f_i + c$. Streaming algorithms that are robust to such changes are accordingly called *turnstile* or *dynamic* in the literature. The setting where only positive updates are permitted is usually referred to as the *cash register* model. The setting where negative updates exist but all elements of f are guaranteed to be nonnegative is called the *strict* turnstile model.

Furthermore, it is often desirable to compare separate streams. Let $\circ$ be the concatenation operator on streams. Given stream σ_1 and σ_2, it is often useful to determine statistics of $f(\sigma_1 \circ \sigma_2)$. For example, consider the problem where σ_1 and σ_2 are streams of IP address, perhaps visitors to different servers or the same server over different time periods. It may prove useful to know, for example, the number of unique visitors between the union of the streams, the ℓ_0 norm of $f(\sigma_1 \circ \sigma_2)$. If we have already accumulated $\mathcal{D}(\sigma_1)$ and $\mathcal{D}(\sigma_2)$, we would prefer to not take another pass over $\sigma_1 \circ \sigma_2$, especially as we may no longer have access to them. For $\mathcal{A}$ a streaming algorithm, we call its data structure $\mathcal{S}$ a *sketch* if there is an operator $\oplus$ such that, for any streams σ_1 and σ_2,

$$\mathcal{S}(\sigma_1) \oplus \mathcal{S}(\sigma_2) = \mathcal{S}(\sigma_1 \circ \sigma_2). \tag{1}$$

This property is useful in many practical settings, as we will describe in greater detail below.

Linear sketches are an important subclass of sketches where $\mathcal{S}$ is a linear function of $f(\sigma)$ of a dimension fixed by n. More specifically, a linear sketch transform is a linear transform $\mathcal{S} : F^n \to F^{\ell(n)}$, where F is the field over which $f(\sigma)$ is a vector and $\ell(n) \ll \min\{m, n\}$ is the dimension of the transform. Typically $\mathcal{S}$ is drawn from a distribution Π over such transforms. We will abuse terminology by referring to sketch distributions, sketch transforms, and sketch data structures as sketches where it is clear which is referenced. In particular, any turnstile streaming algorithm uses a linear sketch as a data structure [47].

Since linear sketches take values in a vector space, we can perform arithmetic on them. For streams σ_1 and σ_2 with frequency vectors f_1 and f_2, scalars a and b, and linear sketch transform S,

$$a\mathcal{S}(f_1) + b\mathcal{S}(f_2) = \mathcal{S}(af_1 + bf_2). \tag{2}$$

Having covered the preliminary concepts, we will now discuss several specific sketches and sketching algorithms that are useful in practice. This is not a comprehensive survey of the topic and we will skip over many technical details. We invite the interested reader to investigate cited sources for greater detail. In particular, we direct the interested reader toward Muthukrishnan's survey of streaming algorithms [53], Mahoney's survey of randomized matrix algorithms [49], Woodruff's excellent book covering sketching in numerical linear algebra [73], and Feigenbaum et al.'s [28] and McGregor's [51] surveys of streaming graph algorithms. Unfortunately, several of these broad sources are now slightly dated and do not include the newest research, particularly sketching graph algorithms. We will address several of these in Sect. 3.3. See the Guha and McGregor's resources from their workshop at KDD2018 for a partial bibliography of recent advances in graph streaming and sketching [33].

2.1 Estimating Distinct Elements

How many distinct elements are present in a stream? We have already used counting distinct IP addresses as a motivating example. This problem, estimating the cardinality of σ when viewing it as a multiset, is a core problem in the streaming literature. We can rephrase it as finding the ℓ_0 norm of $f(\sigma)$, alternatively its 0th frequency moment, given by $\|f(\sigma)\|_0 = F_0 = |\{i \in \{1, 2, \ldots, m\}|f(\sigma)_i \neq 0\}|$. It is known that any data structure that provides relative error guarantees for a multiset with n unique elements requires $\Omega(n)$ space [4]. Consequently, investigators have developed many data structures that provide such relative error guarantees while admitting a small probability of failure, such as PCSA [30], LINEARCOUNTING [72], MINCOUNT [7], LOGLOG [24], Multiresolution Bitmap [26], HYPERLOGLOG [29], and the space-optimal rough-refinement algorithm of [41].

The HYPERLOGLOG sketch is undoubtedly the most popular of these data structures in practice, and has attained widespread adoption [29]. The sketch provides an estimate of the cardinality of a streaming multiset in one pass using only $O(1/\varepsilon^2 \log \log n + \log n)$ bits of memory, where the estimate has standard error 1.04ε. Although there is a known optimal algorithm with space complexity $O(1/\varepsilon^2 + \log n)$ [41] with relative error ε, it is known to be inefficient in practice [67]. Consequently, less asymptotically-optimal HYPERLOGLOG-style sketches tend to be preferred in applications [25].

The HYPERLOGLOG sketch leverages randomness to count distinct elements of a stream. The sketch relies on the key insight that the binary representation of a random machine word of size W starts with $0^{j-1}1$ with probability 2^{-j}. Thus, if the maximum number of leading zeros in a set of random words is $j-1$, then 2^j is a good estimate of the cardinality of the set [30]. However, this estimator is coarse and has high variance. We overcome this by pseudorandomly partitioning the stream into many substreams. Upon reading i we compute $h(i)$, for h a hash function from U to $\{0, 1, \ldots, 2^W - 1\}$. The hash family from which h is drawn should be such that for each element $i \in U$, $h(u)$ is uniformly distributed in $\{0, 1, \ldots, 2^W - 1\}$. This ensures that $h(i)$ serves as a proxy for a uniform random number for each $i \in U$. We maintain $r = 2^p$ registers (S) for integer p, where i is mapped to the register indexed by the first p bits of $h(i)$. We then count the leading number of zeros in the remaining $W - p$ bits of $h(i)$ plus one. If this value is larger than the value stored in the register, then we overwrite it. Once we have read all of σ, S_j holds the maximum number of leading zeros to appear in its substream of hashed values.

To query the cardinality of the stream, we return the bias adjusted harmonic mean of the estimator present in each register,

$$\widetilde{F}_0 = \alpha_r r^2 \left(\sum_{j=0}^{r-1} 2^{-S_j} \right)^{-1}, \tag{3}$$

where α_r is a bias correction term that depends on r given by

$$\alpha_r := \left(r \int_0^\infty \left(\log_2 \left(\frac{2+u}{1+u} \right) \right)^r du \right)^{-1}. \tag{4}$$

Unfortunately, this estimate is known to suffer from bias when F_0 is very small or very large relative to W. The original authors corrected bias on small cardinalities by replaced the estimate with a different estimator LINEARCOUNTING when $\tilde{F}_0$ falls below a threshold [29]. Subsequent work replaces the estimator (3) entirely, with either a more robust bias reduction term in the denominator [64] or a maximum likelihood estimate [25,74]. Subsequent authors corrected the bias on high cardinalities by increasing the machine word size from 32 to 64 bits [36].

After bias correction, (3) has the guarantee that the standard error $|\tilde{F}_0 - F_0| \le 1.04\varepsilon F_0$, where $r = O(1/\varepsilon^2)$. As each register requires only space $O(\log \log n)$, the space complexity of the sketch is thus $O(1/\varepsilon^2 \log \log n + \log n)$.

While there are optimal sketches that offer better space complexity and error guarantees, they tend to be impractical to implement when compared with HYPERLOGLOG, especially because HYPERLOGLOG's PROCESS and OUTPUT subroutines are so fast and easy to implement [36]. Subsequent literature has further improved its efficiency. HYPERLOGLOG++ introduced a sparsification scheme for the register list so that it is compact when few elements have been read [36], while HYPERLOGLOG-TAILCUT significantly reduces the number of bits required for each register [74].

HYPERLOGLOG sketches have seen significant use in applications. For example, as has been used as an example above, HYPERLOGLOG sketches are useful for probabilistically counting the number of unique visitors to a server. Furthermore, while they are not linear sketches, HYPERLOGLOG sketches admit a composition operator of the type (1), which is taking the element-wise maximum of the registers of a collection of sketches sharing the same hash function. This property affords queries of the sort "how many unique IP addresses visited a collection of servers" given a sketch for each server.

One failing of HYPERLOGLOG and other popular cardinality sketches such as MINCOUNT is that they are not robust to negative updates. Linearity is a desirable property, in particular for answering queries of the form "how many unique IP addresses visited *each member of* a collection of servers" given a sketch for each server. So-called Hamming norm sketches address this concern utilizing a dimension-reducing linear transform composed of elements independently sampled from stable distributions [19]. Unfortunately, this reliance upon a very large matrix is a space liability that is only overcome by applying pseudorandom number generators for space bounded computation [57] and generating each element on-the-fly as needed. This results in an impractically slow PROCESS time.

Failing linearity, others have examined instead estimating the *intersection* of accumulated cardinality sketches. A naïve estimator computes estimates of each stream and their union and estimates the intersection using the inclusion-exclusion priciple. However, this can result in negative estimates, and exhibits high variance. Ting shows promising results for a different cardinality sketch

MINCOUNT, which permits combinatorial queries concerning the unions and intersections of MINCOUNT sketches due to their closed union and intersection operations [67]. Cohen et al. and Ertl independently developed maximum likelihood estimators for HYPERLOGLOG intersections that do not yield a closed intersection operator [18,25]. Unfortunately, as might be expected all of these estimators exhibit high variance on small intersections, which limits their utility.

2.2   Approximate Counting

How many times has an element been encountered? Alternately, what are the top k most frequent elements in a collection? Queries of $f(\sigma)_i$ are common in applications, for example: "how many times has an IP address queried this server?" However, storing $f(\sigma)$ is out of question.

Perhaps surprisingly, there are $\widetilde{O}(1)$ sketches that can approximately answer queries of the above type. COUNTMINSKETCH is one such sketch, and perhaps the most popular in applications [20]. The hashing scheme in COUNTMINS-KETCH is in some ways similar to a counting Bloom filter [27], although it differs in key respects. First consider an array of counters S of length r coupled with a hash function h drawn from a 2-universal hash family (e.g. multipy-shift hashing [23]) mapping U onto $\{1, 2, \ldots, r\}$. When $(i, c) \in \sigma$ is observed, we set add c to $S_{h(i)}$. Similarly, upon querying i we return $S_{h(i)}$. However, this method clearly has high variance due to hash function collisions, since $S_h(i) = \sum_{j:h(j)=h(i)} f_j$.

We overcome this variance by parallelizing the approach. The true COUNT-MINSKETCH S consists of k arrays of counters of length r, as well as k hash functions mapping U onto $\{1, 2, \ldots, r\}$. When $(i, c) \in \sigma$ is observed, we add c to $S_{j,h_k(j)}$ for each $j \in \{1, 2, \ldots, k\}$. When we query an accumulated COUNT-MINSKETCH for a particular index i, it returns $\min_{j \in k} S_{j,h_j(i)}$.

Surprisingly enough, this estimate turns out to be quite good in general. For a true frequency vector f, let $\widetilde{f}$ be the hypothetical vector consisting of all frequency estimates for elements of U. Then COUNTMINSKETCH guarantees that for all $i \in \{1, 2, \ldots, m\}$, $\widetilde{f} - f \in [0, \varepsilon(\|f\|_1 - f_i)]$ with probability $1 - \delta$. Here, the accuracy and precision parameters ε and δ are such that $r = O(1/\varepsilon^2)$ and $k = O(\log(1/\delta))$, so that COUNTMINSKETCH has space complexity $\widetilde{O}(1/\varepsilon \log(1/\delta))$. Hence, the size of CountMinSketch is only logrithmically dependent upon the size of its input. It is also worth noting that the hash functions need only be pairwise independent, and so there are practical implementations of COUNTMINSKETCH.

COUNTMINSKETCH is a linear sketch on strict turnstile streams. The slightly more complicated COUNTSKETCH is a linear sketch on arbitrary turnstile streams, although it is slightly less space efficient. We will not go into the details of COUNTSKETCH here, inviting the interested reader to examine [15].

While the returned estimate is biased and may be an overestimate, subsequent modifications seek to correct for this bias by subtracting a bias estimate and taking the median, rather than the min [22], or by taking account of the current estimate when inserting a new observation [31]. In particular, COUNTMINS-KETCH's error guarantees are tightest on heavy hitters - the highest frequency indices of f. Thus, one can probabilistically obtain the heavy hitters of a stream by maintaining a heap while accumulating a COUNTMINSKETCH with high certainty. Charikar et al. provide algorithmic details for such a heap-based heavy hitter recovery algorithm using COUNTSKETCH [15]. If the stream is guaranteed to be a cash register stream, then one could instead use COUNTMINSKETCH for a space savings of $O(1/\varepsilon)$. Although the maximum size of this heap depends upon the distribution of f, so long as this distribution is not nearly uniform the algorithm depends only logarithmically on n. Moreover, as the only data structures are COUNTSKETCH and a heap, the algorithm can be implemented efficiently.

COUNTMINSKETCH and similar variants are widely used to collect frequency statistics in applications. In particular, their linearity makes them suitable for comparing statistics between disparate streams, such as finding heavy hitters - the most frequent elements - as well as changes over time or between streams as a form of anomaly detection. COUNTMINSKETCH has seen use in many different application areas, including network anomaly detection [20], counting $k$mers in genomics [76], and finding corpora statistics in natural language processing [31]. We will apply COUNTMINSKETCH to recovering the heavy hitters of the degree distribution on graphs in Sect. 4.1. COUNTSKETCH-type sketches are also a common subroutine in more complex sketching algorithms.

2.3 Approximating Norms

One of the richest veins in the sketching literature is the approximation of various norms of $f(\sigma)$. Indeed, the HYPERLOGLOG sketch in Sect. 2.1 can be thought of as a non-linear estimate of the ℓ_0 norm of f. Moreover, the ℓ_1 norm of $f(\sigma)$ is trivial to accumulate in a pass over σ. The literature on approximating streaming and sketching ℓ_p norms is much too voluminous to do justice to it here [4, 37, 38, 56]. We will instead discuss one of the early successes in approximating the ℓ_2 norm, sometimes called the TUGOFWAR sketch.

TUGOFWAR is deceptively simple. Consider a single counter x and a hash function h mapping U to $\{-1, 1\}$. While reading $(i, c) \in \sigma$, we simply add $h(i)c$ to x. Surprisingly enough, $|x|$ is an unbiased estimator of $\|f(\sigma)\|_2^2$. Of course, the variance is large.

It is worth noting here that h implicitly defines a linear transformation $s = (h(1), h(2), \ldots, h(m))$ so that $x = \langle s, f \rangle$. If we repeat this operation with r independent hash functions, we have defined a random $r \times m$ matrix S that produces a vector $x = Sf$ when accumulated over σ. It is not immediately obvious that this is an improvement, as the matrix S is still large. Fortunately, the hash function h need only be drawn from a 4-universal hash family (e.g. via tabulation hashing [63]), meaning that we can efficiently compute elements of S as they are needed rather than storing it in memory. In fact, $\frac{1}{\sqrt{r}}\|x\|_2 \in [(1 - \varepsilon)\|f\|_2, (1 + \varepsilon)\|f\|_2]$ with probability greater than $\frac{2}{3}$, where $r = O(1/\varepsilon^2)$. The distribution over projections defined by h and S shares many similarity with the famous Johnson Lindenstrauss Lemma, which has seen extensive application in dimentionality reduction by way of random projections [39]. By repeating this procedure in parallel $t = O(\log(1/\delta))$ times and taking the median, we can guarantee the same error bounds with a $1 - \delta$ probability of failure, using only $\tilde{O}(1/\varepsilon^2 \log(1/\delta))$ space.

The TUGOFWAR is clearly a linear sketch, as it amounts to a linear function on f. Thus, it can be used to answer interesting queries such as finding the norm of the difference between different streamed vectors f_1, f_2 by $Sf_1 - Sf_2 = S(f_1 - f_2)$. Such queries can be useful for detecting when there are large changes in the frequency distribution on, for example, visitors to a server or destinations queried over a proxy. Like COUNTSKETCH, TUGOFWAR and related norm approximation sketches are commonly used as subroutines of more sophisticated sketching algorithms. There exist similar algorithms for estimating ℓ_p norms for other values of p [4,37,38,56], as well as matrix norms [46,48].

3   Advanced Sketching Applications

Section 2 and the references therein provide an incomplete accounting of the history of data sketching on elementary statistics of unstructured data streams. However, structured data such as matrices or graphs arise naturally in many applications. Handling such data in the streaming model often requires a greater degree of finesse than the applications described so far. This section and the included references shall serve as an incomplete overview of more advanced applications of data sketching. The reader may note that many of the sketches described make use of data structures from Sect. 2 as building blocks. Furthermore, many of these algorithms are, by necessity, in the semi-streaming data model. Consequently, they require polyloglinear memory in some dimension of the structured data. This means that, while they improve upon the memory requirements of a RAM algorithm (e.g. quadratic in the case of square matrices), the improvement is not so drastic as seen throughout Sect. 2. Consequently, many of these algorithms have not seen the same degree of adoption in practice.

3.1   Approximate Numerical Linear Algebra

Linear transformations such as that used in TUGOFWAR above that approximately preserves the ℓ_2 norm by a multiplicative factor of ε while performing dimensionality reduction are commonly referred to as $(1 \pm \varepsilon)$-ℓ_2 subspace embeddings, or more concisely just subspace embeddings [55]. Indeed, the implicit transformation defined by the COUNTMINSKETCH and its variants also form subspace embeddings, and are used to great effect in the sublinear approximation of computations on large matrices in the literature [17,73]. Subspace embeddings allow us to approximately perform several linear algebraic operations that might otherwise be intractable to compute naïvely. The general strategy for such algorithms is to apply subspace embeddings to project the largest dimension of all involved matrices and vectors into a much smaller dimensional space, and then solve the original problem on these smaller matrices. Consequently, these methods achieve the most gain on matrices where one dimension is much larger than the other, such as highly overconstrained linear regressions of the sort that often arise in machine learning.

We will here briefly summarize some of these results. Throughout, we will consider that a stream σ accumulates a frequency matrix $M(\sigma)$, a natural generalization of a frequency vector $f(\sigma)$. We will drop parameters where it is clear what is accumulated. For a subspace embedding S of M, we will write SM to denote the matrix accumulated by reading over updates to M in σ and applying the transformation S. We will also take for granted that S can be applied to M in time $O(\mathrm{NNZ}(M))$, where $\mathrm{NNZ}(M)$ is the number of nonzero entries of M. Note below that $\|\cdot\|_F$ is the matrix Frobenius norm, the sum of element-wise squares.

For the purposes of this section we will take as a given that $r \times n$ subspace embeddings S can be computed and stored efficiently without going deeply into detail. Moreover, we will assume that there are distributions Π over such matrices such that for any $n \times d$ matrix A, $S \in_R \Pi$ is an ℓ_2 subspace embedding for A with high probability. In practice, such transformations are defined implicitly by hash functions in the style of the transformations in Sects. 2.2 and 2.3. These hash functions need to be sampled from a k-universal hash family for some small k. A k-universal family of hash functions H mapping $\mathcal{X}$ to $\mathcal{Y}$ have the property that, for any $h \in_R H$ and any $x_1, x_2, \ldots x_k \in \mathcal{X}$, $h(x_1), h(x_2), \ldots, h(x_k)$ is uniformly distributed in $\mathcal{Y}$. Efficient k-universal families are known for most small values of k [63,70]. In particular, the reliance upon k-universal hash families allows the computation of ℓ_2 subspace embeddings without relying upon perfectly random or fully independent hash functions, significantly reducing both the time required to apply transformations as well as their memory footprint.

Consider the problem of multiplying two matrices $A \in \mathbb{R}^{n \times d}$ and $B \subset \mathbb{R}^{n \times d'}$, producing the $d \times d'$ matrix $C = A^T B$. If n is large, then C becomes expensive to compute, even using efficient linear algebra software kernels. Let S be drawn from $r \times n$ subspace embedding distribution. By accumulating SA and SB, we can instead return the product $(SA)^T SB = A^T S^T SB$. Here we project the tall matrices A and B into the lower row dimension $r \ll n$ using the random projection S, and perform the multiplication on these smaller projections. We are guaranteed to obtain the property $\|AB - A^T S^T SB\|_F \leq \varepsilon \|A\|_F \|B\|_F$ with probability $1 - \delta$. This algorithm requires only $\widetilde{\Theta}((d + d')\varepsilon^{-2} \log(1/\delta))$ space where $r = O(\varepsilon^{-2} \log(1/\delta))$. Note that the space depends on the dimensions of A and B that are not sketched. Thus, if the matrices are nearly square, then the space complexity of this algorithm is polylogarithmic in n. Hence, this and the subsequent algorithms we will discuss are properly classified as semi-streaming algorithms.

Another common problem involving tall matrices is linear regression, a subroutine that arises naturally in many analyses including machine learning. Given matrix $A \in \mathbb{R}^{n,d}$ and $b \in \mathbb{R}^n$, we want to find the $x \in \mathbb{R}^d$ that solves the optimization problem $x^* = \mathrm{argmin}_{y \in \mathbb{R}^d} \|Ay - b\|_2$. Although there is a closed-form solution for ℓ_2 linear regression, it can still prove infeasibly expensive in the presence of enough constraints. Similar to the multiplication solution, we can sublinearize this problem by applying an $r \times n$ subspace embedding S to both A and b and solve the embedded problem $\widetilde{x} = \operatorname{argmin}_{y \in \mathbb{R}^d} \|SAy - Sb\|_2$. We are guaranteed to obtain the property $\|SA\widetilde{x} - Sb\|_2 \leq (1 + \varepsilon)\|Ax^* - b\|_2$ with probability $1 - \delta$ using $\widetilde{\Theta}(d^2\varepsilon^{-1}\log(1/\delta))$ space where $r = O(\varepsilon^{-1}\log(1/\delta))$.

We give a final example in the rank-k estimation of a matrix A. In this problem, the object is to find the matrix of rank at most k that most closely approximates A. Call this matrix A_k. If error is measured in the Fröbenius norm, then one can solve this problem by computing the singular value decomposition (SVD) of $A = U\Sigma V^T$ and returning the matrix composed of the top k singular vectors and values, which we will denote $U_k\Sigma_k V_k$. This is a common query, particularly when performing data analysis or dimensionality reduction such as PCA. It is possible to sublinearize this process using mutliple subspace embeddings in a single pass. However, the space complexity of the best known algorithm is not tight, and it is not considered practical [16]. Fortunately, there is a known optimal 2-pass algorithm that we will briefly present. The object of this algorithm is to produce a matrix $\widetilde{A}_k$ that fulfills the guarantee

$$\|A - \widetilde{A}_k\|_F \leq (1 + \varepsilon)\|A - A_k\|_F \tag{5}$$

with probability $1 - \delta$. Such an $\widetilde{A}_k$ is produced as follows, where S is an appropriate $r \times n$ subspace embedding, $r = O(\varepsilon^{-1}\log(1/\delta))$:

1. Accumulate SA in one pass over σ.
2. Compute U^T, an orthonormal basis for the rowspace of SA.
3. Accumulate AU in a second pass over σ.
4. Compute $[AU]_k$, the k-truncated SVD of AU.
5. Return (factored) $[AU]_k U^T$.

An immediate observation of this algorithm is that it is *not* sublinear in n. Indeed, the space complexity of the algorithm is $\tilde{O}(k\varepsilon^{-1}(n+d)\log(1/\delta))$, strictly classifying it as a semi-streaming algorithm. However, if one simply must find a low-rank approximation, or a substitute for the k-truncated SVD of a large matrix A, then this algorithm provably provides a nearly space-optimal solution.

As an addendum to this line of reasoning, often what one needs when computing the k-truncated SVD of a matrix is a decomposition of the principal directions of the action of the matrix when treated as a linear transformation. Properties of the singular vectors of such a matrix are often used in applications to draw conclusions about the data in the matrix. Consequently, in many applications a simpler approximation to the k-truncated SVD may suffice, including those that are not strictly rank k. One such gadget is provided using two subspace embeddings, $r \times n$ S and $d \times r\varepsilon^{-2}$ R, accumulating SA, AR, and SAR in one pass over σ. The matrix $AR(SAR)^+SA$ satisfies the error condition (5), although it may have rank greater than k [16]. Here $M^+ = V\Sigma^{-1}U^T$ is the Moore-Penrose pseudo inverse of the matrix M with SVD $U\Sigma V^T$. The advantage of using the factors AR, $(SAR)^+$, and SA is that they are (roughly) of the form of an SVD and can be computed relatively quickly and in one pass over σ.

3.2   Approximate Norm-Aware Sampling

Frequently one might want to sample some element or elements from a stream. However, it is not immediately clear how to randomly sample over elements in turnstile streams. In general, we want to sample an element from σ that is returned with probability proportional to the norm of the frequency vector f. More precisely, we want to return element i with probability $\frac{|f_i|^p}{\|f\|_p^p}$ for some $p \geq 0$. We call such a sample an ℓ_p sample of f.

Returning a strict ℓ_p sample certainly requires $\Omega(n)$ memory. Fortunately, there are ℓ_p sampling sketches that instead return element i with probability $(1 \pm \varepsilon)\frac{|f_i|^p}{\|f\|_p^p}$ with failure probability $1 - \delta$. The particulars of these sketches are somewhat involved, and so we will not cover them in detail here. We suggest that the interested reader to investigate [5, 40, 52]. Suffice to say that the sketches are linear over turnstile streams. The best known ℓ_1 sampling sketches require $O(\varepsilon^{-1} \log \varepsilon^{-1} \log^2 n \log \delta^{-1})$ space, while the best known ℓ_0 sampling sketches require $O(\log^2 n \log \delta^{-1})$ space [40]. Note that both have only a polylogarithmic dependence on n, and ℓ_0-sampling sketches can achieve arbitrary sampling precision at no additional cost. Interestingly, the main ingredients in these sketches are applications of COUNTSKETCH and TUGOFWAR, which are discussed in Sects. 2.3 and 2.2.

As these sampling sketches are linear, we can perform arithmetic with them. That is, if we have two streams over the same universe with frequency vectors f, f' and S is a sketch matrix drawn from such a sampling distribution, then $Sf + Sf' = S(f + f')$, allowing us to sample probabilistically from $f + f'$. Although these sampling algorithms are certainly difficult to implement and are not as computationally efficient as those already widely adopted in applications, such as those discussed in Sects. 2.1, 2.2, and 2.3, they can provide powerful insights into large data streams. Furthermore, their linearity means that they can be used to sample from immense vectors stored in distributed memory utilizing a MapReduce computation strategy. Each processor that owns part of the vector applies the same sketch transform to it, whereupon all of the sketches are sent to a central processor that adds them together and performs the sampling procedure. In particular, we will see applications of sampling sketches in Sect. 3.3 where they are used to sublinearly answer questions about graph structure, as well as a more practical example in Sect. 4 where they are used in the sampling of random walks on scale-free graphs.

3.3   Approximate Graph Structure

Just like generalizing a stream to update a frequency matrix as discussed in Sect. 3.1, we can also generalize the concept of a data stream as updates to the edges of a graph. There is a large body of work covering streaming graph algorithms where σ is a sequential scan over an edge set. This can also be interpreted as a stream of updates to the graph's adjacency matrix A (or its Laplacian L, or vertex-edge incidence matrix B). Much of the literature on streaming graphs focuses on a less general type of stream, where edges are updated only once and there are no negative updates [28]. See [51] for a good if somewhat outdated overview of this literature. However, many practical problems, including many discussed elsewhere in this document, depend on graphs that are subject to change. In such a stream, every update modifies the weight of an edge in the graph, possibly bringing it into existence or removing it from the graph by increasing its weight from zero or reducing its weight to zero, respectively. Fortunately, recent research [1,2,43] has solved a number of basic graph problems on evolving graphs in the semi-streaming model by way of linear sketches.

We will not go into great detail on the particulars of these algorithms, and will instead explain their general approach. In a recent series of papers, Ahn, Guha, and McGregor used ℓ_0- and ℓ_1-sampling sketches on the columns of the vertex-edge incidence matrix to solve or approximating a number of semi-streaming graph problems [1,2]. Their approach depends on ℓ_p sampling sketches as discussed in Sect. 3.2. Here assume that we have a turnstile stream defining $G = (V, E)$ where $|V| = n$ and $|E| = m$. Consider the vertex-edge incidence matrix B of G, defined as

$$B_{(x,y),z} = \begin{cases} 1 & \text{if } x = z \\ -1 & \text{if } y = z \\ 0 & \text{else.} \end{cases} \tag{6}$$

Each column of B lists a vertex's participation in edges in E. For $x \in V$, let $B_{:,x}$ denote the corresponding column. The graph sparsification algorithms all proceed along the following basic premise. Sample a series of ℓ_0 sampling sketches $S_1, S_2, \ldots, S_t$, for some $t = O(\log n)$. Algorithms depending on ℓ_1 sampling are similar. First, read the stream defining B (which can be adapted from an edge stream) and sketch each column of B using each of $S_1, S_2, \ldots, S_t$. If $x \in V$, then $B_{:,x}$ is a vector whose nonzero entries point to edges of which it is an endpoint. Thus, $S_1(B_{:,x})$ can recover a uniformly sampled neighbor of x, say y, with high probability. Then, $S_2(B_{:,x}) + S_2(B_{:,y}) = S_2(B_{:,x} + B_{:,y})$ can sample a neighbor of the supervertex $(x + y)$, since the row values indexed by the edge (x, y) are cancelled out when adding the two row vectors by the definition of B. New sketches are required for each contraction, as otherwise the samples will not be independent and the guarantees of the sampling fails. The sparisification of the graph so obtained constitutes a random logarithmically-sized subtree of G. This pattern permits the approximation of many of G's structural properties, possibly over several passes, using semi-streaming memory.

Particular applications include answering graph queries such as deciding whether a graph is connected, k-connected, or bipartite, as well as approximating the weight of its minimum spanning tree. Somewhat more memory $\tilde{O}(n^{1+\gamma})$ for γ an accuracy parameter permits the computation of approximate spanning subgraphs that can approximate distances, exactly computing the minimum spanning tree, and approximating the maximum weight matching over several passes [1,2]. A separate $O(\varepsilon^{1,2} n \operatorname{polylog} n)$ algorithm suffices to return $\tilde{U} \subseteq V$ that is has density at least $(1 - \varepsilon)$ of the maximum density subset of V [69].

It is important to note that these algorithms are not especially practical. For many real-world graphs where E is not too much bigger than V, sketching the graphs in this way actually requires more memory to store than just storing G explicitly. Furthermore, the reliance on such a large number of ℓ_p sampling sketches incurs a significant performance overhead. The discovery of a more practical ℓ_p sampling sketch would greatly improve the computational footprint of these algorithms, but it is unlikely that the memory overhead tradeoff will be worthwhile for all but the most dense graphs. A practical implementation of these algorithms using existing tools would likely instead explicitly store sparse representations of $B_{:,x}$ for each $x \in V$ until a given size threshold is reached, at which point a sampling sketch S is initialized based upon the held vector and maintained under subsequent updates. The tree-sampling stage of the algorithms would proceed as normal, augmented with behavior when explict and sketch representations of column vectors are merged. These algorithms are largely included for completeness, as well as to contrast the applications in Sect. 4.

Another application of sketching algorithms to graphs is the approximation of sparse subgraphs that preserve some property of the original graph, especially its Laplacian. The use case for such algorithms is the production of sparse subgraphs that can be used in further analyses while incurring reduced overhead by pruning edges. Algorithms of this type are more practical, as the upfront cost associated with the sketching procedure can yield significant savings on downstream, expensive, measurements so long as some error is tolerable. While a thorough discussion of such algorithms is out of scope for this document, we invite the interested reader to read [43] or Woodruff's treatment in Chap. 5 of [73].

Cardinality sketches also have applications to estimating some graph properties. For example, consider accumulating a HYPERLOGLOG sketch of the adjacency set of each vertex in a graph. Using these sketches, one can estimate the degrees of each vertex. However, one can also combine them to estimate other, less trivial graph properties. Indeed, the HyperANF algorithm uses HYPERLOGLOG sketches to estimate the n-hop or n-th degree neighborhood sizes of all vertices [10]. These quantities are useful for several applications, such as edge prediction in social networks [34] and coarse probabilistic distance calculations [10,54]. By utilizing cardinality sketch intersection estimators, one can also estimate local features such as edge- and vertex-local triangle counts, although the variance inherent in such estimators applied to small intersections implies that estimates will only be reliable on elements with a nontrivial triangles density [62].

Unfortunately, while these cardinality sketch-based approaches are more practical than their sampling-sketch based siblings, they are not linear and so cannot handle negative updates. Of course, implementing them with linear Hamming-norm sketches obviates this problem, but at the cost of once again producing algorithms that are not practicable.

4   Identification of Important Vertices in Massive Graphs

A common task in the analysis of a graph $G = (V, E)$, where $|V| = n$ and $|E| = m$, is the identification of vertices or edges that are "important". Quantifications of importance depend heavily upon the qualities being considered, and preferred methods tend to vary from problem domain to problem domain.

Centrality indices are one major class of importance quantifications. A centrality index is any map C that assigns to every $v \in V$ a nonnegative score $C(v)$. The particulars of C are usually assumed to be conditioned only on the structure of G. Consequently, we can identify the centrality index on G as a function $C_G : V \rightarrow \mathbb{R}_{\geq 0}$. For $x \in V$, we will call $C_G(x)$ the centrality of x in G. Typically, for $x, y \in V$, $C_G(x) > C_G(y)$ implies that x is more important than y in G with respect to the property that C measures. However, such point queries are not particularly informative. In most applications, what is desired are the top k centrality vertices of G, which we will call the vertex heavy hitters. We will generally drop the subscript from C when it is clear from context. It is important to note that if G changes, so does the mapping C. At times, we will write $C(G)$ or $C(V)$ to denote the set of all centrality scores of the vertices in G.

Centrality indices tend toward three distinct types: indices that rely upon distance-based measures, such as degree centrality and closeness centrality, spectral measures, such as eigencentrality and its variants Katz's index and PAGERANK, and path-based measures such as betweenness centrality [11]. Researchers have considered more exotic centrality indices that rely on metadata, such as vertex and edge colorings [42]. Such notions of centrality, while interesting, are out of scope for this document.

Throughout the rest of this section we will discuss to what extent sketching algorithms can improve the scalability of a few popular centrality indices. In general when computing $C(V)$ one is primarily interested in the top k vertices for some given k, as these are the most important vertices with respect to the index in question. Consequently, we can focus on the recovery of these "heavy hitters" instead of attempting to produce a score for each $x \in V$. In the subsequent sections we will investigate approaches to doing so for two popular centrality indices, degree centrality and betweenness centrality.

4.1   Streaming Degree Centrality Heavy Hitters via COUNTSKETCH

Degree centrality is perhaps the oldest and most basic notion of centrality. In an undirected and unweighted graph, the degree centrality of a vertex is simply calculated as its number of neighbors. If the graph is weighted, degree centrality is usually generalized to the sum of the weights of its incident edges. In either case, we will assume the degree centrality of vertex $x \in V$ is equal to the sum of the xth row of the adjacency matrix A. In a directed graph, the indegree (outdegree) centrality of vertex x is the number of incoming (outgoing) edges to (from) x, conventionally corresponding to the xth column (row) of A.

$$DC(x) = |\{(u,v) \in E \mid x \in \{u,v\}\}| = \|A_{x,:}\|_1 = \|A_{:,x}\|_1 \qquad (7)$$

$$IDC(x) = |\{(u,v) \in E \mid x = v\}| = \|A_{x,:}\|_1 \qquad (8)$$

$$ODC(x) = |\{(u,v) \in E \mid x = u\}| = \|A_{:,x}\|_1 \qquad (9)$$

Though simple, degree centrality is still widely used as a benchmark in many applications [11]. Indeed, it is competitive with more sophisticated notions of centrality in some contexts [68]. Moreover, indegree centrality is known to correlate well with PageRank, making it a decent proxy when computing PageRank is not practical [68].

Given an edge stream updating adjacency matrix A, it is a simple to interpolate it as a stream updating a vector v storing the valency of every vertex in the graph. If computing indegree, simply convert an update (x,y,c) to (y,c), where (y,c) is interpreted as "add c to v_y". Outdegree is similar, and if G is undirected one simply applies both updates. This formulation places us in the vector turnstile model. Moreover, the accumulated vector v is such that v_x is exactly the degree centrality of vertex x.

Fortunately, this formulation allows us to directly apply COUNTSKETCH, as discussed in Sect. 2.2 and use its heap accumulation algorithm to approximately obtain the degree centrality heavy hitters of G. So long as G is not nearly regular, which scale-free graphs that arise in most applications are certainly not, the space complexity of this procedure is only logarithmic in n and can be applied quickly to σ in one pass.

4.2 Estimating Betweenness Centrality Heavy Hitters with κ-Path Centrality and Sampling Sketches

Degree centrality is a rather basic measure of vertex importance. What can be done for more complex measures? The betweenness of a vertex is defined in terms of the proportion of shortest paths that pass through it. Thus, a vertex with high betweenness is one that connects many other vertices to each other - such as a boundary vertex connecting tightly-clustered subgraphs. For $x, y, z \in V$, suppose that $\lambda_{y,z}$ is the number of shortest paths from y to z and $\lambda_{y,z}(x)$ is the number of such paths that include x. Then the betweenness centrality of x is calculated as

$$BC(x) = \sum_{\substack{x \notin \{y,z\} \\ \lambda_{y,z} \neq 0}} \frac{\lambda_{y,z}(x)}{\lambda_{y,z}}. \qquad (10)$$

An implementation of betweenness centrality must solve the all pairs all shortest paths problem, which is a resource-hungry process. Moreover, there is no known algorithm for computing the betweenness centrality of a single vertex using less space or time than the best algorithm for computing the betweenness centrality for all of the vertices. The celebrated Brandes algorithm, the best known algorithm for solving betweeness centrality, requires $\Theta(nm)$ time ($\Theta(nm + n^2 \log n)$ for weighted graphs) and $\theta(m)$ space [12].

A significant number of algorithms have arisen that attempt to alleviate this time cost by approximating the betweenness centrality of some or all vertices have been proposed. Some of these approaches depend on adaptively sampling and computing all of the single-source shortest paths of a small number of vertices [6,13], while others sample shortest paths between random pairs of vertices [65,75]. A recent advancement incrementalizes the latter approach to handle evolving graphs [9]. Still other approaches drive down the approximation time by reasoning about sampled hypergraphs [35,75].

Fortunately, researchers have directed much effort in recent years toward maintaining the betweenness centrality of the vertices of evolving graphs [32,71]. The most recent of these approaches keep an eye toward parallelization across computing clusters to maintain scalability [45].

If the evolving graph in question is sufficiently small that storing it in working or even distributed memory is feasible, then existing solutions suffice to solve the problem in a reasonably efficient fashion. However, none of the existing solutions adapt well to the semi-streaming model, as they each require $\Omega(m)$ memory. Indeed, directly approximating betweenness centrality, even for a single vertex, seems likely to be infeasible using $o(m)$ memory.

Often, κ-path centrality is an attempt to obtain the top-k betweenness central vertices by way of relaxing the computations [3]. The κ-centrality of a vertex x is the probability that a random simple path of length at most k over all other source vertices y passes through x. Here a *simple* path is a series of adjacent vertices such that no vertex appears more than once. That is, a simple path has no loops. There is a randomized algorithm for approximating the κ-path centrality of the vertices of a graph that amounts to simulating $T = 2\kappa^2 n^{1-2\alpha} \ln n$ random simple paths over G and maintaining a counter for each vertex. Here α is an accuracy parameter. The authors demonstrate that on several real-world scale-free networks, vertices with high approximate κ-path centrality empirically correlate well with vertices with high betweenness centrality [44]. Moreover, this algorithm is much more efficient than most other attempts to approximate betweenness centrality, as it sidesteps the need to compute any solutions to single source all shortest paths, as it instead samples random walks.

Furthermore, this κ-path centrality algorithm is similar in some respects to those in Sect. 3.3 in that it mostly depends upon sampling. The algorithm is given in Algorithm 2.

The algorithm is mostly the same as the κ-path centrality algorithm, with the exception that sampling is performed approximately using sampling sketches. We can make this algorithm much more practical by distributing it using a Pregel-like model [50]. In a Pregel-like model, vertex adjacency information is distributed among the processors in a cluster. Computation proceeds in a series of passes, where in each pass the processors perform some computations and then communicate. This computation scheme can be made much more efficient in practice by allowing the communication to occur asynchronously [59]. In a data stream setting, each processor is responsible for streams that define the adjacency vectors for some subset of vertices, and each pass corresponds to

---

Algorithm 2. Sublinear κ-Path Centrality

1: $T \leftarrow 2\kappa^2 n^{1-2\alpha} \ln n$
2: $P \leftarrow T$ empty paths
3: $L \leftarrow T$ uniform samples from $\{1, 2, \ldots, \kappa\}$
4: $C \leftarrow n$ empty counters
5: **for** $i \in \{1, 2, \ldots, T\}$ **do**
6:     $P_{i,1} \leftarrow$ a vertex uniformly sampled from V
7: **for** $j \in \{1, 2, \ldots, \kappa - 1\}$ **do**   \\loop corresponds to a pass over σ
8:     **for** $i \in \{1, 2, \ldots, T\}$ **do**
9:       . **if** $j \leq L_i$ **then**
10:         $S \leftarrow$ an ℓ_0-sampling sketch
11:         accumulate $SA_{P_{i,j}}$, ignoring elements in $P_{i,:}$ \\simple paths
12:         $P_{i,j+1} \leftarrow$ sample from $SA_{P_{i,j}}$
13:         $C_{P_{i,j+1}} \leftarrow C_{P_{i,j+1}} + 1$
14: **return** $C_x / 2\kappa n^{-2\alpha} \ln n$ for each $x \in V$

---

processors taking a pass over some of their vertex adjacency streams. Otherwise, the algorithm is mostly the same. The only significant change is that, at the start of the jth iteration of the outer loop, the current holder of $P_{i,:}$ must transmit it to the processor holding $P_{i,j}$. There is also a reduce step wherein the processors sum up C. Note that, instead of performing a naïve sum of C, we could instead communicate each local copy of C to a single processor who treats the incoming communication as a data stream and maintains a k-heap, returning the k-heavy hitters of C if the heavy hitters are all that is desired.

This approach maintains the error guarantees of the original κ-path centrality algorithm at the expense of a small probability of failure, which can be superceded by allowing that multiple passes over streams may be required. It can also be relaxed so that only vertices with vertex adjacency list size over a certain threshold need be sketched. In a scale free graph, most vertices will be small, but some will have degree that is a nontrivial fraction of n [8]. For most practical purposes it is only these large vertices that need be sketched. For a procedure that depends on high betweenness centrality vertices on a very large graph, this procedure will provide an estimate of these heavy hitters using fewer resources. It also avoids the practical problems of storing these large vectors in distributed memory over the compute cluster in delegates [60]. [61] describes and analyzes this and other related algorithms based upon the semi-streaming sampling of random walks.

5 Examples of Sketching Applied to Adaptive Cyber Defense

We will quickly summarize some of the Adaptive Cyber Defenses discussed throughout this chapter, as well as discussing other applications to the work presented in other chapters in this document. We opened Sect. 1 with a high-level discussion of various analyses of network traffic summary statistics and how they can be applied to ACDs. We will now discuss some of these problems in terms of the technologies introduce in Sect. 2.

5.1 Distinct IP Addresses Accessing a Server

HYPERLOGLOG sketches, as discussed in Sect. 2.1, permit a practical solution to the problem of maintaining the number of distinct IP addresses that send and/or receive packets to/from a managed server. By maintaining such a data structure for each managed host over every, say, hour, one can maintain a comprehensive log of the approximate volume of traffic flowing into or out of an enterprise network. Although the HYPERLOGLOG data structure does not permit the recovery of the actual IP addresses, sketches accumulated on different servers can be combined to obtain an approximation of the number of distinct IP addresses that interacted with either server. We can also use estimators of the sketch intersections to estimate the number of distinct IP addresses that visit *both* servers as discussed in Sect. 2.1. Recall, however, that this estimate has high variance when the intersection is small. If the number of distinct elements of both servers and their union is large compared to the intersection (i.e. there is little to no overlap), then the relative error in the approximation of the individual and union values may be much larger than the ground truth intersection. However, if the intersection is large, we will be able to probabilistically recover it albeit with higher variance than the directly approximated elements. This means that the approach is most suitable for identifying and approximating the overlap heavy hitters - those pairs of servers whose IP partner intersections are the largest. Fortunately, for problems such as detecting DDoS attacks or botnets, it is exactly these pairs of servers who share a large number of common communication partners in which we are interested. Maintaining such a list of heavy hitters may prove useful for forensic analysis of such an event, allowing analysts to prioritize the server logs that are most likely to contain the relevant information, as well as contributing to the discovery of such events in the first place by tracking changes.

5.2 Data Flow Volumes

We also discussed the desirability of the maintaining the approximate counts of IP source, destination, and source-destination data flow volume in Sect. 1, which we can approximately accomplish using COUNTMINSKETCH. Say that we want to monitor the volume of traffic flowing over each IP source-destination pair on the boundary between an enterprise network and the Internet. While maintaining a dictionary data structure for each window is clearly intractable, we can instead maintain a logarithmically-sized COUNTMINSKETCH data structure for each window. By accumulating the data structure with a heap as discussed in Sect. 2.2, we can obtain a list of the top k heavy hitters for each time windows. Changes in these lists indicate changes in the underlying dynamics of the network activity, which can include DoS attacks, exfiltration events, or significant changes in the enterprise mission (e.g. new public-facing server launched, etc). Cormode and Muthukrishnan describe many other applications in [21].

Maintaining a norm of the implicit frequency vector defined by each time window can prove useful for such detection schemes as well, in concert with the heavy hitter lists. Significant changes to, say, the ℓ_2 norm of the frequency vector from one time period to another indicates a change in the distribution of flow over communicating pairs, such as an increase in the traffic communicated by a small number of pairs. Combined with the heavy hitters obtained via COUNT-MINSKETCH, norms obtained by TUGOFWAR as discussed in Sect. 2.3 or other norm sketching algorithms can both alert analysts to the presence of anomalous behavior as well as assist them in focusing their attention on likely sources of the anomalous behavior.

5.3   Identifying Attack Patterns

Of course, any ACD that depends upon linear algebraic operations such as linear regression and are also subject to immense scale could benefit from approximate numerical linear algebra tools discussed in Sect. 3.1.

Malicious actors have several atomic cyber attacks at their disposal. A 2014 survey coarsely categorizes cyber attacks into the following four categories:

- Denial of Service (DoS): Attackers issue malicious requests that consume computing or memory resources of a service, degrading the performance of legitimate requests.
- Probing: Attackers scan a network to gather information and identify vulnerable hosts as future targets.
- Remote to Local (User) Attacks (R2L): Attackers compromise a machine remotely by exploiting a software or configuration vulnerability. Attacks of this type are typically employed when attackers do not have an account on the target machine.
- Priviledge Escalation: Attackers exploit vulnerabilities or problematic configurations to gain root access on a machine on which they own an account.

Atomic cyber attacks of the above type are typically single steps in a larger playbook.

The recognition of playbook and cyber-attack patterns or attack paths is a daily task in cyber security operation centers. Although zero-day attacks are difficult to detect, experts can often forensically identify the attack path along which a zero-day exploit occurred, and the automated identification of such paths can help to detect zero-day attacks more quickly [66]. Attack paths are effectively graphs representing possible series of attacks and otherwise benign actions that culminate in an attacker's achieving some goal.

There can be many branching and intersecting paths in such an attack path, and an instance in the wild might even include loops. Figure 1 shows an example attack path relating the insertion of malware. This attacker must perform one of many possible series of atomic actions that result in gaining access to the

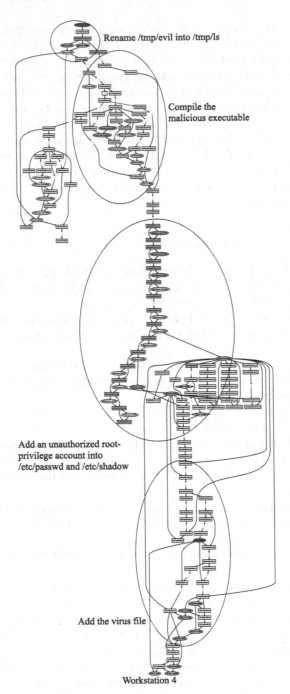

Fig. 1. A sample attack path for malware insertion after the attacker has gained an unprivileged foothold on a network. One of many possible series of atomically benign actions coupled with expressly malicious moves results in the deployment of malware on a targeted system.

target IT system, crafting a malware payload, gaining privileges, and deploying the payload onto a target resource.

The process of performing cyber-attack pattern matching involves inferring discrete events into vertices and edges to build such a graph and identifying when it matches a specified pattern such as Fig. 1. These discrete events are inferred from machine logs of various types, e.g. firewall logs, IDS logs, audit logs, web logs, et cetera. However, it is not always clear how to map these observations onto a given pattern graph. While some of this ambiguity can be handled with supervised machine learning, real-time systems will encounter novel observations and may not have a human in the loop. Consequently, it is necessary to maintain many parallel graph instantiations to handle different interpretations of individual logs. Each of these instantiations is subject to edge updates based upon future observations, including edge deletions if an interpretation is invalidated. Moreover, the log files themselves must often in practice be treated as data streams, as they may be quite large compared to system memory.

Consequently, collecting approximate graph structure of these parallel instantiations may prove to be attractive in practice, particularly if they are sufficiently large and there are sufficiently many that storing them in system memory is inconvenient. Spectral sparsifications of the type discussed in Sect. 3.3 are particularly attractive, particularly if we treat the collection of parallel graphs as a single largely disconnected graph. As the literature on streaming dynamic graphs is still quite young, we expect that more advances relevant to ACD problems will emerge in the future.

5.4   Identifying Important Machines

Consider a simple problem, where a security analyst wants to determine which machines in a enterprise network are the most critical for maintaining mission function. One way of measuring this criticality is to identify it with degree - i.e. the important machines are those with many connections. Such machines might host critical services, e.g. mail servers and databases. One could apply the COUNTSKETCH approach discussed in Sect. 4.1 in order to obtain the degree sketch heavy hitters from a single pass over a description of the network.

Consider that the analyst might want to differentiate between vertices that are connected to mostly user terminals (who have low degree) from those who are also connected to other high-degree vertices, i.e. other backbone services. The latter type of machine is clearly more "core" to the network, and is of more interest to an analyst. Consider the pth neighborhood size of a vertex x, i.e. the number of unique vertices that can be reached by following p edges starting at x. In a security context, ranking vertices by neighborhood size might be a more robust quantification of importance, as it can capture local dependencies that span more than one hop. However, the COUNTSKETCH approach does not generalize to handle the estimation of $p$the neighborhood sizes.

Note that instead of utilizing a COUNTSKETCH instance to estimate G's degree centrality like in Sect. 4.1, we can instead maintain a HYPERLOGLOG instance for each vertex $x \in V$. We will call this data structure DEGREESKETCH.

A DEGREESKETCH instance $\mathcal{D}$ is a distributed dictionary that maps vertices to HYPERLOGLOG sketches, so that $\mathcal{D}[x]$ is the sketch associated with $x \in V$. Upon reading in a description of G, $\mathcal{D}[x]$ can be used to estimate the degree of x. By its nature $\mathcal{D}$ can be easily distributed across a set of networked processors $\mathcal{P}$, where each processor $P \in \mathcal{P}$ is responsible for storing the sketches associated with a partition of V.

DEGREESKETCH gives us a much tighter estimate for the degree of each vertex than the COUNTSKETCH-based approach. However, it also requires memory linear in the size of V. Due to constant associated with the HYPERLOGLOG sketches, this requires more memory than simply maintaining a counter for each vertex, while introducing error! However, the power of DEGREESKETCH comes from the fact that the HYPERLOGLOG sketches are *composable*. Let $\widetilde{\cup}$ be the union operator for HYPERLOGLOG sketches. Given access to E, we can compute a second-degree sketch $\mathcal{D}_2[x]$ via

$$\mathcal{D}_2[x] = \widetilde{\bigcup}_{y:xy \in E} \mathcal{D}[y]. \tag{11}$$

That is, a union of the HYPERLOGLOG sketches for each vertex adjacent to x allows us to directly estimate the second neighborhood size of x. If we have $\mathcal{D}_p$, where $\mathcal{D}_p[x]$ is a HYPERLOGLOG estimating the pth neighborhood size of x, then we can similarly estimate the $(p+1)$th neighborhood size via the sketch produces via

$$\mathcal{D}_{p+1}[x] = \widetilde{\bigcup}_{y:xy \in E} \mathcal{D}_p[y]. \tag{12}$$

Obtaining the pth neighborhood size estimates in this way requires p passes over a stream defining G, and uses only $O\left(\frac{n}{\varepsilon^2} \log \log n + \log n\right)$ memory, where the HYPERLOGLOG sketches have standard error 1.04ε. This approach permits the analyst to estimate the local pth nieghborhood sizes for massive graphs cheaply. Circling back to the original discussion, this approach allows the analyst to cheaply compute identify backbone machines in a network given access to the network's description, even if it is impractically large.

A fully distributed version of this algorithm is presented in full in [61], and is inspired by the ANF [58] and HYPERANF [10] algorithms.

6   Conclusion

We have discussed many of the scalability challenges that arise in the practice of ACDs, particularly at the network level, and provided a survey of the sketching tools available to meet these challenges. We hope that, rather than highlighting a particular technique, this chapter serves as a resource for developing scalable ACD technologies. In particular, we have argued that sketch-based analyses can summarize large batches of data while preserving information relevant both to the automated detection of anomalies as well as information helpful for pointing forensic analysts in the right direction.

In order to meet these efficiency gains, sketches give up random access to the source data and provide only Monte Carlo approximation guarantees on their output. Fortunately, for many ACD applications, such as detecting macro-level changes in network traffic, approximation is generally acceptable. While many of the approaches we have discussed require only a single pass over the data stream, we have also discussed a few problems for which no practicable single-pass algorithm is known, and have instead discussed multi-pass sublinear algorithms. Although these multi-pass algorithms are necessarily much less reactive, they may still prove useful in practice for batch analysis. For example, we have shown that technologies that depend upon the SVD of a very large matrix or approximately finding the top betweenness centrality-indexed vertices in a graph can benefit from sketching algorithms, which still lower the space and sometimes time overhead compared to exact methods.

Though we have discussed many of the prominent practical sketching technologies available in this chapter, there is too large a literature of streaming and sketching algorithms to completely summarize here. The many survey papers we have referenced should serve as good starting places for examining this literature.

As we have discussed, however, not all queries are soluble using data streaming and sketching approaches. Indeed, many seemingly benign queries, such as "does a particular element comprise more than half of a data stream?", "do two sets have a null intersection?", or "does a graph contain at least one triangle?" require space linear in the size of the input. Thus, the techniques discussed in this chapter are no magic bullet for arbitrarily improving the performance of a piece of technology. Care must be taken when selecting the questions to ask of a data set, as the answers may not always be forthcoming.

Acknowledgements. The work presented in this chapter was supported by the Army Research Office under grant W911NF-13-1-0421.

References

1. Ahn, K.J., Guha, S., McGregor, A.: Analyzing graph structure via linear measurements. In: Proceedings of the Twenty-Third Annual ACM-SIAM Symposium on Discrete Algorithms, pp. 459–467. SIAM (2012)
2. Ahn, K.J., Guha, S., McGregor, A.: Graph sketches: sparsification, spanners, and subgraphs. In: Proceedings of the 31st ACM SIGMOD-SIGACT-SIGAI Symposium on Principles of Database Systems, pp. 5–14. ACM (2012)
3. Alahakoon, T., Tripathi, R., Kourtellis, N., Simha, R., Iamnitchi, A.: K-path centrality: a new centrality measure in social networks. In: Proceedings of the 4th Workshop on Social Network Systems, p. 1. ACM (2011)
4. Alon, N., Matias, Y., Szegedy, M.: The space complexity of approximating the frequency moments. J. Comput. Syst. Sci. **58**, 137–147 (1999)
5. Andoni, A., Krauthgamer, R., Onak, K.: Streaming algorithms via precision sampling. In: 2011 IEEE 52nd Annual Symposium on Foundations of Computer Science (FOCS), pp. 363–372. IEEE (2011)

6. Bader, D.A., Kintali, S., Madduri, K., Mihail, M.: Approximating betweenness centrality. In: Bonato, A., Chung, F.R.K. (eds.) WAW 2007. LNCS, vol. 4863, pp. 124–137. Springer, Heidelberg (2007). https://doi.org/10.1007/978-3-540-77004-6_10

7. Bar-Yossef, Z., Jayram, T.S., Kumar, R., Sivakumar, D., Trevisan, L.: Counting distinct elements in a data stream. In: Rolim, J.D.P., Vadhan, S. (eds.) RANDOM 2002. LNCS, vol. 2483, pp. 1–10. Springer, Heidelberg (2002). https://doi.org/10.1007/3-540-45726-7_1

8. Barabási, A.L., Albert, R.: Emergence of scaling in random networks. Science 286(5439), 509–512 (1999)

9. Bergamini, E., Meyerhenke, H., Staudt, C.L.: Approximating betweenness centrality in large evolving networks. In: 2015 Proceedings of the Seventeenth Workshop on Algorithm Engineering and Experiments (ALENEX), pp. 133–146. SIAM (2014)

10. Boldi, P., Rosa, M., Vigna, S.: HyperANF: approximating the neighbourhood function of very large graphs on a budget. In: Proceedings of the 20th International Conference on World Wide Web, pp. 625–634. ACM (2011)

11. Boldi, P., Vigna, S.: Axioms for centrality. Internet Math. 10(3–4), 222–262 (2014)

12. Brandes, U.: A faster algorithm for betweenness centrality. J. Math. Sociol. 25(2), 163–177 (2001)

13. Brandes, U., Pich, C.: Centrality estimation in large networks. Int. J. Bifurc. Chaos 17(07), 2303–2318 (2007)

14. Cárdenas, A.A., Manadhata, P.K., Rajan, S.P.: Big data analytics for security. IEEE Secur. Priv. 11(6), 74–76 (2013)

15. Charikar, M., Chen, K., Farach-Colton, M.: Finding frequent items in data streams. In: Widmayer, P., Eidenbenz, S., Triguero, F., Morales, R., Conejo, R., Hennessy, M. (eds.) ICALP 2002. LNCS, vol. 2380, pp. 693–703. Springer, Heidelberg (2002). https://doi.org/10.1007/3-540-45465-9_59

16. Clarkson, K.L., Woodruff, D.P.: Numerical linear algebra in the streaming model. In: Proceedings of the Forty-First Annual ACM Symposium on Theory of Computing, pp. 205–214. ACM (2009)

17. Clarkson, K.L., Woodruff, D.P.: Low-rank approximation and regression in input sparsity time. J. ACM (JACM) 63(6), 54 (2017)

18. Cohen, R., Katzir, L., Yehezkel, A.: A minimal variance estimator for the cardinality of big data set intersection. In: Proceedings of the 23rd ACM SIGKDD International Conference on Knowledge Discovery and Data Mining, pp. 95–103. ACM (2017)

19. Cormode, G., Datar, M., Indyk, P., Muthukrishnan, S.: Comparing data streams using hamming norms (how to zero in). IEEE Trans. Knowl. Data Eng. 15(3), 529–540 (2003)

20. Cormode, G., Muthukrishnan, S.: An improved data stream summary: the count-min sketch and its applications. J. Algorithms 55(1), 58–75 (2005)

21. Cormode, G., Muthukrishnan, S.: What's hot and what's not: tracking most frequent items dynamically. ACM Trans. Database Syst. (TODS) 30(1), 249–278 (2005)

22. Deng, F., Rafiei, D.: New estimation algorithms for streaming data: count-min can do more (2007)

23. Dietzfelbinger, M., Hagerup, T., Katajainen, J., Penttonen, M.: A reliable randomized algorithm for the closest-pair problem. J, Algorithms 25(1), 19–51 (1997)

24. Durand, M., Flajolet, P.: Loglog counting of large cardinalities. In: Di Battista, G., Zwick, U. (eds.) ESA 2003. LNCS, vol. 2832, pp. 605–617. Springer, Heidelberg (2003). https://doi.org/10.1007/978-3-540-39658-1_55

25. Ertl, O.: New cardinality estimation algorithms for HyperLogLog sketches. arXiv preprint arXiv:1702.01284 (2017)

26. Estan, C., Varghese, G., Fisk, M.: Bitmap algorithms for counting active flows on high speed links. In: Proceedings of the 3rd ACM SIGCOMM Conference on Internet Measurement, pp. 153–166. ACM (2003)

27. Fan, L., Cao, P., Almeida, J., Broder, A.Z.: Summary cache: a scalable wide-area web cache sharing protocol. IEEE/ACM Trans. Netw. (TON) **8**(3), 281–293 (2000)

28. Feigenbaum, J., Kannan, S., McGregor, A., Suri, S., Zhang, J.: On graph problems in a semi-streaming model. Theoret. Comput. Sci. **348**(2–3), 207–216 (2005)

29. Flajolet, P., Fusy, É., Gandouet, O., Meunier, F.: HyperLogLog: the analysis of a near-optimal cardinality estimation algorithm. In: Discrete Mathematics and Theoretical Computer Science. pp. 137–156 (2007)

30. Flajolet, P., Martin, G.N.: Probabilistic counting algorithms for data base applications. J. Comput. Syst. Sci. **31**(2), 182–209 (1985)

31. Goyal, A., Daumé III, H., Cormode, G.: Sketch algorithms for estimating point queries in NLP. In: Proceedings of the 2012 Joint Conference on Empirical Methods in Natural Language Processing and Computational Natural Language Learning, pp. 1093–1103. Association for Computational Linguistics (2012)

32. Green, O., McColl, R., Bader, D.A.: A fast algorithm for streaming betweenness centrality. In: 2012 International Conference on Privacy, Security, Risk and Trust (PASSAT) and 2012 International Conference on Social Computing (SocialCom), pp. 11–20. IEEE (2012)

33. Guha, S., McGregor, A.: Graph streams and sketches: resources (2018). https://people.cs.umass.edu/~mcgregor/graphs/

34. Gupta, P., Goel, A., Lin, J., Sharma, A., Wang, D., Zadeh, R.: WTF: the who to follow service at Twitter. In: Proceedings of the 22nd International Conference on World Wide Web, pp. 505–514. ACM (2013)

35. Hayashi, T., Akiba, T., Yoshida, Y.: Fully dynamic betweenness centrality maintenance on massive networks. Proc. VLDB Endow. **9**(2), 48–59 (2015)

36. Heule, S., Nunkesser, M., Hall, A.: HyperLogLog in practice: algorithmic engineering of a state of the art cardinality estimation algorithm. In: Proceedings of the 16th International Conference on Extending Database Technology, pp. 683–692. ACM (2013)

37. Indyk, P.: Stable distributions, pseudorandom generators, embeddings, and data stream computation. J. ACM (JACM) **53**(3), 307–323 (2006)

38. Indyk, P., Woodruff, D.: Optimal approximations of the frequency moments of data streams. In: Proceedings of the Thirty-Seventh Annual ACM Symposium on Theory of Computing, pp. 202–208. ACM (2005)

39. Johnson, W.B., Lindenstrauss, J.: Extensions of Lipschitz mappings into a Hilbert space. Contemp. Math. **26**(189–206), 1 (1984)

40. Jowhari, H., Sağlam, M., Tardos, G.: Tight bounds for Lp samplers, finding duplicates in streams, and related problems. In: Proceedings of the Thirtieth ACM SIGMOD-SIGACT-SIGART Symposium on Principles of Database Systems, pp. 49–58. ACM (2011)

41. Kane, D.M., Nelson, J., Woodruff, D.P.: An optimal algorithm for the distinct elements problem. In: Proceedings of the Twenty-Ninth ACM SIGMOD-SIGACT-SIGART Symposium on Principles of Database Systems, pp. 41–52. ACM (2010)

42. Kang, C., Kraus, S., Molinaro, C., Spezzano, F., Subrahmanian, V.: Diffusion centrality: a paradigm to maximize spread in social networks. Artif. Intell. **239**, 70–96 (2016)

43. Kapralov, M., Lee, Y.T., Musco, C., Musco, C., Sidford, A.: Single pass spectral sparsification in dynamic streams. SIAM J. Comput. **46**(1), 456–477 (2017)
44. Kourtellis, N., Alahakoon, T., Simha, R., Iamnitchi, A., Tripathi, R.: Identifying high betweenness centrality nodes in large social networks. Soc. Netw. Anal. Min. **3**(4), 899–914 (2013)
45. Kourtellis, N., Morales, G.D.F., Bonchi, F.: Scalable online betweenness centrality in evolving graphs. IEEE Trans. Knowl. Data Eng. **27**(9), 2494–2506 (2015)
46. Li, Y., Nguyen, H.L., Woodruff, D.P.: On sketching matrix norms and the top singular vector. In: Proceedings of the Twenty-Fifth Annual ACM-SIAM Symposium on Discrete Algorithms, pp. 1562–1581. Society for Industrial and Applied Mathematics (2014)
47. Li, Y., Nguyen, H.L., Woodruff, D.P.: Turnstile streaming algorithms might as well be linear sketches. In: Proceedings of the Forty-Sixth Annual ACM Symposium on Theory of Computing, pp. 174–183. ACM (2014)
48. Li, Y., Woodruff, D.P.: Tight bounds for sketching the operator norm, Schatten norms, and subspace embeddings. In: LIPIcs-Leibniz International Proceedings in Informatics, vol. 60. Schloss Dagstuhl-Leibniz-Zentrum fuer Informatik (2016)
49. Mahoney, M.W., et al.: Randomized algorithms for matrices and data. Found. Trends® Mach. Learn. **3**(2), 123–224 (2011)
50. Malewicz, G., et al.: Pregel: a system for large-scale graph processing. In: Proceedings of the 2010 ACM SIGMOD International Conference on Management of Data, pp. 135–146. ACM (2010)
51. McGregor, A.: Graph mining on streams. In: Liu, L., Özsu, M.T. (eds.) Encyclopedia of Database Systems, pp. 1271–1275. Springer, Boston (2009). https://doi.org/10.1007/978-0-387-39940-9_184
52. Monemizadeh, M., Woodruff, D.P.: 1-pass relative-error Lp-sampling with applications. In: Proceedings of the Twenty-First Annual ACM-SIAM Symposium on Discrete Algorithms, pp. 1143–1160. SIAM (2010)
53. Muthukrishnan, S., et al.: Data streams: algorithms and applications. Found. Trends® Theor. Comput. Sci. **1**(2), 117–236 (2005)
54. Myers, S.A., Sharma, A., Gupta, P., Lin, J.: Information network or social network?: the structure of the Twitter follow graph. In: Proceedings of the 23rd International Conference on World Wide Web, pp. 493–498. ACM (2014)
55. Nelson, J., Nguyên, H.L.: OSNAP: faster numerical linear algebra algorithms via sparser subspace embeddings. In: 2013 IEEE 54th Annual Symposium on Foundations of Computer Science (FOCS), pp. 117–126. IEEE (2013)
56. Nelson, J., Nguyn, H.L., Woodruff, D.P.: On deterministic sketching and streaming for sparse recovery and norm estimation. Linear Algebra Appl. **441**, 152–167 (2014)
57. Nisan, N.: Pseudorandom generators for space-bounded computation. Combinatorica **12**(4), 449–461 (1992)
58. Palmer, C.R., Gibbons, P.B., Faloutsos, C.: ANF: a fast and scalable tool for data mining in massive graphs. In: Proceedings of the Eighth ACM SIGKDD International Conference on Knowledge Discovery and Data Mining, pp. 81–90. ACM (2002)
59. Pearce, R.: Triangle counting for scale-free graphs at scale in distributed memory. In: 2017 IEEE High Performance Extreme Computing Conference (HPEC), pp. 1–4. IEEE (2017)
60. Pearce, R., Gokhale, M., Amato, N.M.: Faster parallel traversal of scale free graphs at extreme scale with vertex delegates. In: SC14: International Conference for High Performance Computing, Networking, Storage and Analysis, pp. 549–559. IEEE (2014)

61. Priest, B.W.: Semi-streaming approximation of centrality indices in massive graphs. Ph.D. thesis, Dartmouth College (2019)
62. Priest, B.W., Pearce, R., Sanders, G.: Estimating edge-local triangle count heavy hitters in edge-linear time and almost-vertex-linear space. In: 2018 IEEE High Performance Extreme Computing Conference (HPEC). IEEE (2018)
63. Pătrașcu, M., Thorup, M.: The power of simple tabulation hashing. J. ACM (JACM) **59**(3), 14 (2012)
64. Qin, J., Kim, D., Tung, Y.: LogLog-beta and more: a new algorithm for cardinality estimation based on LogLog counting. arXiv preprint arXiv:1612.02284 (2016)
65. Riondato, M., Kornaropoulos, E.M.: Fast approximation of betweenness centrality through sampling. Data Min. Knowl. Disc. **30**(2), 438–475 (2016)
66. Sun, X., Dai, J., Liu, P., Singhal, A., Yen, J.: Using bayesian networks for probabilistic identification of zero-day attack paths. IEEE Trans. Inf. Forensics Secur. **13**(10), 2506–2521 (2018)
67. Ting, D.: Streamed approximate counting of distinct elements: Beating optimal batch methods. In: Proceedings of the 20th ACM SIGKDD International Conference on Knowledge Discovery and Data Mining, pp. 442–451. ACM (2014)
68. Upstill, T., Craswell, N., Hawking, D.: Predicting fame and fortune: PageRank or indegree. In: Proceedings of the Australasian Document Computing Symposium, ADCS, pp. 31–40 (2003)
69. Vu, H.: Data stream algorithms for large graphs and high dimensional data (2018)
70. Wegman, M.N., Carter, J.L.: New hash functions and their use in authentication and set equality. J. Comput. Syst. Sci. **22**(3), 265–279 (1981)
71. Wei, W., Carley, K.: Real time closeness and betweenness centrality calculations on streaming network data. In: Proceedings of the 2014 ASE Big-Data/SocialCom/Cybersecurity Conference, Stanford University (2014)
72. Whang, K.Y., Vander-Zanden, B.T., Taylor, H.M.: A linear-time probabilistic counting algorithm for database applications. ACM Trans. Database Syst. (TODS) **15**(2), 208–229 (1990)
73. Woodruff, D.P., et al.: Sketching as a tool for numerical linear algebra. Found. Trends® Theor. Comput. Sci. **10**(1–2), 1–157 (2014)
74. Xiao, Q., Zhou, Y., Chen, S.: Better with fewer bits: improving the performance of cardinality estimation of large data streams. In: INFOCOM 2017-IEEE Conference on Computer Communications, pp. 1–9. IEEE (2017)
75. Yoshida, Y.: Almost linear-time algorithms for adaptive betweenness centrality using hypergraph sketches. In: Proceedings of the 20th ACM SIGKDD International Conference on Knowledge Discovery and Data Mining, pp. 1416–1425. ACM (2014)
76. Zhang, Q., Pell, J., Canino-Koning, R., Howe, A.C., Brown, C.T.: These are not the k-mers you are looking for: efficient online k-mer counting using a probabilistic data structure. PLoS ONE **9**(7), e101271 (2014)

Author Index

Printed in the United States
By Bookmasters